AF899115

VOLUME FIFTY SEVEN

ANNUAL REPORTS IN MEDICINAL CHEMISTRY

Coronaviruses and Other Novel Antiviral Targets

ANNUAL REPORTS IN MEDICINAL CHEMISTRY

VOLUME 57

Editor-in-Chief

STEPHEN NEIDLE

School of Pharmacy, University College London, London, United Kingdom

EDITORIAL BOARD MEMBERS
Chair, Editorial Board
STEPHEN NEIDLE

Zhi-shu Huang
Karen E. Lackey
Takuo Tsukuda
Peter Bernstein
Karl-Heinz Altmann
Katherine Seley-Radtke

VOLUME FIFTY SEVEN

Annual Reports in MEDICINAL CHEMISTRY

Coronaviruses and Other Novel Antiviral Targets

Volume Editors

KATHERINE SELEY-RADTKE

Department of Chemistry & Biochemistry,
University of Maryland, Baltimore County,
Baltimore, MD, United States

ELSEVIER

ACADEMIC PRESS

An imprint of Elsevier

Academic Press is an imprint of Elsevier
50 Hampshire Street, 5th Floor, Cambridge, MA 02139, United States
525 B Street, Suite 1650, San Diego, CA 92101, United States
The Boulevard, Langford Lane, Kidlington, Oxford OX5 1GB, United Kingdom
125 London Wall, London, EC2Y 5AS, United Kingdom

First edition 2021

Copyright © 2021 Elsevier Inc. All rights reserved.

No part of this publication may be reproduced or transmitted in any form or by any means, electronic or mechanical, including photocopying, recording, or any information storage and retrieval system, without permission in writing from the publisher. Details on how to seek permission, further information about the Publisher's permissions policies and our arrangements with organizations such as the Copyright Clearance Center and the Copyright Licensing Agency, can be found at our website: www.elsevier.com/permissions.

This book and the individual contributions contained in it are protected under copyright by the Publisher (other than as may be noted herein).

Notices
Knowledge and best practice in this field are constantly changing. As new research and experience broaden our understanding, changes in research methods, professional practices, or medical treatment may become necessary.

Practitioners and researchers must always rely on their own experience and knowledge in evaluating and using any information, methods, compounds, or experiments described herein. In using such information or methods they should be mindful of their own safety and the safety of others, including parties for whom they have a professional responsibility.

To the fullest extent of the law, neither the Publisher nor the authors, contributors, or editors, assume any liability for any injury and/or damage to persons or property as a matter of products liability, negligence or otherwise, or from any use or operation of any methods, products, instructions, or ideas contained in the material herein.

ISBN: 978-0-323-91511-3
ISSN: 0065-7743

For information on all Academic Press publications
visit our website at https://www.elsevier.com/books-and-journals

Publisher: Zoe Kruze
Acquisitions Editor: Jason Mitchell
Developmental Editor: Federico Paulo S. Mendoza
Production Project Manager: Denny Mansingh
Cover Designer: Mark Rogers

Typeset by STRAIVE, India

ELSEVIER Book Aid International
Working together to grow libraries in developing countries
www.elsevier.com • www.bookaid.org

Contents

Contributors

Preface

The goal of this volume is to highlight current efforts across the antiviral field to fight the current pandemic as well as a number of other emerging and reemerging infectious diseases that have plagued the world in the past few decades. Many of us in the antiviral field have been working on coronaviruses as well as many of the other neglected viruses, covered in this volume, for some time. Despite this, we have often struggled to obtain support for our research since a number of funding agencies did not recognize the threat that these viruses pose to spread globally until the stark reality of the SARS-CoV-2 pandemic hit. Fortunately, one of the few positive outcomes of the pandemic is that there is a new focus on the development of small-molecule broad-spectrum inhibitors that could be administered orally, as well as stockpiled to combat future outbreaks. It is critical that we stop being reactive and instead be proactive. As a result, many federal and other funding agencies have now prioritized this approach and have put out calls to support these efforts. The chapters in this volume, although by no means exhaustive, have nicely summarized the efforts by many against SARS-CoV-2, as well as a number of different viral classes that are also desperately in need of effective therapeutics.

In that regard, the first chapter in the volume, "Improving properties of the nucleobase analogs T-705/T-1105 as potential antiviral," was contributed by Prof. Chris Meier and his team, where they have detailed their efforts to design and synthesize prodrugs for favipiravir—an interesting drug that is currently being investigated for use against SARS-CoV-2. Originally developed as an anti-influenza drug, favipiravir is a heterocyclic base that must first be ribosylated and subsequently phosphorylated to form the active triphosphate nucleotide. The triphosphate is then incorporated and serves as both a chain terminator and a lethal mutagenesis inhibitor. Unfortunately, due to the extreme polarity and presence of phosphatases in vivo, it is impossible to deliver any of the phosphate forms of the nucleosides; thus, prodrugs have been used extensively to overcome this limitation.

In an effort to avoid all of the necessary steps to get to the active form, Prof. Meier has been investigating the possibility to administer various prodrugs that would allow for the delivery of mono-, di-, or triphosphate directly. Prof. Meier has been a leader in the nucleoside prodrug field for many years, and the goal of developing a triphosphate prodrug has always been viewed as the holy grail; thus, his recent efforts are very promising and frankly represent a potential game changer for nucleoside therapeutics. In that regard, Prof. Meier and his team detail the design, synthesis, and biological testing for a series of prodrugs of favipiravir. Tapping into the successful cycloSal prodrug moiety, they first describe their routes to make the monophosphate analogs. Then, the team focuses on the development and synthesis of the di-*PPro* and tri-*PPPro* prodrugs. Overall, their thorough coverage of the different synthetic routes, mechanism of action studies, and antiviral activities for the series makes for a nice concise but complete story that is easy for others in the field to follow and, of course, adapt for their own nucleoside analogs.

The second chapter "Anno 2021: Which antivirals for the coming decade?" by Profs. Piet Herdewijn and Erik De Clercq and their team is quite extensive in terms of recent and historical coverage of antivirals. They begin with a brief overview of the current nucleosides being investigated against SARS-CoV-2. These include remdesivir, AT-527, molnupiravir, and favipiravir. In addition to the nucleosides, a protease inhibitor PF-07321332 from Pfizer is also showing promise against the mPro protease. Finally, the viral methyltransferases nsp14 and nsp16 are also coming to focus as potential targets for inhibition, and to date one compound TO507 appears to be gaining traction.

Moving on from SARS-CoV-2, Profs. Herdewijn and De Clercq spend the rest of the chapter covering the various drugs that have been developed or are in development for a number of other classes of viruses. These include well-studied viruses, such as HIV, HCV, and yellow fever virus, as well as respiratory viruses such as influenza, RSV, and rhinoviruses. Finally, they spend some time covering efforts to treat viruses, such as measles, rabies, and polio, and ending up with new coronaviruses that are emerging. The authors have nicely detailed the synthesis and available biological results for these classes of viruses, and, as a result, this represents an excellent resource for researchers looking for a thorough summary of many different diseases.

The third chapter "Broad-spectrum antiviral nucleosides: Our best hope for the future" was contributed by me and two of my graduate students and provides a brief overview of some of the more well-known nucleosides that exhibit activities against a number of different viruses as well as several new ones that are gaining attention. As I mentioned in the beginning of this preface, there is currently a strong push by many government agencies to support the development of broad-spectrum inhibitors that can treat multiple viruses, not only across a family of viruses but also across many families. This would allow for stockpiling so that we would be better prepared for the next pandemic. We have approached our review from a historical, structural, and mechanistic manner, focusing on the key features of the various nucleoside scaffolds and mechanisms of action that have proven successful.

We begin with the early nucleosides that represented the first obligate chain terminators lacking a 3′-hydroxyl group. We also discuss the acyclic nucleosides and phosphonates, which have proved to be a very important class of nucleosides for a number of different viruses. In addition, as mentioned earlier with Prof. Meier's chapter, prodrugs are critical components for antiviral nucleosides to overcome delivery issues as well as to bypass the first phosphorylation step, which is considered by many the rate-limiting step for activation of the nucleoside to its active triphosphate (or diphosphate in the case of phosphonate nucleosides) form. We then spend some time discussing the McGuigan ProTide approach and mechanism, as that component has proven to be invaluable for a number of FDA-approved nucleoside drugs. Next we discuss a number of other types of nucleoside scaffolds that have exhibited several antiviral activities, before moving on to the current efforts against SARS-CoV-2.

In that regard, several of the nucleosides such as favipiravir, molnupiravir, and ribavirin work by lethal mutagenesis in addition to chain termination. Similarly, several other promising nucleosides, including remdesivir, AT-527, galidesivir, and sofosbuvir, also work by multiple mechanisms. The most well known is nonobligate delayed chain termination, but very recent studies have also uncovered new mechanisms of action. For example, remdesivir also works by a new mechanism called translocation, and AT-527 also inhibits the coronavirus NiRAN enzyme, bringing to light new and exciting possibilities for nucleoside drug design.

The fourth chapter "The search for antivirals to treat alphavirus infections" by Prof. Andrea Brancale and Dr. Marcella Bassetto details efforts to target alphaviruses. This particular class of viruses has, in most cases, lacked appropriate or effective treatments for many in the family. These viruses from the Togaviridae genus tend to be most prevalent in third-world countries and are quite debilitating, for example, the chikungunya virus. A number of these viruses are encephalitic and have been associated with horses as well as humans, including Venezuelan equine encephalitis, eastern equine encephalitis, and western equine encephalitis. Other viruses discussed that are lesser known include Sindbis Semliki Forest, Una, Mayaro, Ross River, and O'nyong-nyong viruses. Prof. Brancale and Dr. Bassetto have approached their chapter from the standpoint of drug repurposing, followed by medicinal chemistry strategies, then natural products that have shown promise, followed by future outlooks for finding cures for these deadly diseases.

Similarly, the penultimate chapter "Inhibitors of Ebolavirus targeting innate immune evasion" by Profs. Enzo Tramontano and Angela Corona and their group focuses on another deadly family of viruses—the filoviruses. Ebola has been an extremely difficult virus to target because of the inability to do normal clinical trials and even animal studies, due to ethical issues, and the limited window of opportunity for studying the virus, which ends quickly. Some of the other viruses covered in the chapter include Sudan, Reston, Tai Forest, Bombali, and Bundibugyo viruses.

Initially, remdesivir was being investigated against Ebola and was used in five compassionate cases, but it fell out of favor due to newer more effective treatments. During a recent outbreak, a new vaccine Ervebo was shown to confer immunogenicity against EBOV. As a result, the European Medicines Agency allowed its use during the 2018 outbreak in DRC in an effort to quickly investigate its potential. Erbevo was later approved by the FDA in 2020, and more recently, Inmazeb, a monoclonal antibody cocktail that targets the EBOV glycoprotein, was also approved; however, effective and less cumbersome treatments against Ebola remain elusive.

In that regard, the traditional approach to antiviral drug design involves targeting one or more viral proteins that are essential to the virus's replication. The authors nicely detail the various protein targets and efforts to design drugs to shut down those particular steps with various types of inhibitors as well as drugs like interferon.

Finally, the last chapter in the volume, "Uracil derivatives as non-nucleoside inhibitors of viral infections," from Dr. Anastasia Khandazhinskaya and colleagues provides an exhaustive review of the many different types of uracil-based non-nucleoside inhibitors. Historically, most of these have proven to be highly effective against HIV, and as a result, non-nucleoside reverse transcriptase inhibitors have been included in a number of antiviral drug cocktails. The authors have done an outstanding job of detailing the design rationale, the various synthetic routes, and, importantly, the numerous biological activities for an impressive number of different scaffolds. These non-nucleoside inhibitors have exhibited widespread activities against a number of viruses beyond HIV, which is likely to surprise many. Some of the viruses reported in the chapter include activity against human cytomegalovirus, HCV and HBV, RSV, tickborne encephalitis virus,

SARS-CoV-2, and, lastly, adenoviruses. In short, this is an extremely useful review of this important class of antiviral therapeutics.

On a personal note, I would like to say a big thank you to all of my friends and colleagues, as well as their graduate students and/or postdoctoral researchers, for contributing such excellent chapters for this volume of *Annual Reports in Medicinal Chemistry*. Many of us, including myself, have been extremely busy over the past year and a half working nonstop to come up with new drugs to fight SARS-CoV-2, so I deeply appreciate their willingness and efforts to take time out to participate in this important and timely volume.

Prof. Katherine Seley-Radtke, Editor

CHAPTER ONE

Improving properties of the nucleobase analogs T-705/T-1105 as potential antiviral

Xiao Jia, Benedikt Ganter, and Chris Meier*
Organic Chemistry, Department of Chemistry, Faculty of Mathematics, Informatics and Natural Sciences, Universität Hamburg, Hamburg, Germany
*Corresponding author: e-mail address: chris.meier@chemie.uni-hamburg.de

Contents

1. Introduction

In recent years, RNA viruses have become a significant threat to public health because of apparent sudden epidemic outbreaks and periodic pandemics, which cause considerable medical and socioeconomic burden.[1] These viruses, many of them causing so-called emerging infections, such as Ebola, Influenza, and Zika, are highly transmissible and cause immense

Annual Reports in Medicinal Chemistry, Volume 57
ISSN 0065-7743
https://doi.org/10.1016/bs.armc.2021.08.002
Copyright © 2021 Elsevier Inc.
All rights reserved.

morbidity and mortality.[2–4] The newly emerged coronavirus disease 2019/20 (COVID-19) caused by infection with the severe acute respiratory syndrome coronavirus 2 (SARS-CoV-2) has spread rapidly across the world since December 2019.[5,6] As of September 9, 2021 there have been more than 222 mio. confirmed cases and 4.5 mio. death cases globally related to COVID-19, according to the World Health Organization. Currently, although very efficient vaccines are available, the worldwide number of infections is still rapidly growing, impacting billions of people's lives not only in acute infections but also in long term perspectives as long-COVID.[7] Therefore, developing not only safe and effective vaccines but also drugs against SARS-CoV-2 is urgently needed. First of all, a number of important repurposed antiviral drugs (e.g., Lopinavir, Hydroxychloroquine, and Ritonavir)[8] have been licensed to treat COVID-19. Medicinal chemists in academia and at companies are seeking for new drugs for the treatment of infections by not only SARS-CoV-2 but also for a broad spectrum of RNA viruses.[9]

For several decades, many RNA or DNA polymerase inhibitors based on nucleoside analogs (e.g., HIV-RT and nucleos(*t*)ide reverse transcriptase inhibitors (NRTIs)) are extensively used as the backbone to combat infections caused by HIV, influenza, hepatitis B and hepatitis C virus, herpes virus and recently SARS-CoV-2.[10–15] Mainly these compounds target the virus replication by interfering with the action of the virus specific polymerases. In this respect they either act as obligate or as non-obligate chain terminators which in both cases finally lead to incomplete replication of the viral genome and thus incompetent new virus particles. A further mechanism of action of such compounds is that—if they are not repaired after incorporation—they induce a high number of mutations within the viral genome leading finally to lethal mutagenesis by error catastrophe. In order to combat COVID-19, many approved nucleoside antivirals, such as Remdesivir, Sofosbuvir, and Tenofovir have been used to combat this virus-caused pandemic (repurposing approach).[16,17] Previous studies have demonstrated that the nucleobase analog Favipiravir (6-fluoro-3-hydroxypyrazine-2-carboxamide) is an antiviral agent that inhibits the RNA-dependent RNA polymerase of many RNA viruses in animal infection models,[18,19] including Influenza virus (types A, B and C) infection,[18–22] Lassa virus infection,[23] severe fever with thrombocytopenia syndrome virus (SFTSV) infection,[24] Ebola virus (EBOV) infection,[25] Rabies virus infection,[26] Rift Valley fever virus (RVFV) infection,[27] Western equine encephalitis virus (WEEV) infection,[28] West Nile virus (WNV) infection,[29] Norovirus infection,[30] Crimean-Congo hemorrhagic fever virus (CCHFV) infection,[31] Yellow fever virus disease (YFV) infection,[32] and Hantavirus virus (HRTV) infection.[33] Favipiravir was developed by the Japanese Toyama Chemical Company for the treatment of severe or pandemic influenza[18,21] and was first approved in Japan in 2014 to treat human influenza virus infections.[34] Recently, Favipiravir has been demonstrated to be a potential agent for COVID-19 patients.[35,36] Due to its successful antiviral potency, Favipiravir was first approved for clinical use in China in March 2020 to treat SARS-CoV-2 infections.[35,37]

The antiviral efficacy of many nucleos(*t*)ide polymerase inhibitors is strongly dependent on their intracellular activation by host cellular kinases to yield, via the nucleoside monophosphate (NMP) and nucleoside diphosphate (NDP), ultimately the bioactive nucleoside analog triphosphate (NTP). The nucleobase analog Favipiravir (T-705) first undergoes metabolic activation through phosphoribosylation to form ribofuranosyl-5′-monophosphate (Favipiravir-RMP, T-705-RMP) by the eukaryotic hypoxanthine guanine phosphoribosyltransferase (HGPRT) which is then converted through intracellular phosphorylation into its ribofuranosyl-5′-triphosphate metabolite (Favipiravir-RTP, T-705-RTP) most probably by host cell enzymes (Fig. 1).[22,38] In previous studies the ribofuranosyl-5′-diphosphate metabolite (Favipiravir-RDP, T-705-RDP) was not observed or was formed in amounts below the limit of detection. However, the enzymes involved in the conversion of T-705-RMP into T-705-RDP and T-705-RTP have not been completely identified.[22,38,39]

Earlier, Favipiravir is noted as a promising drug for the treatment of infections caused by different RNA viruses probably due to the inhibition of the RdRP by T-705-RTP in a dose-dependent manner.[18,19,22,40] However, Favipiravir is effective on SARS-CoV-2 and other RNA viruses at high doses only,[41] In addition, it was shown that Favipiravir has a risk for embryotoxicity and teratogenicity.[42] Based on these concerns, the Ministry of Health, Labor and Welfare has formulated strict regulations on its production and clinical use.[42]

Recently, the related pyrazine derivative (T-1105; 3-hydroxy-2-pyrazinecarboxamide), which is the non-fluorinated T-705 analog has also shown promising results in the treatment of influenza virus[43] and other RNA viruses. Also its nucleoside analog form T-1106 (3,4-dihydro-3-oxo-4-β-D-ribofuranosyl-2-pyrazinecarboxamide; Fig. 1) showed potent antiviral activity against RNA viruses without significant toxicity to mammalian cells.[20–22,32,44,45] As T-705, T-1105 needs activation into its triphosphorylated form but proved to be more efficiently activated by human

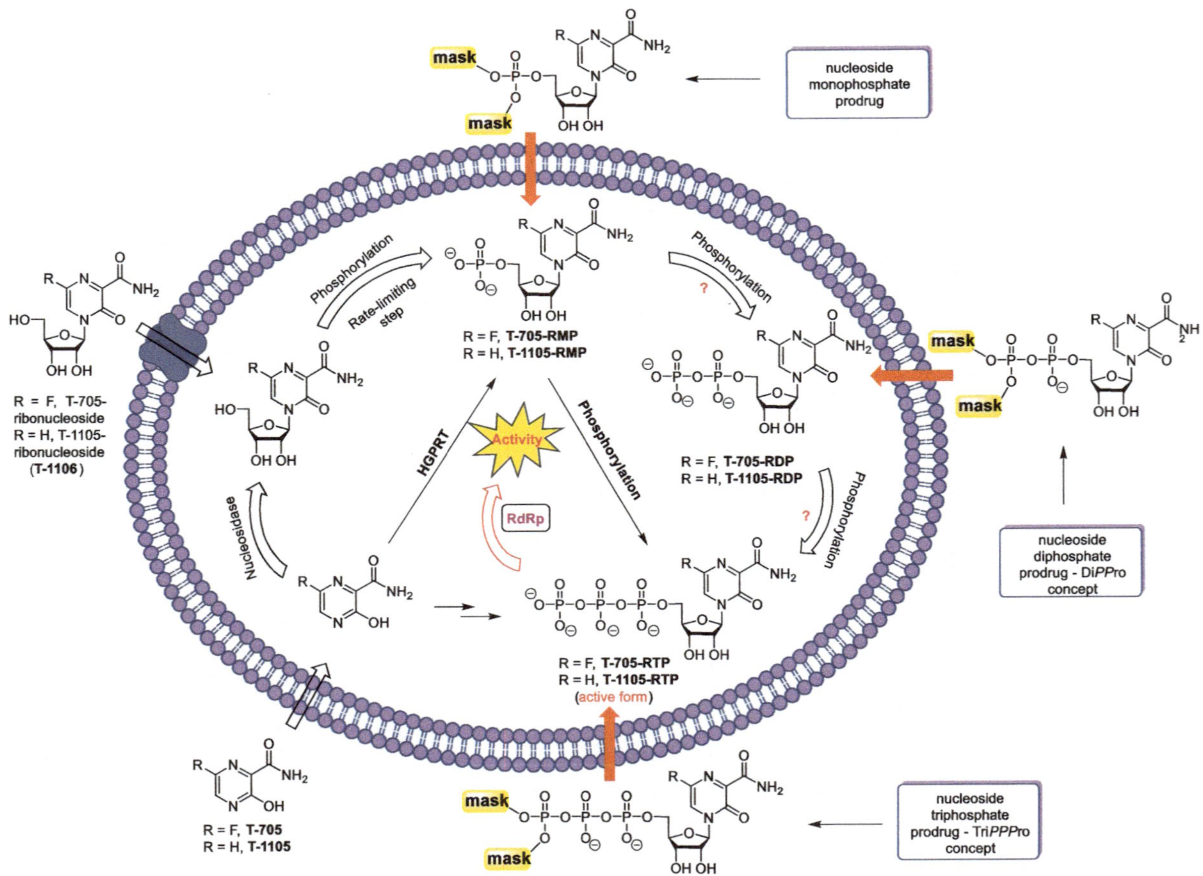

Fig. 1 Mode of action of Favipiravir (T-705) and T-1105 and the corresponding nucleoside prodrug approach.

HGPRT than T-705 in influenza virus infected MDCK cells.[39] The antiviral activity of the nucleoside analog T-1106 was in many cases virtually similar to T-705, although the dose of T-1106 required was lower than that of T-705.

Earlier studies have shown that both T-705 and T-1105 are poor substrates for the HGPRT and thus the overall conversion of T-705 and T-1105 into their corresponding nucleoside triphosphates (T-705-RTP and T-1105-RTP) has been identified as being the limiting factor for the antiviral activity.[39] So, it would be beneficial to use the already phosphorylated metabolites as potential drugs, in the extreme to use the ultimately active triphosphate of the two nucleobase analogs. However, the charged phosphorylated metabolites cannot be considered as viable drug candidates because of their high polarity which prevents cell membrane penetration. To overcome the inefficient phosphorylation of nucleoside analogs T-705 and T-1105 and thus to achieve the higher intracellular concentration of its active metabolites (T-705-RTP and T-1105-RTP), several pronucleotide strategies developed by our group, such as nucleoside monophosphate prodrugs (*cyclo*Sal), nucleoside diphosphate prodrugs (Di*PP*ro-concept), and nucleoside triphosphate prodrugs (Tri*PPP*ro-concept) have been applied to the nucleoside forms of T-705 (Favipiravir) and T-1105 (T-1106; Fig. 1).[46]

This minireview summarizes the application of our mono-, di-, and triphosphate pronucleotide approaches to T-705 and T-1105 and other nucleoside analogs. The synthesis, hydrolysis properties, and structure–activity relationships in the antiviral evaluation of these prodrugs are discussed here briefly.

2. T-705/T-1105 mechanism of action (MoA)

For all positive stranded RNA viruses like influenza, Ebola and also SARS-CoV-2, RdRp's play a crucial role in the viral RNA replication. In case of SARS-CoV-2 this protein is coded nsp12. The role of nsp12 makes it an ideal target for antiviral agents due to the importance for the viral life cycle and the lack of host homologs.

Intracellular nucleoside analog 5′-triphosphates (NA-TPs) are then incorporated by error-prone viral RdRp's into the nascent viral RNA.[47–49] This can lead to a disruption of the RNA synthesis by chain termination, or can lead to lethal mutagenesis. In contrast to other RNA viruses, SARS-CoV-2 has the possibility to remove nucleoside analogs from the viral

RNA after the incorporation by nsp12, which is carried out by the RNA proof-reading exoribonuclease (nsp14-ExoN).[50] This can lead to resistance development of SARS-CoV-2 against nucleoside analogs. For example, it has been shown that nsp14 is capable of excising Ribavirin, a long-known broad-spectrum antiviral nucleoside analog, from the viral genome due to its proof-reading capabilitys.[51,52] Nevertheless several antivirally active nucleoside analogs are currently under investigation as potential anti-SARS-CoV-2 candidates.[17,53,54] Although Favipiravir is currently part of clinical trials for treatment of SARS-CoV-2, the exact MoA is still not finally determined. Several previous studies showed that the antiviral activity of T-705-RTP relies on its mutagenic properties.[55–59]

Moreover, two independent studies were published showing that the incorporation of a single or two consecutive T-705 molecules led to chain termination.[60–62] To get a deeper understanding of the MoA, in a collaborative work with the group of B. Canard, SARS-CoV-2 infected Vero cells were cultivated in the presence (500 μM) or absence of T-705.[63] Subsequently, a deep sequencing of viral RNA was performed. It could be observed, that in the presence of T-705 the total mutations are 3-fold higher and the G—A and C—U transition mutations increased by 12 times compared to the absence of T-705. The increase in mutation frequency has also a measurable antiviral effect on SARS-CoV-2. The viral replication is inhibited with an EC_{50} value of 0.2 mM and the virus-induced cytopathic effect is lowered with an EC_{50} value of 0.1 mM. In SARS-CoV-2 infected cells, the fact that T-705 has an antiviral effect showed first that T-705 manages to avoid removal from the viral RNA by nsp-14. The second conclusion that can be made is that the inhibition of the replication is, at least partially, caused by the lethal mutagenesis induced by T-705. Even more effective as the T-705-RTP was the T-1106 triphosphate analog.

To determine the exact amount of T-705 and T-1105 that are incorporated by the SARS-CoV-2 RdRp into the viral RNA, enzyme assays with the isolated RdRp were performed. These experiments should also help to get a deeper understanding of the MoA. First, the primer dependent activity of the nsp12 using annealed primer-templates (PT) and self-priming hairpin (HP) RNA's was determined. The RNA synthesis of nsp12 alone is not efficient, this result is consistent with prior studies.[64,65] For an efficient RNA synthesis, the cofactors nsp7 and 8 are essential. Together with these cofactors the nsp12 forms a four-component complex containing a nsp7/nsp8 heterodimer and a nsp8 monomer. Interestingly, the addition of further nsp8 to this nsp12:nsp7L8 complex, forming the nsp12:nsp7L8:8 complex further increases the efficiency of the RNA synthesis, both on primer-templates (PT) and self-priming hairpin (HP) RNAs.[63] This nsp12 complex was used for elongation reactions in the presence of T-705-RTP and T-1105-RTP with simultaneously omission of ATP and/or GTP. For both, the PT and HP substrate, multiple incorporation events could be observed. However, T-705-RTP proved less efficient as compared to T-1105-RTP. The lower efficiency of T-705-RTP matches with previous results mentioned (see below) and may be explained by the lower chemical stability of T-705-RTP compared to T-1105-RTP.[66] The experiment also clearly showed that both nucleoside analogs are functioning as purine analogs as they are not incorporated in place of uracil and cytosine.

If the HP complex is used both nucleoside analogs are efficiently incorporated opposite of uracil and cytosine and the formation of full-length products can be observed. These results suggest that lethal mutagenesis is the MoA if HP was used as substrate. If on the other hand PT is used as substrate a different MoA was observed depending on whether the nucleoside analogs are incorporated opposite of uracil und cytosine. Both nucleoside analogs are rapidly incorporated opposite of uracil but further extension is slow and inefficient. Multiple consecutive incorporations even lead to complete stop of the extension. Opposite of cytosine however a marked stall in replication before each incorporation of a nucleoside analog occurred, but after incorporation the elongation proceeds rapidly. In these experiments without adding GTP more full-length products were observed. Previous studies showed for the poliovirus RdRp that these pause events before the incorporation of T-705/T-1105, opposite of cytidine, are caused by backtracking of RdRp, which was attributed the primary cause of the antiviral activity.[67] The results of this work suggest that for nsp12 the MoA of T-705/T-1105 is dictated by the structural and functional properties of the polymerase and the RNA. The proof of not fully elongated RNA strains suggests that chain termination may contribute to the antiviral effect as well as chain termination, although it is not the predominant MoA, because chain termination could only be observed after several consecutive incorporations of the nucleoside analogs. Due to the fact that consecutive incorporation is relatively unlikely and based on the speed and frequency of the incorporation of T-705/T-1105 it is more plausible that the antiviral effect of T-705/T-1105 is caused by its lethal mutagenic effect. This is also consistent with the earlier performed infectious virus experiments. It was also shown that the SARS-CoV-2 polymerase is the fastest viral RdRp known. It is determined that the SARS-CoV-2 nsp12 elongates with a frequency of 600–700

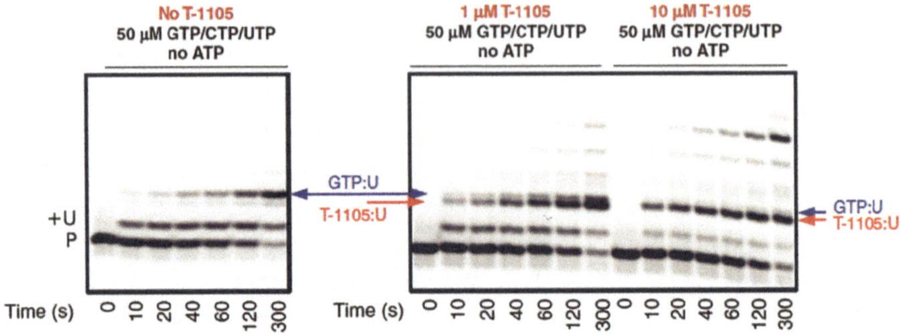

Fig. 2 Elongation reaction with the PT substrate in with each 50 µM of GTP, UTP and CTP in the absence of ATP (left) shows rapid addition of the first uracil followed by slow misincorporation of the GTP:U mismatch. If 1 µM T-1105-RTP it gets incorporated on a similar timescale as the native GTP:U mismatch (right).[63]

nucleotides per second. This is significantly faster compared to picornaviral polymerases (5–20 s^{-1}) or dengue and hepatitis C polymerase (4–18 s^{-1}).[63]

In the no-ATP experiments, GTP:U mismatches were observed, the most common naturally occurring transition mutation, allowing to compare the T-1105 incorporation rate with the appearance of the natural GTP:U transition mutation. If in the elongation reaction 50 µM GTP and 1 µM T-1105-RTP are used the T-1105:U pair occurs 5-fold more often as the GTP:U mismatch. This means that, by taking the concentration into consideration, the T-1105:U pair is by 250-fold more likely to occur then the GTP:U mismatch (Fig. 2). These huge amounts of misincorporations, compared to the natural occurring transition mutations, possibly exceed the proof-reading capacities of nsp14. Even though these results match with the infectious virus experiments, further studies have to examine if nsp14 is capable of excising T-705/T-1105, as it is the case for other nucleoside analogs, or not.[52,68,69]

3. Chemical synthesis of the nucleoside of T-705 and T-1106 as well as their monophosphates

Synthesis of T-705-RMP **3**. The polymerase assays clearly showed the huge antiviral potential of T-705/T-1105 once it gets inside the cell in their phosphorylated nucleoside forms. As described above, one limiting factor of the antiviral activity is the inefficient metabolism of T-705/T-1105 into the corresponding ribonucleoside monophophates by the human HGPRT.[39] A newer study suggests that the metabolism of the RMP into the RDP is a second limiting bottleneck in the enzymatic activation.[39] Additionally, it was reported that the activation heavily depends on the used cell line, which makes the activation by human enzymes even more unreliable.[43] So possibly the antiviral activity could be enhanced by delivering the antiviral active metabolites directly into cells. To achieve this, one aim of our group was to develop reliable synthesis routes to prepare the T-705/T-1105-ribonucleosides which served as starting materials for the synthesis of the phosphorylated metabolites. Moreover, all these compounds then also served for applying several prodrug strategies for intracellular delivery of the corresponding nucleotides (Fig. 1).

T-705-ribonucleoside was synthesized using the Vorbrüggen-coupling method.[66] First, T-705 **1** was silylated by hexamethyldisiloxane (HMDS), the silylated base was then coupled to tetra-*O*-acetyl-β-D-ribofuranose in the presence of tin(*IV*)tetrachloride as Lewis acid. Subsequently the crude product needed to be deacetylated. Unfortunately, it was not possible to use standard deacetylation conditions using either methanolic ammonia, sodium methoxide, or a mixture of triethylamine in water and methanol, because all these conditions led to decomposition of the just formed nucleoside. This decomposition was probably due to the nucleophilicity of the reagents which are typically used in the cleavage of esters. Then, less nucleophilic reagents were used to synthesize T-705-ribonucleoside **2**. Finally, a transesterification reaction using 0.3 equivalents of dibutyltin oxide in methanol gave the desired nucleoside **2** in a yield of 40%.

Following to synthesis of the T-705-ribonucleoside the next aim was to synthesize T-705-ribonucleoside-5′-monophosphate. One of the most common method for the synthesis of 5′-monophosphates is the method of Sowa and Ouchi.[70] Using this method, a tetrachloropyrophosphate pyridinium chloride adduct was generated by adding water and pyridine to phosphorus oxychloride. The tetrachloropyrophosphate pyridinium chloride adduct then reacted with ribonucleosides selectively at the 5′-position. After hydrolysis the 5′-monophosphate was formed together with hydrochloric acid.[70] Usually this aqueous solution was then adjusted to pH 7–8 using solid ammonium carbonate to yield the monophosphate as ammonium salt. However, using these conditions led to decomposition of the T-705-RMP **3**. After modification of the reaction conditions using (i) short reaction times and hydrolysis times and (ii) careful neutralization using aqueous ammonium bicarbonate solution, it was possible to obtain the T-705-RMP **3** in a yield of 50% (Scheme 1).[66] Further studies on the chemical stability showed not only a significant stability of the glycosidic bond but also a chemical instability of the fluorinated heterocycle itself.[66]

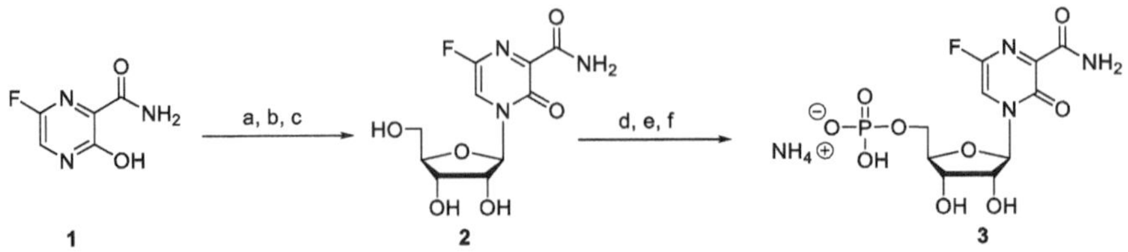

Scheme 1 Synthesis of T-705-ribonucleoside-5′-monophosphate **3**: (A) HMDS, $(NH_4)_2SO_4$, 60 min, 140°C; (B) tetra-*O*-acetyl-β-D-ribofuranose, $SnCl_4$, CH_3CN, 7 h, rt.; (C) Bu_2SnO, MeOH, 24 h, 80°C, overall yield: 40%. (D) phosphorous oxychloride, pyridine, H_2O, 30 min, 0°C. (E) addition of **2**, 20 min, 0°C. (F) ice H_2O, 30 min, aqueous saturated NH_4HCO_3 (pH 7.0), overall yield: 50%.

Synthesis of T-1106-MP **9**. Stability studies showed that T-1106 **5** proved chemically completely stable in contrast to T-705-ribonucleoside **2**.[66] Additionally, Balzarini *et al.* reported that the 3-hydroxyl function of T-705 is critical for their substrate recognition by HGPRT but that the 5-fluoro atom of T-705 is not required for metabolic activation and antiviral activity.[39] Due to these facts the synthetic focus of our group shifted away from T-705 towards T-1105.

T-1106 **5** was synthesized using the Vorbrüggen-coupling method using standard conditions; also the subsequent deacetylation under standard conditions worked without any problems and yielded the desired nucleoside **5** in a yield of 55% (Scheme 2).[66]

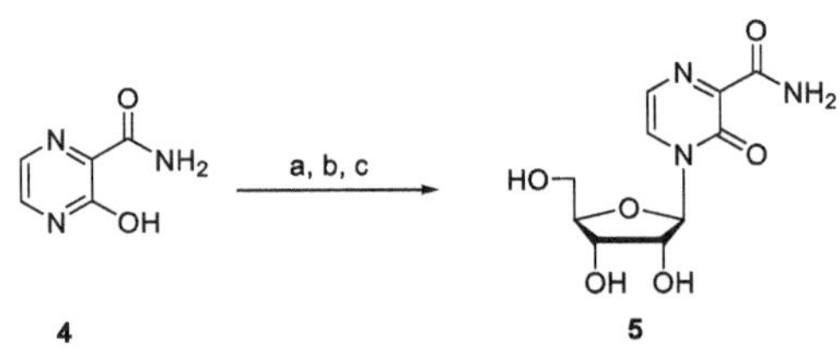

Scheme 2 Synthesis of T-1106 **5**: (A) tetra-*O*-acetyl-β-D-ribofuranose, *N,O*-bis (trimethylsilyl)acetamid, CH_3CN, 30 min, rt.; (B) trimethylsilyl trifluoromethansulfonate, CH_3CN, 44 h, rt.; (C) MeOH, H_2O, Et_3N, 6 h, rt., overall yield: 55%.

The one step phosphorylation method of Sowa and Ouchi[70] to convert T-1106 **5** into the 5′-monophosphate was not successful. Thus, an alternative route was developed to synthesize the monophosphate using phosphoramidite chemistry. First, the 5′-OH group of **5** was dimethoxytrityl (DMTr)-protected followed by the acetylation of the 2′- and 3′-OH groups. Subsequently, the DMTr group was cleaved under acidic conditions to yield compound **7** (65%). The monophosphate **9** was then synthesized using bis(9H-fluoren-9-ylmethyl)-diisopropylamino-phosphite. After basic cleavage of the *O*-acetyl

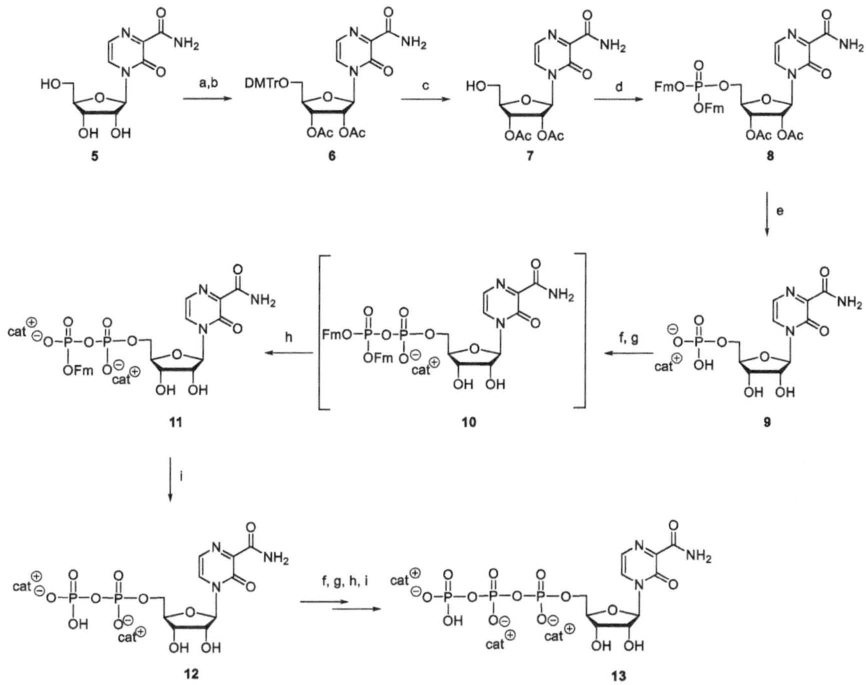

Scheme 3 Synthesis of T-1106-DP **12** and -TP **13**: (A) DMTrCl, C_5H_5N, CH_3CN, 2 h, rt.; (B) TEA, Ac_2O, C_5H_5N, CH_3CN, 16 h, rt., 75% over two steps; (C) CH_3OH, *p*-toluenesulfonic acid, 10 min, rt., 84%; (D) 1. bis(9H-fluoren-9-ylmethyl)-diisopropylaminophosphite (in CH_2Cl_2), DCI, CH_3CN, 15 min, rt.; 2. TBHP, CH_3CN, 10 min, rt.; (E) TEA, CH_3CN, 72 h, rt.; then CH_3OH/H_2O/TEA 7:3:1, 24 h, rt., 72% over steps (D–E); (F) bis(9H-fluoren-9-ylmethyl)-diisopropylamino-phosphite (in CH_2Cl_2), DCI, DMF, 10–20 min, rt.; (G) TBHP, DMF, 5 min, rt.; (H) TEA, CH_3CN, 10–20 min, rt.; (I) TEA, H_2O, CH_3CN, 24–48 h, rt.

and fluorenylmethyl groups, T-1106-MP **9** was synthesized in an overall yield of 47% (Scheme 3).[46]

Synthesis of the T-1106-DP **12** *and -TP* **13**. The T-1106-DP **12** and -TP **13** were synthesized using an iterative strategy starting from the T-1106-MP **9** (Scheme 3).[71] It was not possible to perform a purification on the double Fm-protected diphosphate **10** because it rapidly hydrolyzed during purification. A selective cleavage of one single Fm-group was achieved by the short (5–20 min) treatment of **10** with triethylamine. The isolated intermediate **11** was stable towards hydrolysis and it was possible to separate the monophosphate using RP-chromatography. After an additional deprotection step T-1106-DP **12** was obtained in high purity (>98%) in an overall yield of 53%. The triphosphate **13** was then obtained by repeating the same steps, starting from the diphosphate **12** (purity of >96%; overall yield 25%.

4. *Cyclo*saligenyl (*cyclo*Sal) nucleoside monophosphate prodrug approach

Often the first metabolic phosphorylation of nucleoside analogs into their nucleoside monophosphates (NMPs) has been identified to be the limiting step in the intracellular anabolic phosphorylation cascade. Here, the use of lipophilic NMP pronucleotides may serve as intracellular delivery system for NMPs. In an attempt to obtain a neutral, lipophilic phosphate ester various masking units have been introduced with the aim to neutralize the negative charges at the phosphate or phosphonate metabolites by compensating the hydrophobicity to ensure a cellular uptake.

In our laboratories, one of the most extensively explored type of NMP prodrug approach was developed (*cyclo*Sal-technology).[72] The *cyclo*Sal technology is based on a pH-driven chemical hydrolysis and the phosphate group of the NMP are masked by salicyl alcohols **14** (Scheme 4).

The synthesis of the *cyclo*Sal-pronucleotides has been achieved either by using P(III) or P(V) chemistry (Scheme 4). First salicyl alcohols **14** were synthesized from salicyl aldehydes or salicylic acids.[73] Cyclic saligenylchlorophosphanes **15** were prepared from diols **14** and PCl_3 in yields between 50% and 85% (Method A). Next, compounds **15** were reacted with the nucleoside analog d4T in the presence of diisopropylethylamine (DIPEA)

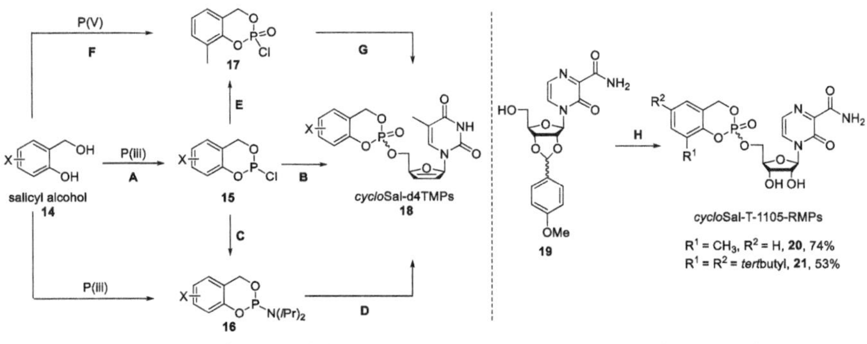

Scheme 4 Synthesis of *cyclo*Sal-d4TMP triesters **18** and T-1106-*cyclo*Sal-prodrugs **20,21**. Method A: PCl_3, pyridine, Et_2O, −10°C, 2 h, 50–85%; Method B: (i) d4T, DIPEA, CH_3CN, −20°C to rt., 1 h; (ii) TBHP, CH_3CN, −20°C to rt., 1 h, 50–73%; Method C: diisopropylamine, Et_2O, 0°C, 30 min, 50–85%; Method D: (i) d4T, pyridinium chloride, tetrazole or imidazolium triflate, CH_3CN, 0°C, 30 min; (ii) TBHP, CH_3CN, rt., 1 h, 70–95%; Method E: O_2, toluene, rt., 16 h, quant; Method F: $POCl_3$, TEA, CH_2Cl_2, −60°C to rt., 61%; Method G: d4T, pyridine, −50°C, 4 h, 83–95%; Method H: (i) *cyclo*Sal-phosphochloridite, TEA, CH_3CN, DMF, 15–20 min, rt.; (ii) TBHP, CH_3CN, DMF, 1–2 h, rt.; (iii) TFA, CH_2Cl_2, TFA, 1 h, rt. *Cyclo*Sal-T-1105-RMP prodrug **21** was obtained in 87% purity.

to generate cyclic phosphite triesters, followed by addition of *tert*butyl hydroperoxide (TBHP) to yield *cyclo*Sal-pronucleotides (exemplified shown here for *cyclo*Sal-d4TMPs **18**). *Cyclo*Sal-pronucleotides such as **18** were obtained as mixtures of diastereomers in reliably yields of 50–73% (Method B). Alternatively, compounds **15** were reacted with diisopropylamine to obtain phosphoramidites **16** (Method C). Then, phosphoramidites **16** were coupled to d4T in the presence of a weak acid and subsequent oxidation with TBHP (Method D). In some cases, the yields were markedly higher (up to 95%) than those obtained according to method B. Using P(V) chemistry, diols **14** were first reacted with $POCl_3$ to form 3-methyl-*cyclo*saligenylphosphorochloridate **17** (Method F).[74] The yields obtained by this method to give the *cyclo*Sal-d4TMPs **18** varied between 83% and 95%. In 2018, this technology has been also applied to T-1105 and T-705 to bypass the rate limiting formation of the phosphorylated nucleobase analog by HGRPT. *Cyclo*Sal-T-1105-RMP prodrugs **20,21** were prepared to deliver the nucleoside monophosphate.[46]

In contrast to other NMP prodrug strategies based on enzymatic activation,[7576–84] *cyclo*Sal pronucleotides enable the intracellular delivery of the target nucleoside monophosphate by a non-enzymatically induced chemical reaction. The *cyclo*Sal-prodrugs were sensitive to basic conditions (Scheme 5).[85] The phenyl phosphate ester bond was preferably cleaved (step a) to form the benzyl phosphate diester **22**, followed by spontaneous cleavage of the benzyl ester bond (step b) releasing the NMP and salicylalcohol **14** (step b). In contrast, the alternative cleavage of the benzyl ester bond in *cyclo*Sal-d4TMP-prodrug **23** is unfavorable (step c). Furthermore, the initial hydrolysis step of these *cyclo*Sal-pronucleotides can be influenced by substituents in the 4-position of the phenol phosphate esters: electron withdrawing groups, such as nitro- or chloro-substituents, result in short half-lives. In the past, a variety of *cyclo*Sal-prodrugs with different nucleoside analogs (d4T, ddA, ddI, BVdU, and ACV) have been reported.[86–90] As expected, T-1106-*cyclo*Sal-prodrugs **20,21** showed a highly selective delivery of T-1105-RMP in phosphate buffer (PBS, pH 7.3) (Fig. 3).[46]

*Cyclo*Sal-phosphate triesters were evaluated for their antiviral activity in HIV-1- and HIV-2-infected wild-type CEM/0 cells and in HIV-2-infected mutant thymidine kinase-deficient CEM/TK^- cells and were compared to the parent nucleoside analogs. Most of the *cyclo*Sal-d4TMP triesters proved at least as active as d4T against HIV-1 and HIV-2 in wild-type CEM/0 cells. Moreover, some *cyclo*Sal phosphate triesters were markedly more active against HIV-2 in CEM/TK^- cells as compared to the parent d4T.[91,92] The antiviral activity of different nucleoside *cyclo*Sal-phosphate triesters

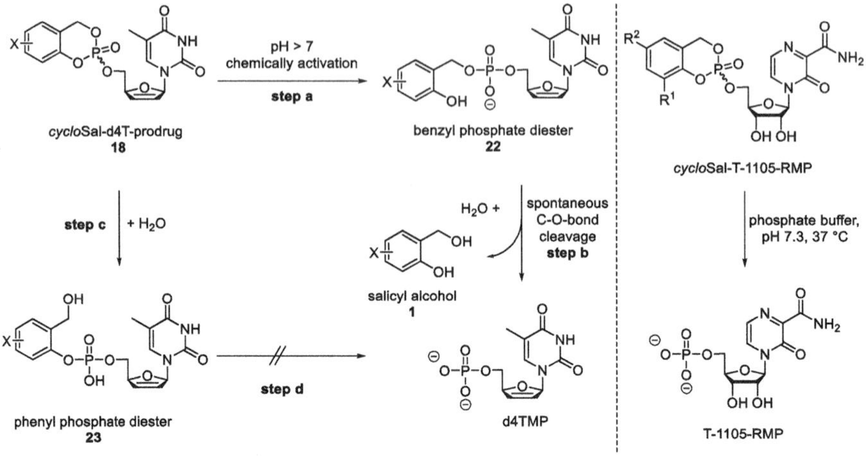

Scheme 5 Hydrolysis pathways of the *cyclo*Sal-d4TMP triesters **18** (left) [85] and T-1106-*cyclo*Sal-prodrugs **20,21** (right).[46]

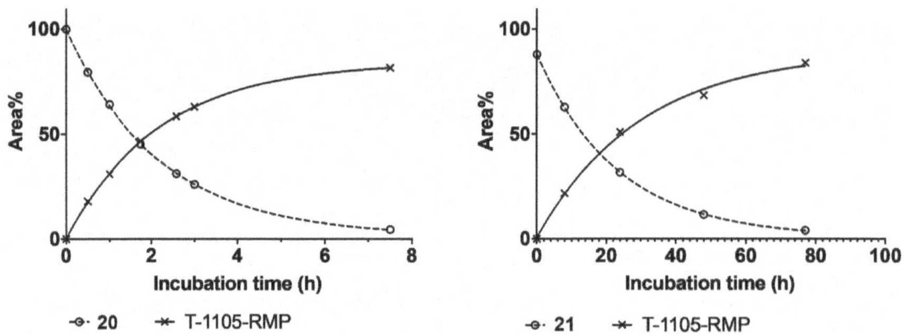

Fig. 3 Chemical hydrolysis of *cyclo*Sal-pronucleotides **20,21** in pH 7.3 phosphate buffer at 37°C.[46]

improved by the introduction of the *cyclo*Sal masks, e.g., *cyclo*Sal-ddAMP (EC_{50} = 0.025 μM) was 100-fold more active as the parent ddA (EC_{50} = 4.3 μM)[93,94] and *cyclo*Sal d4AMP was 600-fold (EC_{50} = 0.05 μM) more potent than d4A (EC_{50} = 30 μM).[88] These results proved (i) the uptake of the *cyclo*Sal-triesters into cells, (ii) the successful intracellular delivery of NMPs, and (iii) that the *cyclo*Sal-concept enables the bypass the first phosphorylation step usually needed for the initial activation of nucleoside analogs. Unexpectedly, the corresponding T-1106-*cyclo*Sal-prodrug **20** showed very poor anti-influenza virus activity (an average EC_{50} = 248 μM in MDCK cells and an average EC_{50} = 232 μM in MDCK-TG^{res} cells, Table 1), probably due to their low chemical stability ($t_{1/2}$ = 1.2 h) and low lipophilicity.

Table 1 Anti-influenza virus activity and cytotoxicity in MDCK and MDCK-TG^{res} cells.[a]

Comp.[b]	MDCK cells						MDCK-TG^{res} cells					
	Cytotoxicity[c]		Antiviral activity (EC_{50} in μM)[b]				Cytotoxicity[c]		Antiviral activity (EC_{50} in μM)[b]			
	MCC	CC_{50}	A/X-31		B/Ned/537/05		MCC	CC_{50}	A/X-31		B/Ned/537/05	
			CPE	MTS	CPE	MTS			CPE	MTS	CPE	MTS
20	>400	>400	280 ± 120	>400	176 ± 42	135 ± 77	>400	>400	256 ± 39	368	200 ± 20	106 ± 47
35	21 ± 5	35 ± 8	5.4 ± 1.0	10 ± 2	7.2 ± 0.5	3.5 ± 0.6	16 ± 0	22 ± 6	6.6 ± 2.5	8.8 ± 2.4	3.4 ± 1.3	2.8 ± 1.2
36a	37 ± 3	>100	1.3 ± 0.2	1.0 ± 0.2	0.81 ± 0.05	0.56 ± 0.08	22 ± 4	44 ± 8	1.3 ± 0.2	0.97 ± 0.13	0.55 ± 0.07	0.39 ± 0.05
36b	21 ± 5	31 ± 0	1.5 ± 0.3	1.6 ± 0.3	0.67 ± 0.12	0.58 ± 0.07	16 ± 0	17 ± 1	1.5 ± 0.3	1.4 ± 0.1	0.66 ± 0.04	0.53 ± 0.17
52a	16 ± 0	30 ± 1	2.6 ± 0.0	3.6 ± 0.5	2.6 ± 0.6	2.4 ± 0.8	13 ± 3	10 ± 0	2.6 ± 0.0	4.5 ± 2.0	1.9 ± 0.3	1.0 ± 0.7
52b	16 ± 0	38 ± 5	6.0 ± 0.4	6.7 ± 2.0	9.9 ± 0.6	6.0 ± 1.9	16 ± 0	10 ± 1	5.6 ± 0.8	5.4 ± 1.3	2.6 ± 0.0	1.5
T-1105	>250	>250	6.4 ± 0.9	5.4 ± 0.6	2.8 ± 0.4	2.1 ± 0.2	>250	>250	>250	>250	>250	>250
T-705	>250	>250	24 ± 1	26 ± 2	9.0 ± 0.8	8.3 ± 1.0	>250	>250	>250	>250	>250	>250
T-1106	>250	>250	58 ± 3	52 ± 3	24 ± 2	19 ± 2	>250	>250	53 ± 4	51 ± 3	20 ± 3	19 ± 2

[a]Data are the means ± SEM of 2–5 independent tests. MDCK: Madin-Darby canine kidney cells; MDCK-TG^{res}: 6-thioguanine-resistant MDCK cell line lacking functional HGPRT.
[b]Antiviral activity is expressed as the EC_{50}, defined as the compound concentration producing 50% inhibition of virus replication, as estimated by microscopic scoring of the CPE or by measuring cell viability in the formazan-based MTS assay.
[c]Cytotoxicity is expressed as the MCC, the compound concentration producing minimal changes in cell morphology, as estimated by microscopy; or CC_{50}, 50% cytotoxic concentration estimated by MTS cell viability assay.

Although the half-lives for T-1106-*cyclo*Sal-prodrug **21** ($t_{1/2}$ = 16 h) was found to be significantly higher by almost a factor of 13, again no anti-influenza virus activity was detected.[46] Because the *cyclo*Sal-pronucleotide system gave nice improvements in antiviral activity and the cleavage proceeded selectively to yield the monophosphates, the failure of by-passing the enzyme responsible for the ribophosphorylation of T-1105 may also point to a further bottleneck in the activation leading to the triphosphate: the conversion of the monophosphate analog into its diphosphate derivative. This might be an explanation, why the delivery of T-1105-RMP was not sufficient to improve the antiviral activity.

5. Nucleoside diphosphate prodrugs

Despite these obvious advantages, not in all cases the *cyclo*Sal-technology was successful. As just mentioned above, T-1105 *cyclo*Sal-pronucleotides **20**,**21** showed no or very poor antiviral activity in MDCK and MDCK-TG^{res} cells.

The same happened when 3′-deoxy-3′-azidothymidine (AZT) was used.[95–97] Here, the *cyclo*Sal-approach failed to improve the antiviral activity in thymidine kinase-deficient cells.[97] One reason might be a rapid dephosphorylation of AZTMP by 3′,5′-(deoxy)nucleotidase present in the cells.[98,99] More importantly, in the metabolism of AZT, not the formation of the monophosphate derivative (AZTMP) is critical but the formation of the corresponding nucleoside diphosphate AZTDP by the host cell enzyme thymidylate kinase (TMPK).[100,101] Consequently, it was expected that the development of nucleoside diphosphate prodrug systems would help to overcome these problems.

Earlier, very few approaches have been developed for nucleoside diphosphate prodrugs, probably because of the inherent instability of the phosphate anhydride bond. In 1990, Hostetler *et al.* designed several potential NDP-diglyceride prodrugs.[102–105] This approach was successfully used for different nucleosides and improved antiviral activities compared to the parent nucleosides, e.g., for 3′-deoxythymidine (3dT, up to 37-fold). Later, Huynh-Dinh *et al.* developed the acylphosphate concept based on a selective cleavage of these mixed anhydrides releasing NDPs in aqueous buffer.[106–108] However, in contrast to the NDP-diglyceride approach, no improvement of the antiviral activity was detected as compared to the parent nucleoside.

In order to improve the stability of the masked pyrophosphate unit, we decided to design a pronucleotide approach based on masking the negatively charged β-phosphate group only, namely, the Di*PP*ro-approach.

5.1 Nucleoside diphosphate prodrugs based on *cyclo*Sal-approach

Initially the previously developed *cyclo*Sal-technology was applied to AZTDP, (*cyclo*Sal-AZTDP prodrugs **24**; Scheme 6, upper section). The nucleoside monophosphate was prepared from AZT by the Sowa–Ouchi method.[70] After the coupling reaction of *cyclo*Sal-phosphorochloridate **17** (P(V)-reagent) and AZTMP, *cyclo*Sal-AZTDP prodrugs **24** were prepared using phosphorochloride chemistry. It was expected that such a construct would allow a rapid conversion of the Di*PP*ro-AZT prodrugs **24** into AZTDP. The *cyclo*Sal-AZTDPs **24** were incubated in PBS (pH 7.3) to study their chemical stability and to identify the formed hydrolysis products. To our surprise, only a very small amount of AZTDP (1%, Scheme 6, pathway a) but a predominate formation of AZTMP (99%, Scheme 6, pathway b) was detected in these studies. The undesired pyrophosphate cleavage leading to AZTMP and *cyclo*Sal-phosphate were caused by nucleophiles attacking the β-phosphate group. Consequently, the *cyclo*Sal-principle is unsuitable to bypass the second intracellular phosphorylation step.

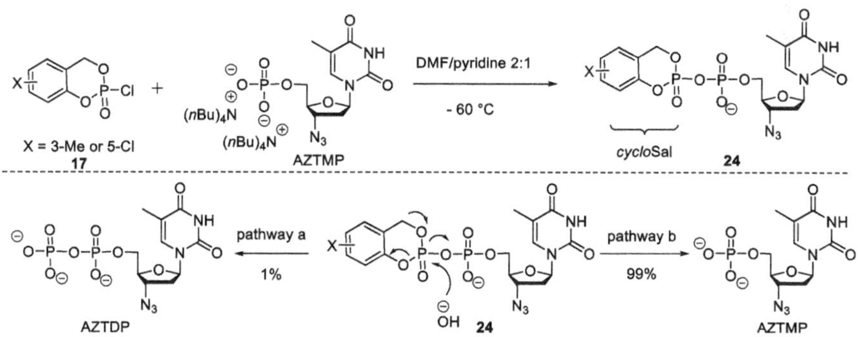

Scheme 6 Synthesis and hydrolysis of *cyclo*Sal-AZTDPs **24**.

5.2 The symmetric Di*PP*ro-nucleoside diphosphate prodrugs

Guided by the results from *cyclo*Sal-AZTDP prodrugs **24**, we aimed to design a new approach based on an enzymatically triggered pronucleotide delivery system[109] that has two masking groups at the β-phosphate. In 2008, we reported on the synthesis of the first examples of the bioreversible protection of nucleoside diphosphates using the phosphoramidite chemistry.[110] The cleavage of the phenyl ester moiety within the masking units was induced by chemical or enzymatic means to form NDP in CEM cell extracts. Later, we developed a series of Di*PP*ro-d4TDPs **28** bearing different lipophilic acyl chain lengths to study their hydrolysis properties in chemical or biological media and described structure–activity relationships in the antiviral evalution.[111–115] The synthesis of Di*PP*ro-d4TDPs **28** is summarized in Scheme 7. The method was applied to different nucleoside analogs (e.g., AZT, d4U, *carba*-isoddA) and the overall yields varied between 10% and 72%.[110–113,115]

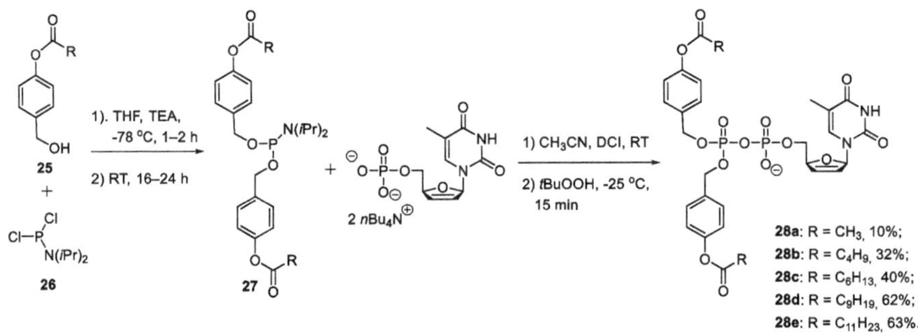

Scheme 7 Synthesis of the Di*PP*ro-d4TDPs **28**.

For the synthesis, the required nucleoside monophosphates were prepared from the nucleoside analogs as described before.[70] In contrast to the synthesis of *cyclo*Sal-nucleoside diphosphate prodrugs **24**, first bis(tetra-*n*-butylammonium)nucleoside monophosphates were coupled with a reactive P(III)-reagent comprising the lipophilic bis(acyloxybenzyl) groups to form P(III)–P(V) intermediates. Subsequently, the intermediates were oxidized with *tert*-butylhydroperoxide (TBHP) to form the corresponding diphosphates. As shown in Scheme 7, bis(4-acyloxylbenzyl)phosphoramidites **27** were prepared from benzyl alcohols **25** and (*N*,*N*-diisopropyl)dichlorophosphine **26** at low temperatures. Purified Di*PP*ro-d4TDPs **28** were obtained as brown syrups by reversed-phase (rp) column chromatography. It should be noted that the reaction could be easily followed by RP-HPLC: i) NMP was completely consumed, ii) the P(III)–P(V) intermediates were formed within few minutes and then converted to give the products.[111]

Stability studies. The Di*PP*ro-d4TDPs **28** were investigated in PBS (pH 7.3) and CEM cell extracts to study their chemical and biological stability, respectively, and more importantly to study the delivery of the corresponding d4TDPs. Two hydrolysis pathways of Di*PP*ro-d4TDPs **28** were proposed (Scheme 8).

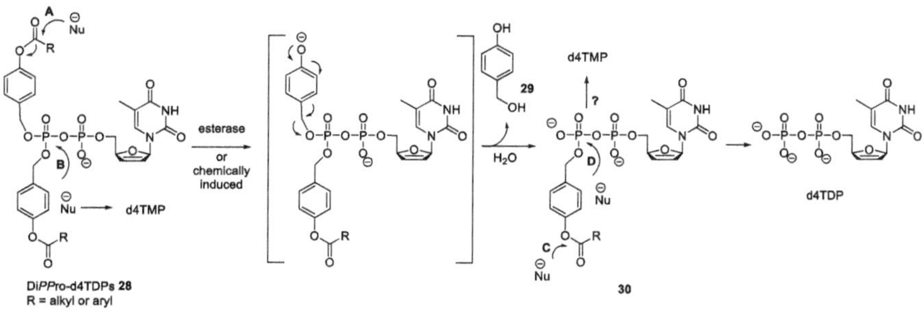

Scheme 8 Hydrolysis mechanism of Di*PP*ro-d4TDPs **28**.

In the case of Di*PP*ro-d4TDPs **28** in PBS, the chemical stability increased with the increase of the alkyl chain lengths attached to the 4-acceptor-substituted benzyl esters. The half-lives of Di*PP*ro-d4TDPs **28** were found to be between 10 h and 100 h, indicating that Di*PP*ro-d4TDPs **28** seem to be quite stable under chemical hydrolysis condition. In PBS, the starting compounds **28** disappeared and the expected mono-masked intermediates **30** were detected (Scheme 8; pathway **A**). In these studies, the ester (acceptor unit) group is converted into a strong donor hydroxy group, thus a spontaneous cleavage of the benzyl-C-O-bond occurred subsequently, leading to the formation of intermediates **30**. The half-lives of intermediates **30** (280–1400 h) were found to be much higher than for starting compounds **28**, potentially caused by repulsive interaction between the two negative charges of the intermediates and the incoming nucleophiles. Additionally, when Di*PP*ro-d4TDPs **28** were almost consumed, an increase of the d4TDP concentration and a small amount of d4TMP were observed in these studies indicating that the hydrolysis of compounds **30** mainly followed pathway **C**. Furthermore, d4TMP concentrations increased at the beginning of the hydrolytic process, probably due to the anhydride bond breakage between the α-phosphate and β-phosphate (Scheme 8; pathway **B**).

Di*PP*ro-d4TDPs **28** were investigated in human $CD4^+$ T-lymphocyte CEM cell extracts *to identify the hydrolysis products. The half-lives of* Di*PP*ro-d4TDPs **28** were in the range of 0.05–21 h and found to be markedly lower than the half-lives of prodrugs measured in PBS (up to 200-fold), which indicates an enzymatic hydrolysis reaction. In the case of Di*PP*ro-d4TDPs **28**, the cleavage and the formation of intermediates **30** were observed. Former studies have shown that intermediates **30** were readily cleaved and hydrolyzed in CEM cell extracts, thus a predominate formation of d4TDP was observed and no further increase of compounds **30** concentrations were detected. It was impossible to quantify the concentrations of d4TDP released in CEM cell extracts because this compound was further dephosphorylated by enzymes, such as phosphatases, to form d4TMP.

Di*PP*ro-d4TDPs **28** were investigated for their antiviral activity in HIV-1 and HIV-2-infected wild-type CEM/0 and in HIV-2-infected mutant thymidine-kinase-deficient cells (CEM/TK^-). The parent d4T was used as reference compound. D4T showed no antiviral activity in thymidine kinase (TK)-deficient cells, because the first phosphorylation step to yield the monophosphate metabolite catalyzed by TK is the metabolism-limiting step. Most of the Di*PP*ro-d4TDPs **28** showed similar activities against HIV-1 and HIV-2 than the parent d4T in wild-type (CEM/0) cell cultures. In contrast, all Di*PP*ro-d4TDP prodrugs **28** were highly active in CEM/TK^- cell cultures. The antiviral activity determined for prodrugs **28a–d** (AB:C1-C9) increased with increasing alkyl chain lengths due to their advantageous permeability. Di*PP*ro-d4TDP **28d** (AB-C9) was the most active compound of all the reported prodrugs: The inhibition of HIV-1 and HIV-2 replication by Di*PP*ro-d4TDP **28d** ($EC_{50}=0.08$ μM/HIV-1; $EC_{50}=0.32$ μM/HIV-2) improved by 11-fold and 7-fold, respectively as compared to nucleoside d4T in wild-type (CEM/0) cells and the activity was improved 1570-fold in CEM/TK^- cell cultures. The antiviral activity observed in the wild-type (CEM/0) cell cultures **28d** ($EC_{50}=$ 0.32 μM/HIV-2) was completely retained in CEM/TK^- cell cultures ($EC_{50}=0.11$ μM/HIV-2). It was concluded that these Di*PP*ro-prodrugs **28** were successfully taken-up across the cell membranes and delivered phosphorylated metabolites, most probably d4TDPs.

5.3 Non-symmetric Di*PP*ro-nucleoside diphosphate prodrugs

Taking all previous results into account, we next developed a second generation of Di*PP*ro-NDP compounds as shown in compounds **34** (Scheme 9). The second generation compounds comprised two *different* acyloxybenzyl moieties linked to the β-phosphate of the nucleoside diphosphate. One of the bioreversible groups is bearing a short alkyl chain carboxylic acid ester (low lipophilicity) while the second group comprises a long alkyl residue carboxylic acid ester or a substituted benzoic acid ester (high lipophilicity). It was expected that such a design would allow a rapid conversion of the Di*PP*ro-compounds **34** into the mono-masked intermediates **30** and therefore avoid the unwanted phosphoanhydride bond hydrolysis in the prodrugs leading to NMP. The second masking unit would be subsequently cleaved slowly to form the NDP. We expected that such a design would achieve a highly selective conversion of the Di*PP*ro-prodrugs into nucleoside diphosphates.

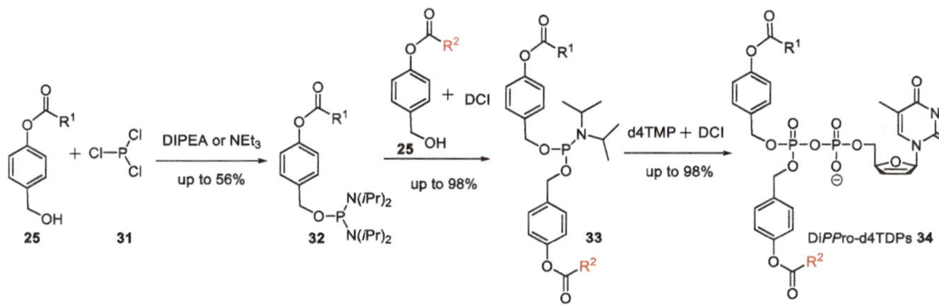

Scheme 9 Synthesis of non-symmetric Di*PP*ro-nucleoside diphosphates **34**.

Di*PP*ro-d4TDPs **34** were made according to the phosphoramidite route as summarized above.[110,112,115] First, bis(di*iso*propylamino)phosphoramidites **32** were prepared from 4-acyloxybenzyl alcohols **25** and phosphorus trichloride **31** in the presence of *N,N*-di*iso*propylamine (DIPA). Then, compounds **32** were reacted with the second 4-acyloxybenzyl alcohols **25** to form non-symmetric phosphoramidites **33**. Finally, phosphoramidites **33** were coupled with d4TMP in an acid-activated reaction followed by oxidation (Scheme 9).

Di*PP*ro-d4TDPs **34** were studied with regard to their stabilities in different media and their hydrolysis products were analyzed. Interestingly, in all cases, both intermediates **30a** (short acyl residues) and intermediates **30b** (long acyl residues) were formed, indicating that both masking groups of Di*PP*ro-d4TDPs **34** were involved in the hydrolysis (pathways **A** and **B**; Scheme 10). From analyzing the amounts of the formed products (intermediates **30**, d4TDP, and d4TMP) the following conclusions can be taken: (i) Di*PP*ro-d4TDPs **34** were found to be more readily cleaved to form intermediates **30b**, (ii) the intermediates **30b** bearing long alkanoyl ester moieties proved to be more stable than intermediates **30a** bearing short alkyl moieties, and (iii) an increase of d4TDP concentration and a very small amount of

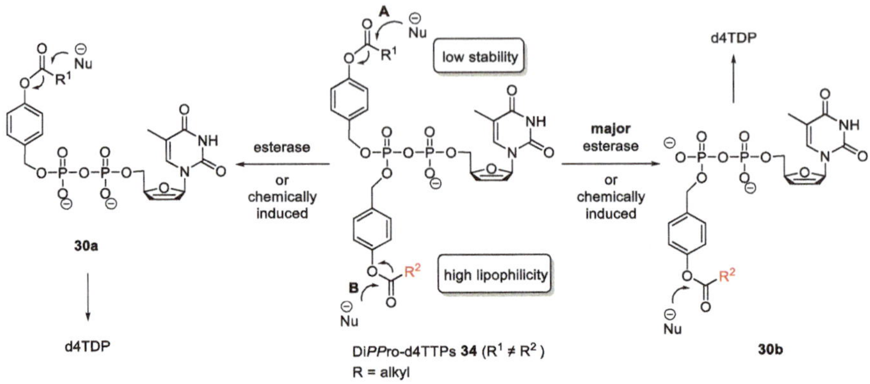

Scheme 10 Hydrolysis pathways of non-symmetric Di*PP*ro-d4TDPs **34**.

d4TMP were detected. Furthermore, the chemical stability of Di*PP*ro-d4TDPs **34** was found to be higher as compared to those of the symmetric Di*PP*ro-d4TDPs **28**. For example, the half-lives for Di*PP*ro-d4TDPs **34** (acyloxybenzyl (AB)-C1;AB-C9) ($t_{1/2}=40$ h)[113] improved by 4-fold compared to **34a** (AB-C1;AB-C1) ($t_{1/2}=10$ h).[110]

We evaluated the stability of Di*PP*ro-d4TDPs **34** in CEM cell extracts as well. As expected, the stability of Di*PP*ro-d4TDPs **34** in cell extracts ($t_{1/2}=0.03$–1.91 h) was significantly lower than the stability in PBS (up to 1450-fold). The masking groups in Di*PP*ro-d4TDPs **34** were selectively cleaved to form intermediates **30a** and intermediates **30b**, respectively. In the case of Di*PP*ro-d4TDPs **34** (AB:C4;AB:R^2), the formation of the C_4H_9-intermediate and the long alkyl chain acyl group (R^2) intermediates were detected. However, for Di*PP*ro-d4TDPs **34** (AB-C1;AB-R^2), a highly selective cleavage of the short biodegradable moiety (AB:C1) led to the formation of intermediates **30b** (AB:R^2) in cell extracts. Di*PP*ro-d4TDPs **34** were also rapidly hydrolyzed by PLE and delivered d4TDP within a few minutes (0.14–0.19 min). However, for Di*PP*ro-d4TDP **34** (AB-C1;AB-C9), both alkanoyl esters were cleaved at almost identical rates. In contrast, a large amount of C_4H_9-intermediate was observed in the hydrolysis of Di*PP*ro-d4TDP **34** (AB-C4;AB-C9) with PLE.

In antiviral assays, the inhibition of the replication of HIV-1 and HIV-2 by Di*PP*ro-d4TDPs **34** was similar, or slightly better, compared to their parent nucleoside d4T in wild-type CEM/0 cells. In some cases, the antiviral activity for Di*PP*ro-d4TDPs **34** increased with the higher lipophilicity associated with the introduced masking units of the Di*PP*ro-prodrugs. The high antiviral activity of Di*PP*ro-d4TDP **34** (AB-C4;AB-C11)

(EC_{50} = 0.13 μM) was retained in thymidine-deficient CEM cells (TK^-) and compared to the parent nucleoside d4T (EC_{50} = 150 μM) the anti-HIV activity was improved by 1153-fold.

5.4 Nucleoside diphosphate prodrugs: T-1106-Di*PP*ro-prodrugs

We applied the bis(acyloxybenzyl)-masking approach to nucleoside analogs T-705 and T-1106.[46] Starting from T-705-RMP and T-1105-RMP, T-705-Di*PP*ro-prodrug **35** ($R^1 \neq R^2$) and T-1106-Di*PP*ro-prodrugs **36** ($R^1 = R^2$ and $R^1 \neq R^2$) were successfully prepared using the previously reported phosphoramidite protocol, respectively.[110,113] The cleavage of Di*PP*ro-prodrugs **35,36** was further investigated in PBS (pH 7.3), with PLE (PBS, pH 7.3), and in MDCK cell extracts.

In PBS (pH 7.3), T-705-Di*PP*ro-compound **35** (AB-C4;AB-C14) was fully decomposed after 30 h. However, the expected intermediates were *not* observed. We assumed that the nucleophilic displacement of the fluorine atom by water took place which led finally to the destruction of the nucleobase analog. Furthermore, T-705-Di*PP*ro-compound **35** (AB-C4; AB-C14) was rapidly hydrolyzed with PLE and delivered the expected T-705-intermediates (with R either C_4H_9 or $C_{14}H_{29}$) much faster compared to the chemical hydrolysis. Compared to its T-705 counterpart, T-1106-Di*PP*ro-prodrugs **36** (for example **36a**, $t_{1/2}$ = 30 h) showed much higher chemical stability. The chemical hydrolysis of T-1106-Di*PP*ro-prodrug **36a** (AB-C9;AB-C9) released mono-masked T-1106-intermediate **37** (AB-C9), T-1105-RDP, and T-1105-RMP (Fig. 4A, condition *without* esterase). With PLE, almost no formation of T-1105-RMP and the unknown by-product was observed. The starting material disappeared and the mono-masked T-1106-intermediate **37** (AB-C9) was formed. Subsequently, T-1106-intermediate **37** (AB-C9) was fully converted into the final metabolite, T-1105-RDP (Fig. 4A, condition *with* esterase). Furthermore, T-1106-Di*PP*ro-prodrugs **36a,b** were efficiently converted into the corresponding mono-masked intermediates **37** in extracts from MDCK and MDCK TG^{res} cells. Subsequently, T-1105-RDP was also detected as intermediate metabolite that was further dephosphorylated to T-1105-RMP (Fig. 5). T-1106-Di*PP*ro-prodrugs **36a,b** activation and release of T-1105-RDP was found to proceed faster for symmetrically masked T-1106-Di*PP*ro-compound **36a** (AB-C9;AB-C9) than for the non-symmetrically masked T-1106-Di*PP*ro-compound **36b** (AB-C4; AB-C14).

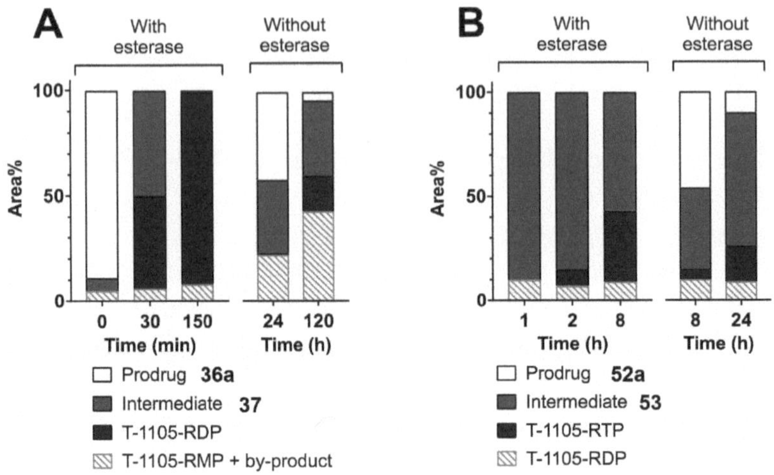

Fig. 4 Metabolic profiles of T-1106-Di*PP*ro-prodrug **36a** (A) and T-1106-Tri*PPP*ro-compound **52a** (B). T-1106-prodrugs **36a** (**A**) and **52a** (**B**) were incubated at 450 μM in the absence or presence of PLE (3 U per mL).[46]

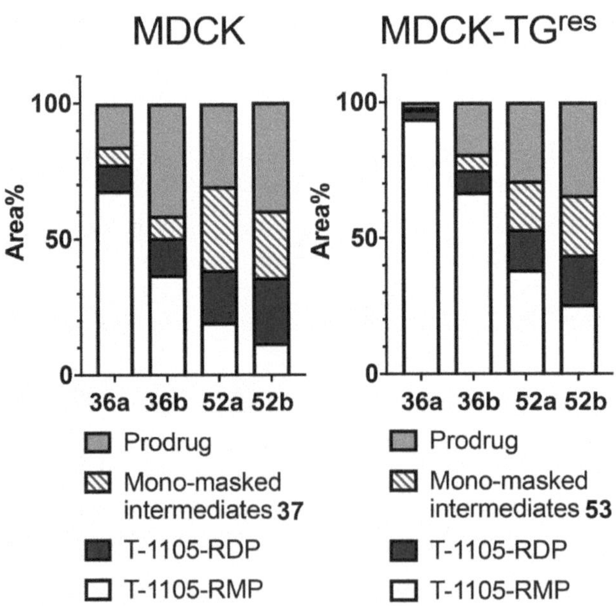

Fig. 5 Metabolic conversion of T-1106-Di*PP*ro-prodrugs **36** and T-1106-Tri*PPP*ro-prodrugs **52** in MDCK and MDCK-TG^{res} cell extracts. Prodrugs **36,52** (500 μM) were incubated in cell extract and after 2 h, proteins were precipitated.[46]

As expected, the initial cleavage step of the hydrolysis mechanism happened similarly to the cleavage pathways for compounds **28,34**, respectively. The hydrolysis pathways are shown in Scheme 11. The studies showed that: (i) a minor pathway involved the cleavage of the phosphoanhydride bond in T-1106-Di*PP*ro-prodrugs **36**, leading to T-1105-RMP, (ii) the major pathway involved an enzymatic trigger mechanism leading to T-1105-RDP via a mono-masked intermediate **37**, and (iii) with PLE, a large amount of T-1105-RDP and almost no T-1105-RMP was detected.[46]

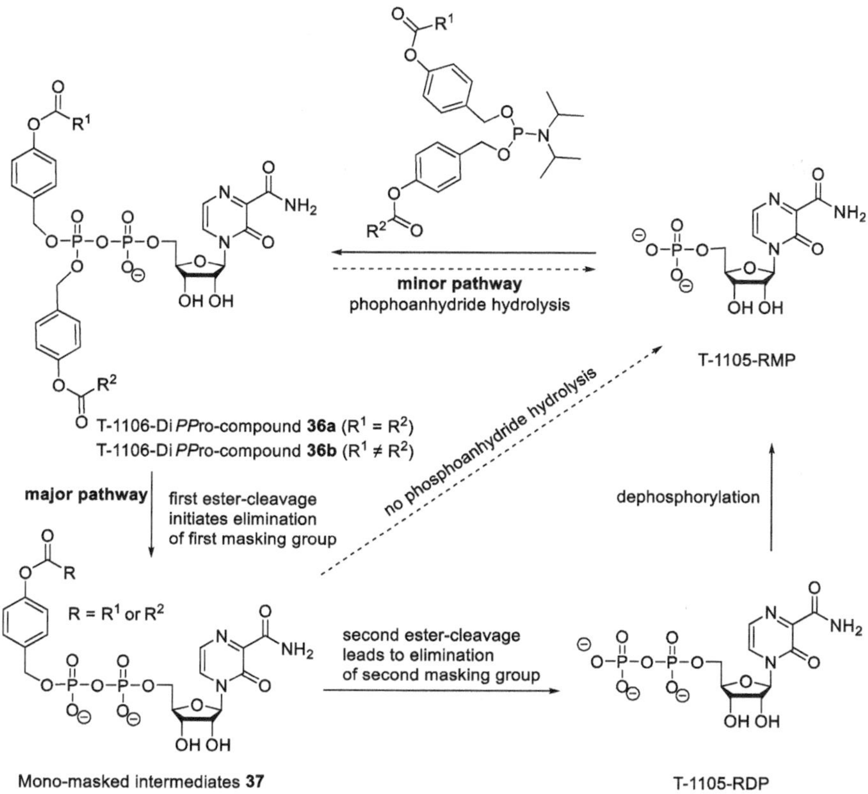

Scheme 11 Synthesis and hydrolysis mechanism of T-1106-Di*PP*ro-prodrugs **36.**

T-705-Di*PP*ro-prodrug **35** and T-1106-Di*PP*ro-prodrugs **36** were evaluated for their anti-influenza activity in MDCK and MDCK-T^{Gres} cells. As shown in Table 1, T-705-Di*PP*ro-compound **35** (AB-C4;AB-C14) had an average EC_{50} of 6.5 μM, which is 3-fold more active than the nucleobase T-705 (an average EC_{50} = 16.8 μM). As for T-1105 ribonucleoside

(T-1106), the anti-influenza activity of the T-1106-Di*PP*ro-prodrugs **36a,b** was 35 to 42-fold higher as compared to the parent nucleoside T-1106 (an average $EC_{50}=38.2\ \mu M$), 4 to 5-fold higher than the nucleobase T-1105 (an average $EC_{50}=4.3\ \mu M$), and 6 to 7-fold higher than its T-705 counterpart **35** (AB-C4;AB-C14), respectively.

Compared to the non-symmetric T-1106-Di*PP*ro-compound **36b** (AB-C4;AB-C14) ($EC_{50}=1.1\ \mu M$), the symmetric T-1106-Di*PP*ro-compound **36a** (AB-C9;AB-C9) had similar (an average $EC_{50}=$ 0.91 μM) anti-influenza activity. More importantly, the antiviral activity observed in MDCK-TG^{res} cell cultures (an average $EC_{50}=0.91\ \mu M$) was completely retained in the case of the lipophilic T-1106-Di*PP*ro-compound **36a** (AB-C9;AB-C9) in HGPRT-deficient MDCK-TG^{res} cell cultures (an average $EC_{50}=0.8\ \mu M$), whereas T-1106 lacked any relevant anti-influenza activity ($EC_{50} > 250\ \mu M$). In antiviral assays, good anti-influenza activity of the T-1106-Di*PP*ro-prodrugs **36a** (AB-C9;AB-C9),**36b** (AB-C4; AB-C14) was observed in MDCK-TG^{res} cells with 5 to 7-fold improved activity as compared to the T-705-Di*PP*ro-compound **35** (AB-C4; AB-C14). Thus, T-1106-Di*PP*ro-compounds **36** provided a successful and selective delivery of T-1105-RDP inside cells and improved the anti-influenza activity and this Di*PP*ro-system proved to be highly successful in the esterase-triggered release of T-1105-RDP. As for T-705, the anti-influenza activity of T-705-Di*PP*ro-compound **35** (an average EC_{50} of 6.5 μM, 6-fold less active than T-1106-Di*PP*ro-compound **36b**) is probably limited by the chemical instability of its fluorinated nucleobase.

6. Nucleoside triphosphate prodrugs

As summarized above, we have developed a concept to deliver NDPs inside cells using lipophilic but still partially charged Di*PP*ro-compounds.[110–115] However, the released NDPs still need further intracellular phosphorylation into their antivirally active triphosphate forms by cellular kinases to interact with viral polymerases. Moreover, it was reported that even the last phosphorylation might be problematic, e.g., the conversion of FTC-DP to form FTC-TP by NDPK.[116] As a consequence, the development of nucleoside triphosphate (NTP) prodrugs is still highly interesting and desirable because this enables: i) the bypass of *all* steps of the intracellular phosphorylation and ii) the delivery of NTPs would in principle maximize the intracellular concentration of the NTPs.

In an attempt to develop NTP prodrugs, the following challenges have to solved: (i) the highly negatively charged triphosphate unit requires appropriate masking, (ii) *two* reactive phosphate anhydrides linkages in the triphosphate moieties should be kept intact, (iii) the high sensitivity for enzymatic dephosphorylation of NTPs should be prevented, and (iv) a suitable chemical synthesis has to be established. Only very few NTP prodrugs have been reported in the past.[117,118] These nucleoside triphosphate prodrugs involved the attachment of only an alkyl or acyl moiety (high lipophilicity) to the γ-phosphate moiety.[109,117–121] Unfortunately, no increase of antiviral activity was observed with these nucleoside triphosphate prodrugs as compared to their parent nucleosides.[121]

6.1 Symmetric Tri*PP*Pro-compounds—Acyloxybenzyl (AB)-nucleoside triphosphate prodrugs

Our group is working intensively on the development of NTP prodrug systems. Encouraged by the positive results of the Di*PP*ro-prodrugs, we developed a novel nucleoside triphosphate prodrug concept in which two biodegradable masking units were attached to the γ-phosphate moiety, namely the Tri*PPP*ro-approach (Scheme 12; upper section).[122] Later, we report on the application of the Tri*PPP*ro-concept on several nucleoside analogs to demonstrate the strength of the strategy (Scheme 12; lower section).[123] For the synthesis of Tri*PPP*ro-prodrugs **38**, we developed two different synthesis routes: the phosphoramidite route and the *H*-phosphonate route.[122,123]

In the initial studies, Tri*PPP*ro-d4TTPs **38** bearing two biodegradable masking groups (alkyl, alkoxyl, and aminoalkyl chains) attached to the γ-phosphate group of d4TTP were prepared using the phosphoramidite route with modest to good chemical yields (26–70%). The P(III) chemistry was similar to our previously reported phosphoramidite chemistry used for the synthesis of Di*PP*ro-prodrugs.[110,112] The difference was that the P(III)-compounds **27** were reacted with d4TDP instead of d4TMP, as the nucleotide components. For this, phosphoramidites **27** were prepared in high yields.[110,112] D4TDP was synthesized using the *cyclo*Sal-technique (55% yield). Then, compounds **27** were mixed with d4TDP in the presence of dicyanoimidazol (DCI). After oxidation, the target products were purified by automated column RP-18 chromatography.

Additionally, a further method (*H*-phosphonate route) was developed. The chemistry was based on a coupling reaction of pyrophosphate **41** (P(V)-reagent) and NMPs. *H*-Phosphonate **40** was easily prepared from

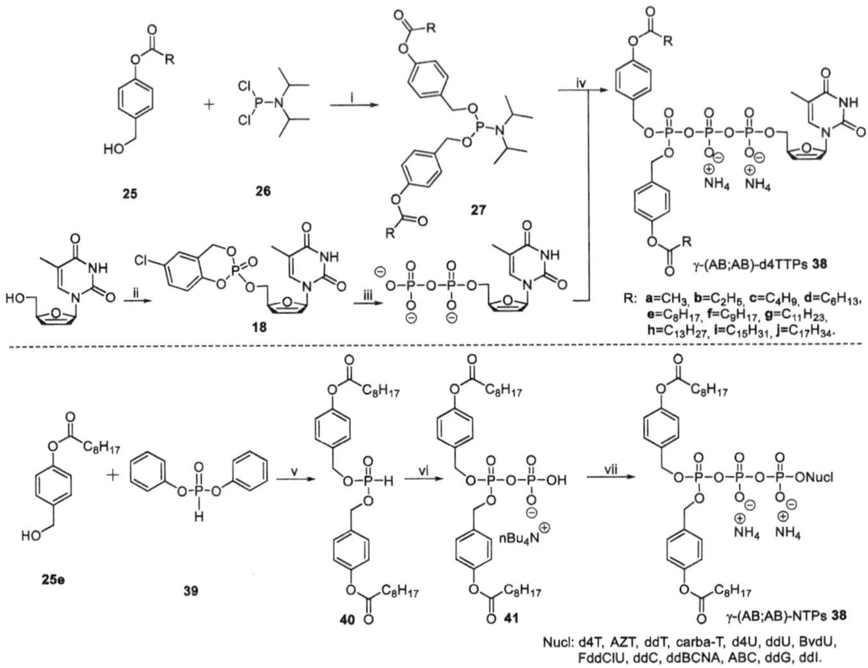

Scheme 12 Synthesis of Tri*PPP*ro-prodrugs **38** using via P(III) and P(V) chemistry. (i) Triethylamine, THF, 0 C-rt, 20 h; (ii) 1. 5-chlorosaligenylchlorophosphite, *N*,*N*-diisopropylethylamine, CH_3CN, −20 C-rt, 3 h, 2. *t*-BuOOH in *n*-decane, 0 C-rt, 30 min; (iii) $(H_2PO_4)Bu_4N$, DMF, rt., 20 h; (iv) 1. DCI, CH_3CN, rt., 1 min, 2. *t*-BuOOH in *n*-decane, 0 C-rt,15 min; (v) pyridine, 38°C, 2 h; (vi) 1. NCS, CH_3CN, rt., 2 h, 2. (H_2PO_4) Bu_4N, CH_3CN, rt., 1 h (vii) 1. TFAA, Et_3N, CH_3CN, 0 C, 10 min, 2. 1-methylimidazole, Et_3N, CH_3CN, 0 C-rt, 10 min, 3. NMP, rt., 1–3 h.

4-acyloxybenzyl alcohol **25e** and diphenyl hydrogen phosphonate (DPP) **39**. The *H*-phosphonate **40** was reacted with *N*-chlorosuccinimide (NCS) to generate a phosphorochloridate which was then reacted with tetra-*n*-butylammonium phosphate to yield pyrophosphate **41** in almost quantitative yields. Due to its high chemical instability, compound **41** was quickly purified by extraction (CH_2Cl_2/H_2O) and immediately used in the next step. Finally, compound **41** was activated with trifluoroacetic acid anhydride (TFAA) and *N*-methylimidazole,[124,125] then reacted with NMPs to give Tri*PPP*ro-NTPs **38** (n-Bu_4N^+ form). The target Tri*PPP*ro-NTPs **38** (NH_4^+ form) were obtained in yields between 7% and 71% using the above separation method. As compared to the phosphoramidite pathway, the *H*-phosphonate route offers some advantages: (i) NMPs are generally easier to prepare as the corresponding NDPs,

(ii) the diphosphate bond between the α- and β-phosphates was formed without oxidation step, (iii) the method was more tolerant to the used solvents, and (iv) the chemical stability of compound **40** was higher than that of the phosphoramidites **27**.

In order to study the hydrolysis properties and the delivery mechanism of Tri*PPP*ro-compounds **38**, the mono-masked acyloxybenzyl (AB)-NTP derivatives **43** were synthesized as well (Scheme 13). First, 4-acyloxybenzyl alcohols **25** reacted with 5-nitrosaligenylchlorophosphite **15** to synthesis benzyl-(5-nitro-*cyclo*Sal)-phosphate triesters **42** in high yields (up to 89%). Then, the mono-masked triphosphates **43** were prepared from 5-nitro-*cyclo*Sal-triesters **42** and d4TDP in yields of 26–30%.

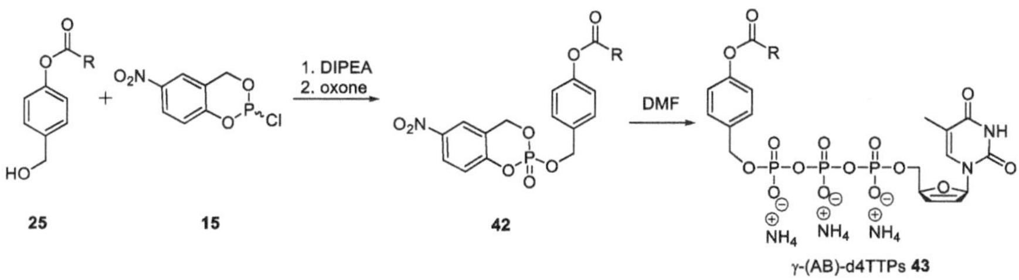

Scheme 13 Synthesis of mono-masked NTPs **43**.

In PBS, generally the half-lives of γ-(AB;AB)-d4TTPs **38a-h** increased with increasing acyl chain lengths (AB:C1-C13). Tri*PPP*ro-d4TTPs **38** hydrolyzed following similar cleavage pathways as described for Di*PP*ro-d4TDPs **28** (Scheme 8). In all cases to some extent, d4TDP was formed as well and a very low amount of d4TMP was observed in addition to the expected predominate formation of γ-(AB)-d4TTPs **43** and d4TTP, indicating that Tri*PPP*ro-d4TTPs **38** hydrolyzed mainly followed pathway **A** ($\mathbf{A^1}$ and $\mathbf{A^2}$; Scheme 14).

Furthermore, after complete conversion of the starting Tri*PPP*ro-d4TTPs **38**, no further increase of amounts of d4TDP and d4TMP was observed. It was concluded that d4TDP and d4TMP were formed from Tri*PPP*ro-d4TTPs **38** by a nucleophilic attack at the γ-phosphate (pathway **B**) or β-phosphate (pathway **C**) moiety, respectively. As observed for Di*PP*ro-d4TDPs **28**,[112] the stability of the intermediately formed γ-(AB)-d4TTPs **43** ($t_{1/2}$=95–637 h)[122] was higher than the corresponding Tri*PPP*ro-d4TTPs **38** ($t_{1/2}$=17–99 h),[122] which was in agreement with the results obtained from the studies of prodrugs **28** in PBS. Interestingly, Tri*PPP*ro-NTPs **38** were also hydrolyzed to form γ-(AB)-NTPs as **43**

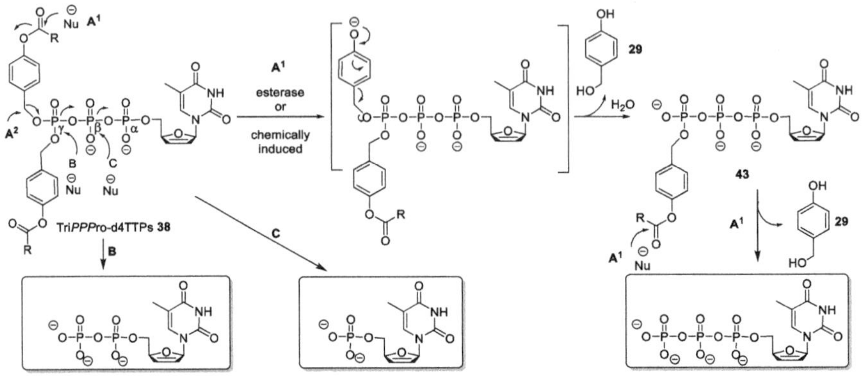

Scheme 14 Hydrolysis mechanism of Tri*PPP*ro-prodrugs **38** in PBS (pH 7.3) (shown for d4T as an example).

and then released the corresponding NTPs in PBS. The hydrolysis pathways leading to the different phosphorylated nucleotide metabolites are shown in Scheme 14.

The Tri*PPP*ro-concept is designed to be cleaved by esterases/lipases. As compared to the chemical hydrolyses, Tri*PPP*ro-compounds **38** were rapidly hydrolyzed with PLE and formed intermediates **43** demonstrating the enzymatic cleavage. The cleavage of intermediates **43** was again much slower than for Tri*PPP*ro-prodrugs **38**. Again, Tri*PPP*ro-compounds **38** were also readily cleaved to form intermediates **43** in CEM/0 cell extracts and then delivered the corresponding NTPs. In addition to NTPs (e.g., d4TTP, $t_{1/2}$=38 min)[122] also a high amount of NDPs (e.g., d4TDP, $t_{1/2}$=59 h)[122] were detected as well, probably due to the presence of dephosphorylating enzymes in cell extracts.

To confirm the Tri*PPP*ro-concept, primer extension assays were performed with HIV's reverse transcriptase (HIV-RT) after hydrolysis of γ-(AB-C8;AB-C8)-d4TTP. As expected, an immediate DNA chain termination was observed after incorporation of d4TMP, indicating that d4TTP (substrate for RT) was released from γ-(AB-C8;AB-C8)-d4TTP with PLE. Additionally, the thymidine-bearing γ-(AB-C8;AB-C8)-TTP was investigated in the same way. The hydrolysis of γ-(AB-C8;AB-C8)-TTP was performed and then the hydrolysis mixture was added to the polymerase assay. Here, the expected canonical incorporation of TMP was observed because the delivered TTP was accepted by HIV-RT as a substrate.

Tri*PPP*ro-prodrugs **38** were tested against HIV replication in infected wild-type (CEM/0) and TK-deficient (CEM/TK^-) cell cultures.

Tri*PPP*ro-d4TTPs **38** showed virtually similar activities against HIV-1 and HIV-2 as compared to the parent d4T in infected wild-type (CEM/0) cells. Importantly, good to very good antiviral activity of Tri*PPP*ro-d4TTPs **38** ($R \geq C_8H_{17}$) was detected in CEM/TK$^-$ cells with 15–535-fold improved activity as compared to the nucleoside. It was concluded from these studies that Tri*PPP*ro-d4TTPs **38** with biodegradable masking groups efficiently enter cells and deliver d4TTP, thereby bypass all intracellular phosphorylation steps. Again, Tri*PPP*ro-NTPs **38** (AB:C8;AB:C8) bearing different nucleoside analogs were active against HIV-1 and HIV-2 in infected wild-type (CEM/0) cell cultures, more importantly, they remained active in CEM/TK$^-$ cells as well.[123]

6.2 Non-symmetric Tri*PPP*ro-nucleoside triphosphate prodrugs

The first generation Tri*PPP*ro-prodrugs **38** comprised two *identical* biodegradable masking moieties and enabled the delivery of NTPs inside cells. In the case of Tri*PPP*ro-d4TTPs **38**, the desired cleavage of the phenol ester dominated, consequently, the prodrug mainly yielded the corresponding intermediates **43** and then released d4TTP (Scheme 14). From the results summarized for Tri*PPP*ro-d4TTPs **38**, the stability and antiviral activity were related to the alkyl chain length attached to the acyl residues. The shorter alkyl chain length led to lower stability of Tri*PPP*ro-d4TTPs **38** (AB:C1-C6) with no antiviral activity. In addition, Tri*PPP*ro-d4TTPs **38** (AB:C8-C15) comprising longer alkyl residues in the ester moiety exhibited higher chemical and biological stability and also higher antiviral activity. Compared to acyloxy ester Tri*PPP*ro-compounds, the alkyl carbonate functional groups (ACB) increase the stability of these prodrugs. These observations formed the basis for our research to conduct a study on a series of NTP prodrugs bearing two *different* biodegradable masking groups ($R^1 \neq R^2$). Non-symmetric Tri*PPP*ro-prodrugs bearing two *different* biodegradable masking groups (AB or ACB) are shown in Scheme 15.[126–128]

In 2020, we first described the non-symmetric γ-(AB;ACB)-d4TTPs (Scheme 15) with the aim to deliver d4TTP with high selectivity in PBS and by enzyme-triggered reactions in CEM cell extracts.[126] Studies have shown that the introduction of these different groups led to the selective formation of γ-(ACB)-d4TTPs **51** by chemical hydrolysis and in particular in cell extracts. Subsequently, the use of the Tri*PPP*ro-concept on a series of approved and also on so-far non-active nucleoside analogs was described to demonstrate the general applicability and the great potential that this approach

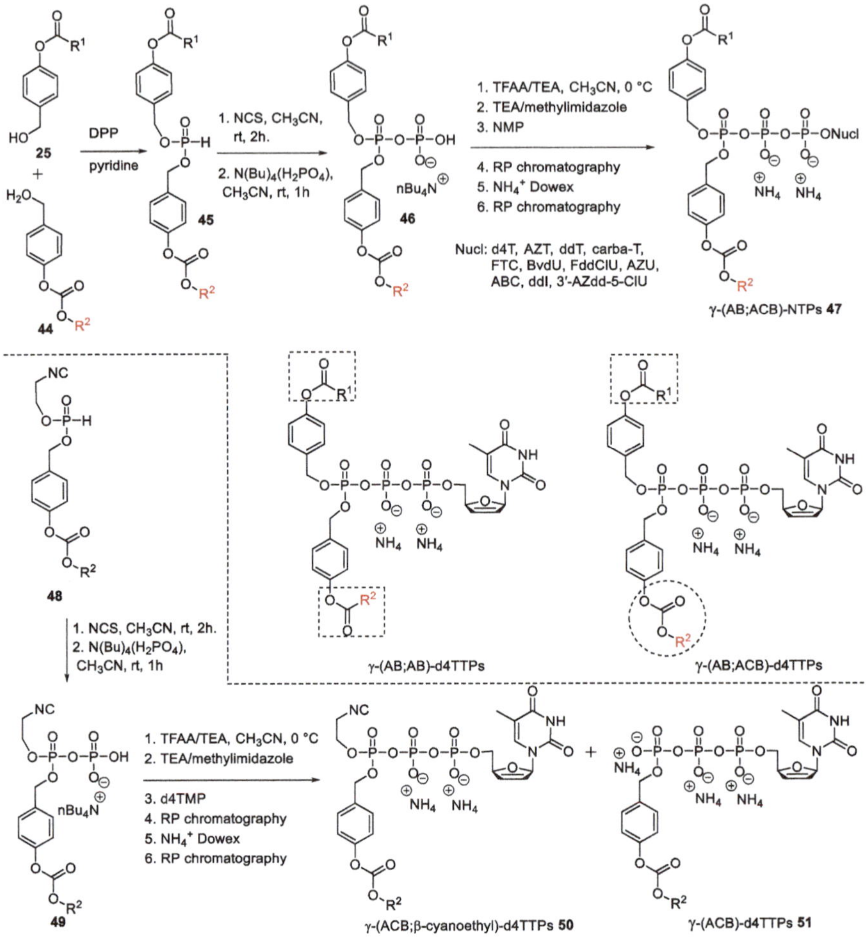

Scheme 15 Synthesis of Tri*PPP*ro-prodrugs **47** (d4T as an example) and intermediates **51** via *H*-phosphonate route.

promises (γ-(AB;ACB)-NTPs **47**; Scheme 15).[127] For comparison, we also made and studied Tri*PPP*ro-prodrugs (γ-(AB;AB)-d4TTPs) comprising two different AB moieties.[128]

6.3 Synthesis of non-symmetric Tri*PPP*ro-prodrugs 47 and γ-(ACB)-d4TTPs 51

Tri*PPP*ro-prodrugs **47** were preferably prepared using the *H*-phosphonate chemistry (up to 85% yield).[126–128] As shown in Scheme 15, the non-symmetric *H*-phosphonates **45** were stepwise activated with trifluoroacetic acid anhydride (TFAA) and *N*-methylimidazole and reacted with d4TMP to yield Tri*PPP*ro-prodrugs **47**.[123] In addition to the γ-(AB;ACB)-d4TTPs **47**, the mono-masked intermediates **51** were synthesized as well (Scheme 15).[126] *H*-Phosphonates **48** were synthesized from DPP, 3-hydroxypropionitrile, and 4-alkoxycarbonyloxybenzyl alcohols **44**. *H*-Phosphonates **48** were converted into pyrophosphates **49** and then reacted with d4TMP to form γ-(ACB;β-cyanoethyl)-d4TTPs **50** (n-Bu_4N^+ form). It was found that the crude product was hydrolyzed during the ion-exchange to yield the mixture of γ-(ACB;β-cyanoethyl)-d4TTPs **50** (NH_4^+ form, yields of 52–63%) and γ-(ACB)-d4TTPs **51** (NH_4^+ form, yields of 10–23%). We assumed that the β-cyanoethyl protecting group was cleaved by the known β-elimination leading to γ-(ACB)-d4TTPs **51**.

As before, Tri*PPP*ro-prodrugs **47** were studied in PBS (25 mM, pH 7.3), pig liver esterase (PLE), and CEM cell extracts. The calculated half-lives of prodrugs **47** reflect the removal of the AB group and/or the ACB group to yield the corresponding intermediates, γ-(AB)-d4TTPs **43** and γ-(ACB)-d4TTPs **51**, respectively. Possible hydrolysis pathways of Tri*PPP*ro-d4TTPs **47** are summarized in Scheme 16.

Generally, the chemical stability of Tri*PPP*ro-d4TTPs **47** (AB-C2;ACB: C9-C16) ($t_{1/2}$=25–83 h) increased with increasing alkyl chain lengths. However, half-lives of more lipophilic compounds **47** (AB-C3-C9;ACB: C12-C18) remained in the same range ($t_{1/2}$=74–90 h).[126] The half-lives of the carbonate intermediates **51** were significantly higher than those of

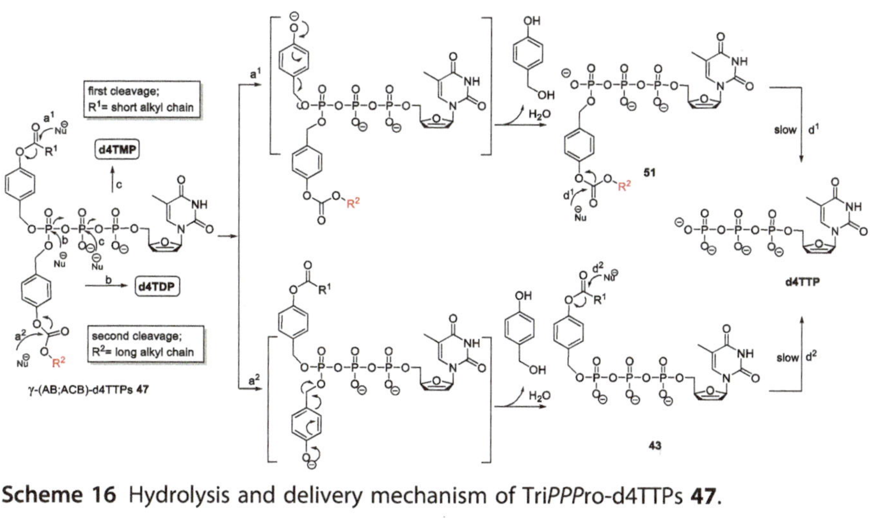

Scheme 16 Hydrolysis and delivery mechanism of Tri*PPP*ro-d4TTPs **47**.

γ-(AB;ACB)-d4TTPs **47**. Moreover, chemical stabilities of γ-alkyl carbonate bearing γ-(ACB)-d4TTPs **51** were higher than the corresponding ester intermediates γ-(AB)-d4TTPs **43**, which was in full agreement with the results observed from the studies of γ-(AB;ACB)-d4TTPs **47** in PBS. The half-life for γ-(ACB-C16)-d4TTP ($t_{1/2} > 1600$ h)[126] was almost 3-fold higher as compared to the studies of γ-(AB-C17)-d4TTP ($t_{1/2}=$ 580 h)[122] described before. In the case of γ-(AB-C11;ACB-C6)-d4TTP, the γ-(AB-C11)-d4TTP/γ-(ACB-C6)-d4TTP ratio was found to be 1:3. In addition, the half-life of γ-(AB-C11)-d4TTP ($t_{1/2}=460$ h)[122] was also found to be lower than that for γ-(ACB-C6)-d4TTP ($t_{1/2}=540$ h).[126]

As an example, when γ-(AB-C2;ACB-C16)-d4TTP[126] was totally consumed after incubation for 50 days in PBS, both intermediates γ-(ACB-C16)-d4TTP (mainly) and γ-(AB-C2)-d4TTP (trace) were formed although in very different concentration and thus, the hydrolysis proceeded highly selective. Both masking groups of γ-(AB;ACB)-d4TTPs **47** were involved in the chemical hydrolysis pathway a^1 and pathway a^2 (Scheme 16). Before complete consumption of the starting γ-(AB-C2;ACB-C16)-d4TTP, an increase of d4TTP and γ-(ACB-C16)-d4TTP concentrations and a very small amount of γ-(AB-C2)-d4TTP were observed, which supports that the γ-(AB-C2)-d4TTP was prone to hydrolysis to form d4TTP.

With PLE, Tri*PP*Pro-compounds **47** were rapidly hydrolyzed ($t_{1/2}=0.17$–13.8 h).[126–128] In the case of the enzymatically catalyzed hydrolysis of γ-(AB-C4;ACB-C16)-NTPs **47**, both hydrolysis intermediates γ-(ACB-C16)-NTPs and γ-(AB-C4)-NTPs were formed in markedly different amounts. Both intermediates first accumulated and later were slowly hydrolyzed to yield NTPs. As in PBS, the highly selective cleavage of the AB group (e.g., AB-C2) led to the formation of γ-(ACB-C16)-NTPs. In contrast, for the hydrolysis of γ-(AB-C4;AB-C15)-d4TTP, both possible intermediates γ-(AB-C14)-d4TTP and γ-(AB-C15)-d4TTP were formed in almost identical amounts.

As before, Tri*PP*Pro-compounds **47** were quickly hydrolyzed in CEM/0 cell extracts and delivered the two intermediates. However, the stabilities of γ-(AB;ACB)-d4TTPs were in the range of 0.8–6.4 h without showing a clear trend. Interestingly, our studies have shown that γ-(AB;ACB)-d4TTPs were readily cleaved to form intermediates γ-(ACB)-d4TTPs **51** but in low concentrations in CEM/0 cell extracts. In addition, a small amount of d4TTP and a large amount of d4TDP were detected due to the fast dephosphorylation of the d4TTP ($t_{1/2}=38$ min)[122] by phosphorylases/kinases present in the cell extracts. These results were in line with the previous results of the symmetric Tri*PPP*ro-prodrugs **38**. Moreover, in case of γ-(AB-C11;ACB-C6)-d4TTP, a predominant formation of γ-(ACB-C6)-d4TTP was observed, the ratio of γ-(AB-C11)-d4TTP and γ-(ACB-C6)-d4TTP was 1:10 after 8 h incubation in CEM/0 cell extracts. This was in full agreement with the results obtained from the studies of this compound in PBS that an almost selective cleavage took place in cell extracts with the AB-moiety being cleaved first. More importantly, it was proven that NTPs were also successfully released from the Tri*PPP*ro-NTPs **47** (AB-C4;ACB:C16) in cell extracts.[127] In all cases, the enzymatic cleavage also took place for the mono-masked intermediates γ-(AB-C4)-NTPs and γ-(ACB-C6)-NTPs, respectively. Again, NTPs were detected in low concentrations only because of the fast enzymatic dephosphorylation by phosphorylases/kinases (main product) and ultimately NMPs. In contrast to γ-(AB-C2; ACB-C16)-d4TTP, no selective cleavage of the different AB-masks of γ-(AB;AB)-d4TTPs was observed in the hydrolysis in CEM cell extracts.[128]

Prodrugs **47** and intermediates **51** were evaluated for HIV replication inhibition in HIV-1/2-infected wild-type CEM/0 cells and in HIV-2-infected mutant thymidine kinase-deficient (CEM/TK$^-$) cells, and the different nucleoside analogs as reference compounds. For γ-(AB-C2; ACB-C16)-d4TTP (EC_{50} = 0.027 μM/HIV-1, EC_{50} = 0.0048 μM/HIV-2), the antiviral activity in this infected cell line improved by 16-fold and 65-fold, respectively, as compared to d4T (EC_{50} = 0.43 μM/HIV-1, EC_{50} = 0.31 μM/HIV-2).[126] More importantly, also high activities were obtained depending on the lipophilicity of the γ-(AB;ACB)-d4TTPs against HIV-2 in this thymidine kinase-deficient (CEM/TK$^-$) cell model. With γ-(AB-C2;ACB-C16)-d4TTP (EC_{50} = 0.11 μM/HIV-2) the antiviral activity in this infected cell line was improved by a 282-fold as compared to d4T (EC_{50} = 31.05 μM/HIV-2). However, the antiviral activity detected in CEM/0 cell cultures decreased in the case of γ-(AB-C2;ACB-C16)-d4TTP in CEM/TK$^-$ cells (23-fold), potentially due to the instability of the these prodrugs in addition to a fast cleavage of the bioreversible AB- or ACB-moiety. We speculated that γ-(AB;ACB)-d4TTPs **47** were rapidly hydrolyzed in cell extracts and d4TTP was present in insufficient concentrations (fast dephosphorylation) to exhibit anti-HIV activity in cells. Moreover, it cannot be excluded that the antiviral activity observed in the infected CEM/TK$^-$ deficient cells was at least in part due to the formation of d4TDP and d4TMP. In the hydrolysis of γ-(AB-C2;ACB-C16)-d4TTP in cell extracts, a large amount of d4TDP and some d4TMP were detected.[126]

Interestingly, the intermediate γ-(ACB-C16)-d4TTP (EC_{50} = 1.46 μM/HIV-2) was also potent in CEM/TK^- cell cultures, indicating that even *one* long aliphatic chain in the ACB-units provides enough lipophilicity to enable a cellular uptake of the aliphatic γ-(AB;ACB)-d4TTPs.

More interestingly, the high potential of the Tri*PPP*ro-approach was also demonstrated by the nucleoside triphosphate prodrugs **47** bearing different nucleoside analogs.[127] These prodrugs proved also highly potent in CEM/TK^- cells, whereas their parent nucleosides lacked any relevant antiviral activity in wild-type CEM/0 cell model. It was confirmed that these Tri*PPP*ro-NTPs **47** were also taken-up by the cells and delivered intracellularly a phosphorylated form of the parent nucleosides, most likely NTPs. Therefore, it was confirmed that the Tri*PPP*ro-strategy enabled the intracellular delivery of NTPs, and the concept could be used to convert inactive nucleoside analogs into powerful biologically active metabolites.

6.4 Nucleoside triphosphate prodrugs: T-1106-Tri*PPP*ro-prodrugs

Encouraged by the previously obtained results, we synthesized two T-1106-Tri*PPP*ro prodrugs **52**, the symmetrical T-1106-Tri*PPP*ro prodrug **52a** (AB-C9;AB-C9) and the non-symmetrical T-1106-Tri*PPP*ro prodrug **52b** (AB-C4;AB-C14), with the aim of achieving metabolic bypass and superior antiviral potency.[46] T-1106-Tri*PPP*ro-prodrugs **52** were prepared using the phosphoramidite protocol (Scheme 17).[110,113] The conversion of T-1106-RDP to T-1106-Tri*PPP*ro prodrugs **52a,b** resulted in yields of 27% and 44%. Due to the already mentioned stability problems, until now we were unable to synthesize the T-705-RTP prodrug. T-1106-Tri*PPP*ro prodrugs **52** were also studied with regard to their chemical and biological stabilities and their hydrolysis products in PBS (pH 7.3, Fig. 4B, condition without esterase), PLE (Fig. 4B, condition with esterase), and crude enzyme preparations (i.e., MDCK and MDCK TG^{res} cell extracts) (Fig. 5). Additionally, these prodrugs were evaluated for their anti-influenza virus activity in wild-type and HGPRT-deficient cells.

Similar to the cleavage pathway for Tri*PPP*ro-d4TTPs **38**, T-1106-Tri*PPP*ro prodrug **52a** (AB-C9;AB-C9) was hydrolyzed to give the mono-masked intermediate **53** (AB-C9) with some concomitant cleavage to T-1105-RTP and subsequently T-1105-RDP (Fig. 4B). Interestingly, the mono-masked triphosphate intermediates **53** proved to be more stable than the corresponding mono-masked T-1106-Di*PP*ro-compounds **37** (Fig. 4A and B). However, when T-1106-Tri*PPP*ro-prodrugs **52a,b** were exposed

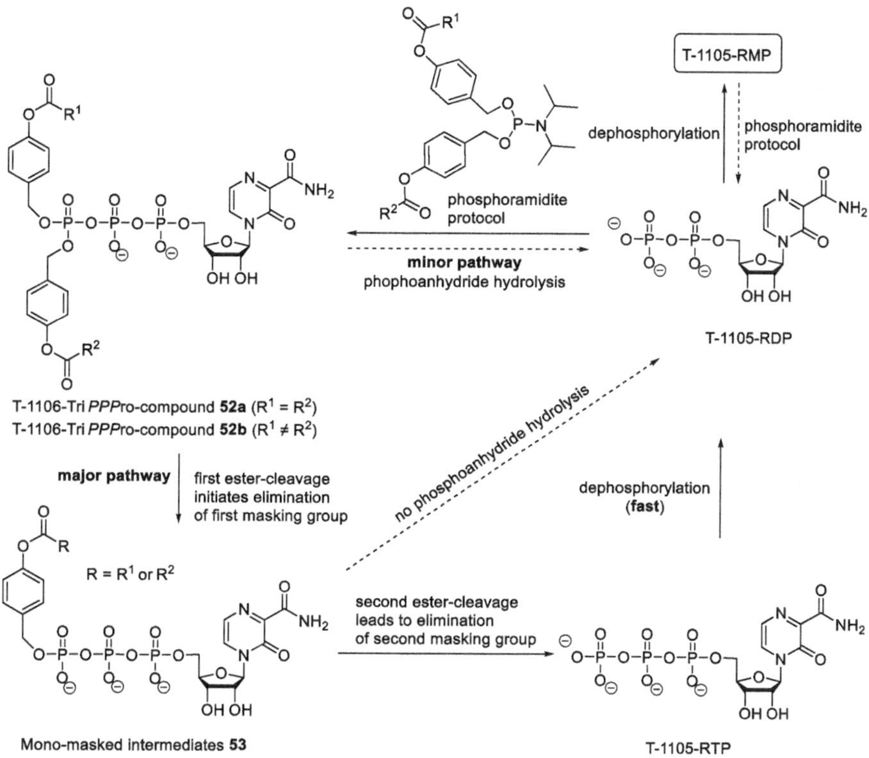

Scheme 17 Synthesis and hydrolysis mechanism of T-1106-Tri*PPP*ro-prodrugs **52**.

to crude enzyme preparations (i.e., MDCK or MDCK TG^{res} cell extracts), no T-1105-RTP was observed, probably due to its fast dephosphorylation after initial formation (Fig. 5). Compared to the T-1106-Di*PP*ro-prodrugs **36**, after 2 h incubation, a higher amount of T-1105-RDP was detected from the corresponding T-1106-Tri*PPP*ro-prodrugs **52a,b** (Fig. 5, compare dark gray stacks). As shown in Fig. 5, the stability of the T-1106-Tri*PPP*ro-prodrugs **52a,b** was found to be lower than the corresponding T-1106-Di*PP*ro-prodrugs **36a,b**, respectively. In addition, the enzymatic cleavage of the symmetric T-1106-Tri*PPP*ro-prodrug **52a** proceeded faster than the non-symmetric T-1106-Tri*PPP*ro-prodrug **52b**.

The anti-influenza activity of T-1106-Tri*PPP*ro-prodrugs **52a,b** in MDCK and MDCK-TG^{res} cells was lower as compared to T-1106-Di*PP*ro-prodrugs **36a,b** (Table 1). T-1106-Tri*PPP*ro-prodrugs **52a** (AB-C9;AB-C9) and **52b** (AB-C4;AB-C14) had EC_{50} values of 2.8 μM and 7.2 μM, respectively. We assumed that the difference in antiviral activity may be caused by the differences in cell membrane permeability. Moreover, this ranking in antiviral activity corresponds to the ranking in activation rate (Fig. 5). Interestingly, T-1106-Tri*PP*Pro prodrugs **52a,b** also retained their anti-influenza activity in HGPRT-deficient MDCK-TG^{res} cells. For instance, T-1106-Tri*PP*Pro-prodrugs **52a** (AB-C9;AB-C9) had an average EC_{50} value of 2.8 μM in MDCK cells versus 2.5 μM in MDCK-TG^{res} cells (Table 1). Therefore, the T-1106-Tri*PP*Pro-prodrugs **52** were successfully released T-1105-RTP, thus bypassing their HGPRT and kinase dependence (Fig. 1). The T-1106-Tri*PP*Pro-prodrugs **52** enabled the bypass the second step of the intracellular phosphorylation in cells, i.e., phosphorylation of T-1105-RMP to T-1105-RDP. We proved that the conversion of T-1106-RDP into T-1106-RTP was not limiting. This was shown in enzymatic studies.

7. Conclusion

The results summarized in this minireview show that nucleoside analogs still have enormous potential to be used as antivirals mainly as inhibitors of the viral polymerases as described here with the example of T-1105-RTP and the RNA-dependent-RNA-polymerase of SARS-CoV-2. Although this polymerase has a proof-reading system, the incorporated T-1105-RMP at least in part escapes its repair after incorporation. With the already known broad-spectrum antiviral activity of Favipiravir (T-705) this may also lead in the case of T-1105 or even other derivatives thereof to broad spectrum antivirals. Moreover, the application of nucleotide delivery approaches may contribute significantly to the improvement of antivirally active nucleoside analogs. This was already accepted in the past for the use of nucleoside monophosphate prodrugs such as the phosphoramidate system, but here we described results that in other cases the intracellular delivery of nucleoside monophosphates is not sufficient for the improvement of the antiviral activity due to metabolic hurdles that appear after the conversion into the monophosphate. In such cases nucleoside di- or even nucleoside triphosphate prodrug systems can help to overcome these limitations. This has been shown here by the failure of the application of the *cyclo*Sal-nucleoside monophosphate delivery approach and the use of the Di*PP*Pro- or the Tri*PP*Pro-nucleoside di- or triphosphate compounds. The chemical synthesis routes towards these latter compounds are now robust and can be applied to a large variety of nucleoside analogs. We are convinced that the further development of these prodrug systems of higher phosphorylated nucleosides will have an important impact in nucleoside drug development.

References

1. *Annual Review of Diseases Prioritized Under the Research and Development Blueprint*; WHO Research and Development Blueprint, 2018.
2. Iuliano, A. D.; Roguski, K. M.; Chang, H. H. Estimates of Global Seasonal Influenza-Associated Respiratory Mortality: A Modelling Study. *Lancet* **2018**, *391*, 1285–1300.
3. Kaner, J.; Schaack, S. Understanding Ebola: The 2014 Epidemic. Glob. Health 2016, 12, 53–60.
4. Xia, H. J.; Xie, X. P.; Shan, C.; Shi, P. Y. Potential Mechanisms for Enhanced Zika Epidemic and Disease. *Acs Infect. Dis.* **2018**, *4*, 656–659.
5. Wu, F.; Zhao, S.; Yu, B.; Chen, Y. M.; Wang, W.; Song, Z. G.; Hu, Y.; Tao, Z. W.; Tian, J. H.; Pei, Y. Y.; Yuan, M. L.; Zhang, Y. L.; Dai, F. H.; Liu, Y.; Wang, Q. M.; Zheng, J. J.; Xu, L.; Holmes, E. C.; Zhang, Y. Z. A New Coronavirus Associated with Human Respiratory Disease in China. *Nature* **2020**, *579*, 265–269.
6. Zhu, N.; Zhang, D. Y.; Wang, W. L.; Li, X. W.; Yang, B.; Song, J. D.; Zhao, X.; Huang, B. Y.; Shi, W. F.; Lu, R. J.; Niu, P. H.; Zhan, F. X.; Ma, X. J.; Wang, D. Y.; Xu, W. B.; Wu, G. Z.; Gao, G. G. F.; Tan, W. J.; Coronavirus, C. N. A Novel Coronavirus from Patients with Pneumonia in China, 2019. *N. Engl. J. Med.* **2020**, *382*, 727–733.
7. WHO. Coronavirus (COVID-19) Dashboard, 2021. https://www.who.int/emergencies/diseases/novel-coronavirus-2019 (accessed September 9, 2021).
8. Wu, R.; Wang, L.; Kuo, H. D.; Shannar, A.; Peter, R.; Chou, P. J.; Li, S.; Hudlikar, R.; Liu, X.; Liu, Z.; Poiani, G. J.; Amorosa, L.; Brunetti, L.; Kong, A. N. An Update on Current Therapeutic Drugs Treating COVID-19. *Curr. Pharmacol. Rep.* **2020**, 1–15.
9. Chaudhuri, S.; Symons, J. A.; Deval, J. Innovation and Trends in the Development and Approval of Antiviral Medicines: 1987-2017 and Beyond. *Antivir. Res.* **2018**, *155*, 76–88.
10. Burton, J. R.; Everson, G. T. HCV NS5B Polymerase Inhibitors. *Clin. Liver Dis.* **2009**, *13*, 453–465.
11. Cihlar, T.; Ray, A. S. Nucleoside and Nucleotide HIV Reverse Transcriptase Inhibitors: 25 Years after Zidovudine. *Antivir. Res.* **2010**, *85*, 39–58.
12. Jordheim, L. P.; Durantel, D.; Zoulim, F.; Dumontet, C. Advances in the Development of Nucleoside and Nucleotide Analogues for cancer and Viral Diseases. *Nat. Rev. Drug Discov.* **2013**, *12*, 447–464.
13. Deval, J. Antimicrobial Strategies Inhibition of Viral Polymerases by 3′-Hydroxyl Nucleosides. *Drugs* **2009**, *69*, 151–166.
14. Wang, G. Y.; Wan, J. G.; Hu, Y. J.; Wu, X. Y.; Prhavc, M.; Dyatkina, N.; Rajwanshi, V. K.; Smith, D. B.; Jekle, A.; Kinkade, A.; Symons, J. A.; Jin, Z. N.; Deval, J.; Zhang, Q. L.; Tam, Y.; Chanda, S.; Blatt, L.; Beigelman, L. Synthesis and Anti-Influenza Activity of Pyridine, Pyridazine, and Pyrimidine C-Nucleosides as Favipiravir (T-705) Analogues. *J. Med. Chem.* **2016**, *59*, 4611–4624.
15. Atta, M. G.; De Seigneux, S.; Lucas, G. M. Clinical Pharmacology in HIV Therapy. *Clin. J. Am. Soc. Nephrol.* **2019**, *14*, 435–444.
16. Li, G. D.; De Clercq, E. Therapeutic Options for the 2019 Novel Coronavirus (2019-nCoV). *Nat. Rev. Drug Discov.* **2020**, *19*, 149–150.
17. Elfiky, A.; Ribavirin, A. Remdesivir, Sofosbuvir, Galidesivir, and Tenofovir Against SARS-CoV-2 RNA Dependent RNA Polymerase (RdRp): A Molecular Docking Study. *Life Sci.* **2020**, *253*, 117592.
18. Furuta, Y.; Gowen, B. B.; Takahashi, K.; Shiraki, K.; Smee, D. F.; Barnard, D. L. Favipiravir (T-705), a Novel Viral RNA Polymerase Inhibitor. *Antivir. Res.* **2013**, *100*, 446–454.

19. Furuta, Y.; Takahashi, K.; Shiraki, K.; Sakamoto, K.; Smee, D. F.; Barnard, D. L.; Gowen, B. B.; Julander, J. G.; Morrey, J. D. T-705 (Favipiravir) and Related Compounds: Novel Broad-Spectrum Inhibitors of RNA Viral Infections. *Antivir. Res.* **2009**, *82*, 95–102.
20. Takahashi, K.; Furuta, Y.; Fukuda, Y.; Kuno, M.; Kamiyama, T.; Kozaki, K.; Nomura, N.; Egawa, H.; Minami, S.; Shiraki, K. In Vitro and In Vivo Activities of T-705 and Oseltamivir against Influenza Virus. *Antivir. Chem. Chemother.* **2003**, *14*, 235–241.
21. Furuta, Y.; Takahashi, K.; Fukuda, Y.; Kuno, M.; Kamiyama, T.; Kozaki, K.; Nomura, N.; Egawa, H.; Minami, S.; Watanabe, Y.; Narita, H.; Shiraki, K. In Vitro and In Vivo Activities of Anti-Influenza Virus Compound T-705. *Antimicrob. Agents Chemother.* **2002**, *46*, 977–981.
22. Furuta, Y.; Takahashi, K.; Kuno-Maekawa, M.; Sangawa, H.; Uehara, S.; Kozaki, K.; Nomura, N.; Egawa, H.; Shiraki, K. Mechanism of Action of T-705 Against Influenza Virus. *Antimicrob. Agents Chemother.* **2005**, *49*, 981–986.
23. Oestereich, L.; Rieger, T.; Ludtke, A.; Ruibal, P.; Wurr, S.; Pallasch, E.; Bockholt, S.; Krasemann, S.; Munoz-Fontela, C.; Gunther, S. Efficacy of Favipiravir Alone and in Combination with Ribavirin in a Lethal, Immunocompetent Mouse Model of Lassa Fever. *J. Infect. Dis.* **2016**, *213*, 934–938.
24. Tani, H.; Fukuma, A.; Fukushi, S.; et al. Efficacy of T-705 (Favipiravir) in the Treatment of Infections with Lethal Severe Fever with Thrombocytopenia Syndrome Virus. *mSphere* **2016**, *1*, e00061–15.
25. Oestereich, L.; Ludtke, A.; Wurr, S.; Rieger, T.; Munoz-Fontela, C.; Gunther, S. Successful Treatment of Advanced Ebola Virus Infection with T-705 (Favipiravir) in a Small Animal Model. *Antivir. Res.* **2014**, *105*, 17–21.
26. Yamada, K.; Noguchi, K.; Komeno, T.; Furuta, Y.; Nishizono, A. Efficacy of Favipiravir (T-705) in Rabies Postexposure Prophylaxis. *J. Infect. Dis.* **2016**, *213*, 1253–1261.
27. Delang, L.; Guerrero, N. S.; Tas, A.; Querat, G.; Pastorino, B.; Froeyen, M.; Dallmeier, K.; Jochmans, D.; Herdewijn, P.; Bello, F.; Snijder, E. J.; de Lamballerie, X.; Martina, B.; Neyts, J.; van Hemert, M. J.; Leyssen, P. Mutations in the Chikungunya Virus Non-structural Proteins Cause Resistance to Favipiravir (T-705), a Broad-Spectrum Antiviral. *J. Antimicrob. Chemother.* **2014**, *69*, 2770–2784.
28. Julander, J. G.; Smee, D. F.; Morrey, J. D.; Furuta, Y. Effect of T-705 Treatment on Western Equine Encephalitis in a Mouse Model. *Antivir. Res.* **2009**, *82*, 169–171.
29. Morrey, J. D.; Taro, B. S.; Siddharthan, V.; Wang, H.; Smee, D. F.; Christensen, A. J.; Furuta, Y. Efficacy of Orally Administered T-705 Pyrazine Analog on Lethal West Nile Virus Infection in Rodents. *Antivir. Res.* **2008**, *80*, 377–379.
30. Scharton, D.; Bailey, K. W.; Vest, Z.; Westover, J. B.; Kumaki, Y.; Van Wettere, A.; Furuta, Y.; Gowen, B. B. Favipiravir (T-705) Protects against Peracute Rift Valley Fever Virus Infection and Reduces Delayed-Onset Neurologic Disease Observed with Ribavirin Treatment. *Antivir. Res.* **2014**, *104*, 84–92.
31. Oestereich, L.; Rieger, T.; Neumann, M.; Bernreuther, C.; Lehmann, M.; Krasemann, S.; Wurr, S.; Emmerich, P.; de Lamballerie, X.; Olschlager, S.; Gunther, S. Evaluation of Antiviral Efficacy of Ribavirin, Arbidol, and T-705 (Favipiravir) in a Mouse Model for Crimean-Congo Hemorrhagic Fever. *PLoS Negl. Trop. Dis.* **2014**, *8*, e2804.
32. Julander, J. G.; Shafer, K.; Smee, D. F.; Morrey, J. D.; Furuta, Y. Activity of T-705 in a hamster Model of Yellow Fever Virus Infection in Comparison with that of a Chemically Related Compound, T-1106. *Antimicrob. Agents Chemother.* **2009**, *53*, 202–209.
33. Safronetz, D.; Falzarano, D.; Scott, D. P.; Furuta, Y.; Feldmann, H.; Gowen, B. B. Antiviral Efficacy of Favipiravir against Two Prominent Etiological Agents of Hantavirus Pulmonary Syndrome. *Antimicrob. Agents Chemother.* **2013**, *57*, 4673–4680.

34. De Clercq, E.; Li, G. D. Approved Antiviral Drugs over the Past 50 Years. *Clin. Microbiol. Rev.* **2016**, *29*, 695–747.
35. Cai, Q. X.; Yang, M. H.; Liu, D. J.; Chen, J.; Shu, D.; Xia, J. X.; Liao, X. J.; Gu, Y. B.; Cai, Q.; Yang, Y.; Shen, C. G.; Li, X. H.; Peng, L.; Huang, D. L.; Zhang, J.; Zhang, S. R.; Wang, F. X.; Liu, J. Y.; Chen, L.; Chen, S. Y.; Wang, Z. Q.; Zhang, Z.; Cao, R. Y.; Zhong, W.; Liu, Y. X.; Liu, L. Experimental Treatment with Favipiravir for COVID-19: An Open-Label Control Study. *Engineering* **2020**, *6*, 1192–1198.
36. Yamamura, H.; Matsuura, H.; Nakagawa, J.; et al. Effect of Favipiravir and an Anti-Inflammatory Strategy for COVID-19. *Crit. Care* **2020**, *24*, 413.
37. Chen, C.; Zhang, Y.; Huang, J.; Yin, P.; Cheng, Z.; Wu, J.; Chen, S.; Zhang, Y.; Chen, B.; Lu, M.; Luo, Y.; Ju, L.; Zhang, J.; Wang, X. Favipiravir Versus Arbidol for COVID-19: A Randomized Clinical Trial. *medRxiv* **2020**, https://doi.org/10.1101/2020.03.17.20037432.
38. Smee, D. F.; Hurst, B. L.; Egawa, H.; Takahashi, K.; Kadota, T.; Furuta, Y. Intracellular Metabolism of Favipiravir (T-705) in Uninfected and Influenza A (H5N1) Virus-Infected Cells. *J. Antimicrob. Chemother.* **2009**, *64*, 741–746.
39. Naesens, L.; Guddat, L. W.; Keough, D. T.; van Kuilenburg, A. B. P.; Meijer, J.; Vande Voorde, J.; Balzarini, J. Role of Human Hypoxanthine Guanine Phosphoribosyltransferase in Activation of the Antiviral Agent T-705 (Favipiravir). *Mol. Pharmacol.* **2013**, *84*, 615–629.
40. Yu, X. J.; Liang, M. F.; Zhang, S. Y.; Liu, Y.; Li, J. D.; Sun, Y. L.; Zhang, L. H.; Zhang, Q. F.; Popov, V. L.; Li, C.; Qu, J.; Li, Q.; Zhang, Y. P.; Hai, R.; Wu, W.; Wang, Q.; Zhan, F. X.; Wang, X. J.; Kan, B.; Wang, S. W.; Wan, K. L.; Jing, H. Q.; Lu, J. X.; Yin, W. W.; Zhou, H.; Guan, X. H.; Liu, J. F.; Bi, Z. Q.; Liu, G. H.; Ren, J.; Wang, H.; Zhao, Z.; Song, J. D.; He, J. R.; Wan, T.; Zhang, J. S.; Fu, X. P.; Sun, L. N.; Dong, X. P.; Feng, Z. J.; Yang, W. Z.; Hong, T.; Zhang, Y.; Walker, D. H.; Wang, Y.; Li, D. X. Fever with Thrombocytopenia Associated with a Novel Bunyavirus in China. *N Engl J Med* **2011**, *364*, 1523–1532.
41. Kaptein, S. J. F.; Jacobs, S.; Langendries, L.; Seldeslachts, L.; ter Horst, S.; Liesenborghs, L.; Hens, B.; Vergote, V.; Heylen, E.; Barthelemy, K.; Maas, E.; De Keyzer, C.; Bervoets, L.; Rymenants, J.; Van Buyten, T.; Zhang, X.; Abdelnabi, R.; Pang, J.; Williams, R.; Thibaut, H. J.; Dallmeier, K.; Boudewijns, R.; Wouters, J.; Augustijns, P.; Verougstraete, N.; Cawthorne, C.; Breuer, J.; Solas, C.; Weynand, B.; Annaert, P.; Spriet, I.; Velde, G. V.; Neyts, J.; Rocha-Pereira, J.; Delang, L. Favipiravir at High Doses Has Potent Antiviral Activity in SARS-CoV-2-Infected Hamsters, Whereas Hydroxychloroquine Lacks Activity. *Proc Natl Acad Sci USA* **2020**, *117*, 26955–26965.
42. Nagata, T.; Lefor, A. K.; Hasegawa, M.; Ishii, M. Favipiravir: A New Medication for the Ebola Virus Disease Pandemic. *Disaster Med. Public Health Prep.* **2015**, *9*, 79–81.
43. Huchting, J.; Vanderlinden, E.; Van Berwaer, R.; Meier, C.; Naesens, L. Cell Line-Dependent Activation and Antiviral Activity of T-1105, the Non-fluorinated Analogue of T-705 (Favipiravir). *Antivir. Res.* **2019**, *167*, 1–5.
44. Gowen, B. B.; Wong, M. H.; Jung, K. H.; Sanders, A. B.; Mendenhall, M.; Bailey, K. W.; Furuta, Y.; Sidwell, R. W. In Vitro and In Vivo Activities of T-705 against Arenavirus and Bunyavirus Infections. *Antimicrob. Agents Chemother.* **2007**, *51*, 3168–3176.
45. Sidwell, R. W.; Barnard, D. L.; Day, C. W.; Smee, D. F.; Bailey, K. W.; Wong, M. H.; Morrey, J. D.; Furuta, Y. Efficacy of Orally Administered T-705 on Lethal Avian Influenza a (H5N1) Virus Infections in Mice. *Antimicrob. Agents Chemother.* **2007**, *51*, 845–851.

46. Huchting, J.; Vanderlinden, E.; Winkler, M.; Nasser, H.; Naesens, L.; Meier, C. Prodrugs of the Phosphoribosylated Forms of Hydroxypyrazinecarboxamide Pseudobase T-705 and its de-Fluoro Analogue T-1105 as Potent Influenza Virus Inhibitors. *J. Med. Chem.* **2018**, *61*, 6193–6210.
47. Eriksson, B.; Helgstrand, E.; Johansson, N. G.; Larsson, A.; Misiorny, A.; Noren, J. O.; Philipson, L.; Stenberg, K.; Stening, G.; Stridh, S.; Oberg, B. Inhibition of Influenza-Virus Ribonucleic-Acid Polymerase by Ribavirin Triphosphate. *Antimicrob. Agents Chemother.* **1977**, *11*, 946–951.
48. Reguera, J.; Gerlach, P.; Cusack, S. Towards a Structural Understanding of RNA Synthesis by Negative Strand RNA Viral Polymerases. *Curr. Opin. Struct. Biol.* **2016**, *36*, 75–84.
49. Ortin, J.; Martin-Benito, J. The RNA Synthesis Machinery of Negative-Stranded RNA Viruses. *Virology* **2015**, *479*, 532–544.
50. Ogando, N. S.; Zevenhoven-Dobbe, J. C.; Meer, Y. V. D.; Bredenbeek, P. J.; Posthuma, C. C.; Snijder, E. J.; Gallagher, T. The Enzymatic Activity of the nsp14 Exoribonuclease Is Critical for Replication of MERS-CoV and SARS-CoV-2. *J. Virol.* **2020**, *94*, e01246–20.
51. Robson, F.; Khan, K. S.; Le, T. K.; Paris, C.; Demirbag, S.; Barfuss, P.; Rocchi, P.; Ng, W.-L. Coronavirus RNA Proofreading: Molecular Basis and Therapeutic Targeting. *Mol. Cell* **2020**, *79*, 710–727.
52. Ferron, F.; Subissi, L.; Silveira De Morais, A. T.; Le, N. T. T.; Sevajol, M.; Gluais, L.; Decroly, E.; Vonrhein, C.; Bricogne, G.; Canard, B.; Imbert, I. Structural and Molecular Basis of Mismatch Correction and Ribavirin Excision from Coronavirus RNA. *PNAS* **2018**, *115*, E162.
53. Sheahan, T. P.; Sims, A. C.; Zhou, S.; Graham, R. L.; Pruijssers, A. J.; Agostini, M. L.; Leist, S. R.; Schäfer, A.; Dinnon, K. H.; Stevens, L. J.; Chappell, J. D.; Lu, X.; Hughes, T. M.; George, A. S.; Hill, C. S.; Montgomery, S. A.; Brown, A. J.; Bluemling, G. R.; Natchus, M. G.; Saindane, M.; Kolykhalov, A. A.; Painter, G.; Harcourt, J.; Tamin, A.; Thornburg, N. J.; Swanstrom, R.; Denison, M. R.; Baric, R. S. An Orally Bioavailable Broad-Spectrum Antiviral Inhibits SARS-CoV-2 in Human Airway Epithelial Cell Cultures and Multiple Coronaviruses in Mice. *Sci. Transl. Med.* **2020**, *12*, eabb5883.
54. Lo, H. S.; Hui, K. P. Y.; Lai, H.-M.; Khan, K. S.; Kaur, S.; Huang, J.; Li, Z.; Chan, A. K. N.; Cheung, H. H.-Y.; Ng, K.-C.; Wang Ho, J. C.; Chen, Y. W.; Ma, B.; Cheung, P. M.-H.; Shin, D.; Wang, K.; Lee, M.-H.; Selisko, B.; Eydoux, C.; Guillemot, J.-C.; Canard, B.; Wu, K.-P.; Liang, P.-H.; Dikic, I.; Zuo, Z.; Chan, F. K. L.; Hui, D. S. C.; Mok, V. C. T.; Wong, K.-B.; Ko, H.; Aik, W. S.; Chan, M. C. W.; Ng, W.-L. Simeprevir Potently Suppresses SARS-CoV-2 Replication and Synergizes with Remdesivir. *ACS Cent. Sci.* **2021**, 7, 792–802.
55. Goldhill, D. H.; Langat, P.; Xie, H.; Galiano, M.; Miah, S.; Kellam, P.; Zambon, M.; Lackenby, A.; Barclay, W. S. J. J. O. V. Determining the Mutation Bias of Favipiravir in Influenza Virus Using Next-Generation Sequencing. *J. Virol.* **2019**, *93*, e01217–18.
56. Baranovich, T.; Wong, S.-S.; Armstrong, J.; Marjuki, H.; Webby, R. J.; Webster, R. G.; Govorkova, E. A. J. J. O. V. T-705 (Favipiravir) Induces Lethal Mutagenesis in Influenza A H1N1 Viruses In Vitro. *J. Virol.* **2013**, *87*, 3741–3751.
57. Escribano-Romero, E.; Jimenez de Oya, N.; Domingo, E.; Saiz, J. C. J. A. A. Chemotherapy. Extinction of West Nile virus by Favipiravir Through Lethal Mutagenesis. *Antimicrob. Agents Chemother.* **2017**, *61*, e01400–17.
58. De Ávila, A. I.; Gallego, I.; Soria, M. E.; et al. Lethal Mutagenesis of Hepatitis C Virus Induced by Favipiravir. *Plos One* **2016**, *11*, e0164691.
59. Arias, A.; Thorne, L.; Goodfellow, I. J. E. Favipiravir Elicits Antiviral Mutagenesis During Virus Replication In Vivo. *eLife* **2014**, *3*, e03679.

60. Sangawa, H.; Komeno, T.; Nishikawa, H.; Yoshida, A.; Takahashi, K.; Nomura, N.; Furuta, Y. Mechanism of Action of T-705 Ribosyl Triphosphate against Influenza Virus RNA Polymerase. *Antimicrob. Agents Chemother.* **2013**, *57*, 5202–5208.
61. Jin, Z. N.; Smith, L. K.; Rajwanshi, V. K.; Kim, B.; Deval, J. The Ambiguous Base-Pairing and High Substrate Efficiency of T-705 (Favipiravir) Ribofuranosyl 5′-Triphosphate Towards Influenza A Virus Polymerase. *PLoS One* **2013**, *8*, e68347.
62. Abdelnabi, R.; de Morais, A. T. S.; Leyssen, P.; et al. Understanding the Mechanism of the Broad-Spectrum Antiviral Activity of Favipiravir (T-705): Key Role of the F1 Motif of the Viral Polymerase. *J. Virol.* **2017**, *91*, e00487–17.
63. Shannon, A.; Selisko, B.; Le, N. T. T.; et al. Rapid Incorporation of Favipiravir by the Fast and Permissive Viral RNA Polymerase Complex Results in SARS-CoV-2 Lethal Mutagenesis. *Nat. Commun.* **2020**, *11*, 4682.
64. Te Velthuis, A. J.; Arnold, J. J.; Cameron, C. E.; Van Den Worm, S. H.; Snijder, E. J. J. N. A. R. The RNA Polymerase Activity of SARS-Coronavirus nsp12 Is Primer Dependent. *Nucleic Acids Res.* **2010**, *38*, 203–214.
65. Cheng, A.; Zhang, W.; Xie, Y.; Jiang, W.; Arnold, E.; Sarafianos, S. G.; Ding, J. J. V. Expression, Purification, and Characterization of SARS Coronavirus RNA Polymerase. *Virology* **2005**, *335*, 165–176.
66. Huchting, J.; Winkler, M.; Nasser, H.; Meier, C. Synthesis of T-705-Ribonucleoside and T-705-Ribonucleotide and Studies of Chemical Stability. *ChemMedChem* **2017**, *12*, 652–659.
67. Dulin, D.; Arnold, J. J.; van Laar, T.; Oh, H. S.; Lee, C.; Perkins, A. L.; Harki, D. A.; Depken, M.; Cameron, C. E.; Dekker, N. H. Signatures of Nucleotide Analog Incorporation by an RNA-Dependent RNA Polymerase Revealed Using High-Throughput Magnetic Tweezers. *Cell Rep.* **2017**, *21*, 1063–1076.
68. Eckerle, L. D.; Lu, X.; Sperry, S. M.; et al. High Fidelity of Murine Hepatitis Virus Replication Is Decreased in nsp14 Exoribonuclease Mutants. *J. Virol.* **2007**, *81*, 12135–12144.
69. Smith, E. C.; Blanc, H.; Vignuzzi, M.; et al. Coronaviruses Lacking Exoribonuclease Activity Are Susceptible to Lethal Mutagenesis: Evidence for Proofreading And Potential Therapeutics. *PLOS Pathog.* **2013**, *9*, e1003565.
70. Sowa, T.; Ouchi, S. Facile Synthesis of 5′-Nucleotides by Selective Phosphorylation of a Primary Hydroxyl Group of Nucleosides with Phosphoryl Chloride. *Bull. Chem. Soc. Jpn.* **1975**, *48*, 2084–2090.
71. Cremosnik, G. S.; Hofer, A.; Jessen, H. J. J. A. C. I. E. Iterative Synthesis of Nucleoside Oligophosphates with Phosphoramidites. *Angew. Chem., Int. Ed.* **2014**, *53*, 286–289.
72. Meier, C. 2-Nucleos-5'-O-yl-4H-1,3,2-Benzodioxaphosphinin-2-Oxides—A New Concept for Lipophilic, Potential Prodrugs of Biologically Active Nucleoside Monophosphates. *Angew. Chem., Int. Ed.* **1996**, *35*, 70–72.
73. Meier, C.; De Clercq, E.; Balzarini, J. Nucleotide Delivery from *cyclo*Saligenyl-3'-Azido-3'-Deoxythymidine Monophosphates (*cyclo*Sal-AZTMP). *Eur. J. Org. Chem.* **1998**, 837–846.
74. Meier, C.; Ludek, O.; Krämer, T.; et al. Divergent Synthesis and Biological Evaluation of Carbocyclic α-, Iso- and 3′-Epi-Nucleosides and Their Lipophilic Nucleotide Prodrugs. *Synthesis* **2006**, *8*, 1313–1324.
75. Meier, C.; Lorey, M.; De Clercq, E.; Balzarini, J. *Cyclo*Sal-2',3'-Dideoxy-2', 3'-Didehydrothymidine Monophosphate (*cyclo*Sal-d4TMP): Synthesis and Antiviral Evaluation of a New d4TMP Delivery System. *J. Med. Chem.* **1998**, *41*, 1417–1427.
76. McGuigan, C.; Cahard, D.; Sheeka, H. M.; DeClercq, E.; Balzarini, J. Aryl Phosphoramidate Derivatives of d4T Have Improved Anti-HIV Efficacy in Tissue Culture and May Act by the Generation of a Novel Intracellular Metabolite. *J. Med. Chem.* **1996**, *39*, 1748–1753.

77. Balzarini, J.; Karlsson, A.; Aquaro, S.; Perno, C. F.; Cahard, D.; Naesens, L.; DeClercq, E.; McGuigan, C. Mechanism of Anti-HIV Action of Masked Alaninyl d4TMP Derivatives. *Proc. Natl. Acad. Sci. U. S. A.* **1996**, *93*, 7295–7299.
78. Balzarini, J.; Egberink, H.; Hartmann, K.; Cahard, D.; Vahlenkamp, T.; Thormar, H.; DeClercq, E.; McGuigan, C. Antiretrovirus Specificity and Intracellular Metabolism of 2′,3′-Didehydro-2′,3′-Dideoxythymidine (Stavudine) and Its 5′-Monophosphate Triester Prodrug So324. *Mol. Pharmacol.* **1996**, *50*, 1207–1213.
79. Girardet, J. L.; Perigaud, C.; Aubertin, A. M.; Gosselin, G.; Kirn, A.; Imbach, J. L. Increase of the Anti-HIV Activity of d4T in human T-Cell Culture by the Use of the Sate Pronucleotide Approach. *Bioorg. Med. Chem. Lett.* **1995**, *5*, 2981–2984.
80. Farquhar, D.; Khan, S.; Srivastva, D. N.; Saunders, P. P. Synthesis and Antitumor Evaluation of Bis[(Pivaloyloxy)Methyl] 2′-Deoxy-5-Fluorouridine 5′-Monophosphate (FdUMP)—A Strategy to Introduce Nucleotides into Cells. *J. Med. Chem.* **1994**, *37*, 3902–3909.
81. Farquhar, D.; Chen, R.; Khan, S. 5′-[4-(Pivaloyloxy)-1,3,2-Dioxaphosphorinan-2-Yl]-2′-Deoxy-5-Fluorouridine—A Membrane-Permeating Prodrug of 5-Fluoro-2′-Deoxyuridylic Acid (FdUMP). *J. Med. Chem.* **1995**, *38*, 488–495.
82. Briggs, A. D.; Camplo, M.; Freeman, S.; Lundstrom, J.; Pring, B. G. Acyloxymethyl and 4-Acyloxybenzyl Diester Prodrugs of Phosphonoformate. *Tetrahedron* **1996**, *52*, 14937–14950.
83. Lefebvre, I.; Perigaud, C.; Pompon, A.; Aubertin, A. M.; Girardet, J. L.; Kirn, A.; Gosselin, G.; Imbach, J. L. Mononucleoside Phosphotriester Derivatives with S-Acyl-2-Thioethyl Bioreversible Phosphate-Protecting Groups—Intracellular Delivery of 3′-Azido-2′,3′-Dideoxythymidine 5′-Monophosphate. *J. Med. Chem.* **1995**, *38*, 3941–3950.
84. Puech, F.; Gosselin, G.; Lefebvre, I.; Pompon, A.; Aubertin, A. M.; Kirn, A.; Imbach, J. L. Intracellular Delivery of Nucleoside Monophosphates through a Reductase-Mediated Activation Process. *Antivir. Res.* **1993**, *22*, 155–174.
85. Wagner, C. R.; Chang, S. L.; Griesgraber, G. W.; Song, H.; McIntee, E. J.; Zimmerman, C. L. Antiviral Nucleoside Drug Delivery via Amino Acid Phosphoramidates. *Nucleosides Nucleotides Nucleic Acids* **1999**, *18*, 913–919.
86. Meier, C.; Lomp, A.; Meerbach, A.; Wutzler, P. *Cyclo*Sal-BVDUMP Pronucleotides: How to Convert an Antiviral-Inactive Nucleoside Analogue into a Bioactive Compound against EBV. *J. Med. Chem.* **2002**, *45*, 5157–5172.
87. Meier, C. *Cyclo*Sal Phosphates as Chemical Trojan Horses for Intracellular Nucleotide and Glycosylmonophosphate Delivery—Chemistry Meets Biology. *Eur. J. Org. Chem.* **2006**, *5*, 1081–1102.
88. Meier, C. *Cyclo*Sal-Pronucleotides—Design of Chemical Trojan Horses. *Mini-Rev. Med. Chem.* **2002**, *2*, 219–234.
89. Meier, C.; Habel, L.; Haller-Meier, F.; Lomp, A.; Herderich, M.; Klocking, R.; Meerbach, A.; Wutzler, P. Chemistry and Anti-Herpes Simplex Virus Type 1 Evaluation of *cyclo*Sal-Nucleotides of Acyclic Nucleoside Analogues. *Antivir. Chem. Chemother.* **1998**, *9*, 389–402.
90. Meier, C.; Balzarini, J. Application of the *cyclo*Sal-Prodrug Approach for Improving the Biological Potential of Phosphorylated Biomolecules. *Antivir. Res.* **2006**, *71*, 282–292.
91. Meier, C.; Ruppel, M. F. H.; Vukadinovic, D.; Balzarini, J. Second Generation of *cyclo*Sal-Pronucleotides with Esterase-Cleavable Sites: The "Lock-in"-Concept. *Nucleosides Nucleotides Nucleic Acids* **2004**, *23*, 89–115.
92. Gisch, N.; Balzarini, J.; Meier, C. Enzymatically Activated *cyclo*Sal-d4T-Monophosphates: The Third Generation of *cyclo*Sal-Pronucleotides. *J. Med. Chem.* **2007**, *50*, 1658–1667.

93. Meier, C.; Knispel, T.; DeClercq, E.; Balzarini, J. ADA-Bypass by Lipophilic *cyclo*Sal-ddAMP pro-Nucleotides a Second Example of the Efficiency of the *cyclo*Sal-Concept. *Bioorg. Med. Chem. Lett.* **1997**, 7, 1577–1582.
94. Meier, C.; Knispel, T.; De Clercq, E.; Balzarini, J. *Cyclo*Sal-Pronucleotides of 2′,3′-Dideoxyadenosine and 2′,3′-Dideoxy-2′,3′-Didehydroadenosine: Synthesis and Antiviral Evaluation of a Highly Efficient Nucleotide Delivery System. *J. Med. Chem.* **1999**, *42*, 1604–1614.
95. Lorey, M.; Meier, C.; DeClercq, E.; Balzarini, J. New Synthesis and Antitumor Activity of *cyclo*Sal-Derivatives of 5-Fluoro-2′-Deoxyuridinemonophosphate. *Nucleosides Nucleotides Nucleic Acids* **1997**, *16*, 789–792.
96. Lorey, M.; Meier, C.; DeClercq, E.; Balzarini, J. *Cyclo*-Saligenyl-5-Fluoro-2′-Deoxyuridinemonophosphate (*cyclo*Sal-FdUMP)—A New Prodrug Approach for FdUMP. *Nucleosides Nucleotides Nucleic Acids* **1997**, *16*, 1307–1310.
97. Meier, C.; DeClercq, E.; Balzarini, J. *Cyclo*-Saligenyl-3′-Azido-2′,3′-Dideoxy-Thymidinemonophosphate (*cyclo*Sal-AZTMP)—A New pro-Nucleotide Approach. *Nucleosides Nucleotides Nucleic Acids* **1997**, *16*, 793–796.
98. Balzarini, J.; Naesens, L.; Aquaro, S.; Knispel, T.; Perno, C. F.; De Clercq, E.; Meier, C. Intracellular Metabolism of *cyclo*Saligenyl 3’-Azido-2’,3’-Dideoxythymidine Monophosphate, a Prodrug of 3’-Azido-2’,3’-Dideoxythymidine (Zidovudine). *Mol. Pharmacol.* **1999**, *56*, 1354–1361.
99. Mazzon, C.; Rampazzo, C.; Scaini, M. C.; Gallinaro, L.; Karlsson, A.; Meier, C.; Balzarini, J.; Reichard, P.; Bianchi, V. Cytosolic and Mitochondrial Deoxyribonucleotidases: Activity with Substrate Analogs, Inhibitors and Implications for Therapy. *Biochem. Pharmacol.* **2003**, *66*, 471–479.
100. Furman, P. A.; Fyfe, J. A.; Stclair, M. H.; Weinhold, K.; Rideout, J. L.; Freeman, G. A.; Lehrman, S. N.; Bolognesi, D. P.; Broder, S.; Mitsuya, H.; Barry, D. W. Phosphorylation of 3′-Azido-3′-Deoxythymidine and Selective Interaction of the 5′-Triphosphate with Human-Immunodeficiency-Virus Reverse-Transcriptase. *Proc. Natl. Acad. Sci. U. S. A.* **1986**, *83*, 8333–8337.
101. Balzarini, J.; Herdewijn, P.; Declercq, E. Differential Patterns of Intracellular Metabolism of 2′,3′-Didehydro-2′,3′-Dideoxythymidine and 3′-Azido-2′, 3′-Dideoxythymidine, 2 Potent Anti-Human Immunodeficiency Virus Compounds. *J. Biol. Chem.* **1989**, *264*, 6127–6133.
102. Hostetler, K. Y.; Stuhmiller, L. M.; Lenting, H. B. M.; Vandenbosch, H.; Richman, D. D. Synthesis and Antiretroviral Activity of Phospholipid Analogs of Azidothymidine and Other Antiviral Nucleosides. *J. Biol. Chem.* **1990**, *265*, 6112–6117.
103. Vanwijk, G. M. T.; Hostetler, K. Y.; Vandenbosch, H. Lipid Conjugates of Antiretroviral Agents—Release of Antiretroviral Nucleoside Monophosphates by a Nucleoside Diphosphate Diglyceride Hydrolase Activity from Rat-Liver Mitochondria. *Biochim. Biophys. Acta* **1991**, *1084*, 307–310.
104. Hostetler, K. Y.; Richman, D. D.; Carson, D. A.; Stuhmiller, L. M.; Vanwijk, G. M. T.; Vandenbosch, H. Greatly Enhanced Inhibition of Human-Immunodeficiency-Virus Type-1 Replication in Cem and Ht4-6c Cells by 3′-Deoxythymidine Diphosphate Dimyristoylglycerol, a Lipid Prodrug of 3′-Deoxythymidine. *Antimicrob. Agents Chemother.* **1992**, *36*, 2025–2029.
105. Hostetler, K. Y.; Parker, S.; Sridhar, C. N.; Martin, M. J.; Li, J. L.; Stuhmiller, L. M.; Vanwijk, G. M. T.; Vandenbosch, H.; Gardner, M. F.; Aldern, K. A.; Richman, D. D. Acyclovir Diphosphate Dimyristoylglycerol—A Phospholipid Prodrug with Activity against Acyclovir-Resistant Herpes-Simplex Virus. *Proc. Natl. Acad. Sci. U. S. A.* **1993**, *90*, 11835–11839.

106. Bonnaffe, D.; Dupraz, B.; Ughetto-Monfrin, J.; et al. Synthesis of Acyl Pyrophosphates - Application to the Synthesis of Nucleotide Lipophilic Prodrugs. *Tetrahedron Lett.* **1995**, *36*, 531–534.
107. Bonnaffe, D.; Dupraz, B.; Ughetto-Monfrin, J.; et al. Synthesis of Nucleotide Lipophilic Prodrugs Containing 2 Inhibitors Targeted against Different Phases of the HIV Replication Cycle. *Nucleosides, Nucleotides Nucleic Acids* **1995**, *14*, 783–787.
108. Bonnaffe, D.; Dupraz, B.; Ughetto-Monfrin, J.; et al. Potential Lipophilic Nucleotide Prodrugs: Synthesis, Hydrolysis, and Antiretroviral Activity of AZT and d4T Acyl Nucleotides. *J. Org. Chem.* **1996**, *61*, 895–902.
109. Thomson, W.; Nicholls, D.; Irwin, W. J.; et al. Synthesis, Bioactivation and Anti-HIV Activity of the Bis(4-Acyloxybenzyl) and Mono(4-Acyloxybenzyl) Esters of the 5′-Monophosphate of AZT. *J. Chem. Soc. Perkin Trans.* **1993**, *11*, 1239–1245.
110. Jessen, H. J.; Schulz, T.; Balzarini, J.; Meier, C. Bioreversible Protection of Nucleoside Diphosphates. *Angew. Chem., Int. Ed.* **2008**, *47*, 8719–8722.
111. Pertenbreiter, F.; Balzarini, J.; Meier, C. Nucleoside Mono- and Diphosphate Prodrugs of 2', 3'-Dideoxyuridine and 2', 3'-Dideoxy-2', 3'-Didehydrouridine. *ChemMedChem* **2015**, *10*, 94–106.
112. Schulz, T.; Balzarini, J.; Meier, C. The Di*PP*ro Approach: Synthesis, Hydrolysis, and Antiviral Activity of Lipophilic d4T Diphosphate Prodrugs. *ChemMedChem* **2014**, *9*, 762–775.
113. Weinschenk, L.; Schols, D.; Balzarini, J.; Meier, C. Nucleoside Diphosphate Prodrugs: Nonsymmetric Di*PP*ro-Nucleotides. *J. Med. Chem.* **2015**, *58*, 6114–6130.
114. Meier, C.; Jessen, H. J.; Schulz, T.; Weinschenk, L.; Pertenbreiter, F.; Balzarini, J. Rational Development of Nucleoside Diphosphate Prodrugs: Di*PP*ro-Compounds. *Curr. Med. Chem.* **2015**, *22*, 3933–3950.
115. Weinschenk, L.; Gollnest, T.; Schols, D.; Balzarini, J.; Meier, C. Bis(Benzoyloxybenzyl)-Di*PP*ro Nucleoside Diphosphates of Anti-HIV Active Nucleoside Analogues. *ChemMedChem* **2015**, *10*, 891–900.
116. Zhu, Y. L.; Dutschman, G. E.; Liu, S. H.; Bridges, E. G.; Cheng, Y. C. Anti-Hepatitis B Virus Activity and Metabolism of 2',3'-Dideoxy-2',3'-Didehydro-beta-L(−)-5-Fluorocytidine. *Antimicrob. Agents Chemother.* **1998**, *42*, 1805–1810.
117. Kreimeyer, A.; Andre, F.; Gouyette, C.; Huynh-Dinh, T. Transmembrane Transport of Adenosine 5'-Triphosphate Using a Lipophilic Cholesteryl Derivative. *Angew. Chem., Int. Ed.* **1998**, *37*, 2853–2855.
118. Bonnaffe, D.; Dupraz, B.; Ughettomonfrin, J.; Namane, A.; Dinh, T. H. Synthesis of Acyl Pyrophosphates—Application to the Synthesis of Nucleotide Lipophilic Prodrugs. *Tetrahedron Lett.* **1995**, *36*, 531–534.
119. Vanwijk, G. M. T.; Hostetler, K. Y.; Kroneman, E.; Richman, D. D.; Sridhar, C. N.; Kumar, R.; Vandenbosch, H. Synthesis and Antiviral Activity of 3′-Azido-3′-Deoxythymidine Triphosphate Distearoylglycerol—A Novel Phospholipid Conjugate of the Anti-HIV Agent AZT. *Chem. Phys. Lipids* **1994**, *70*, 213–222.
120. Bonnaffe, D.; Dupraz, B.; Ughettomonfrin, J.; Namane, A.; Dinh, T. H. Synthesis of Nucleotide Lipophilic Prodrugs Containing 2 Inhibitors Targeted against Different Phases of the HIV Replication Cycle. *Nucleosides Nucleotides Nucleic Acids* **1995**, *14*, 783–787.
121. Bonnaffe, D.; Dupraz, B.; UghettoMonfrin, J.; Namane, A.; Henin, Y.; Dinh, T. H. Potential Lipophilic Nucleotide Prodrugs: Synthesis, Hydrolysis, and Antiretroviral Activity of AZT and d4T Acyl Nucleotides. *J. Organomet. Chem.* **1996**, *61*, 895–902.
122. Gollnest, T.; de Oliveira, T. D.; Schols, D.; Balzarini, J.; Meier, C. Lipophilic Prodrugs of Nucleoside Triphosphates as Biochemical Probes and Potential Antivirals. *Nat. Commun.* **2015**, *6*, 8716.

123. Gollnest, T.; de Oliveira, T. D.; Rath, A.; Hauber, I.; Schols, D.; Balzarini, J.; Meier, C. Membrane-Permeable Triphosphate Prodrugs of Nucleoside Analogues. *Angew. Chem. Int. Ed.* **2016**, *55*, 5255–5258.
124. Mohamady, S.; Jakeman, D. L. An Improved Method for the Synthesis of Nucleoside Triphosphate Analogues. *J. Organomet. Chem.* **2005**, *70*, 10588–10591.
125. Mohamady, S.; Taylor, S. D. General Procedure for the Synthesis of Dinucleoside Polyphosphates. *J. Organomet. Chem.* **2011**, *76*, 6344–6349.
126. Jia, X.; Schols, D.; Meier, C. Anti-HIV-Active Nucleoside Triphosphate Prodrugs. *J. Med. Chem.* **2020**, *63*, 6003–6027.
127. Jia, X.; Schols, D.; Meier, C. Lipophilic Triphosphate Prodrugs of Various Nucleoside Analogues. *J. Med. Chem.* **2020**, *63*, 6991–7007.
128. Zhao, C. L.; Jia, X.; Schols, D.; Balzarini, J.; Meier, C. γ-Non-Symmetrically Dimasked Tri*PP*Pro-Prodrugs as Potential Antiviral Agents against HIV. *ChemMedChem* **2020**, *16*, 499–512.

CHAPTER TWO

Anno 2021: Which antivirals for the coming decade?

Elisabetta Groaz[a,b,*], Erik De Clercq[c], and Piet Herdewijn[a]
[a]Medicinal Chemistry, Rega Institute for Medical Research, KU Leuven, Leuven, Belgium
[b]Department of Pharmaceutical and Pharmacological Sciences, University of Padova, Padova, Italy
[c]Department of Microbiology, Immunology and Transplantation, Rega Institute for Medical Research, KU Leuven, Leuven, Belgium
*Corresponding author: e-mail addresses: elisabetta.groaz@kuleuven.be; elisabetta.groaz@unipd.it

Contents

1. Introduction

Interest in antiviral research has changed focus within the scientific community over the years. After the pioneering work in the 1960s–1970s on herpesviruses that led to the identification of acyclovir as the gold standard for the treatment of herpes simplex virus (HSV) infections, the field subsequently boomed in the 1980s–1990s after the discovery of the immunodeficiency virus (HIV) as the causative agent of acquired immunodeficiency syndrome (AIDS). In the early 2000s, direct acting antivirals (DAAs) were developed for the treatment of hepatitis C (HCV), following

Annual Reports in Medicinal Chemistry, Volume 57
ISSN 0065-7743
https://doi.org/10.1016/bs.armc.2021.09.004
Copyright © 2021 Elsevier Inc. All rights reserved.

a tremendous medicinal chemistry effort that culminated in the discovery of sofosbuvir in 2007. Over a period of approximately 50 years, the focus has shifted from DNA viruses to retroviruses, and further onto RNA viruses as targets for drug discovery. At this moment, worldwide efforts are ongoing to discover new drugs, including nucleoside analogues, against the newly emerged severe acute respiratory syndrome coronavirus type 2 (SARS-CoV-2). An important question to answer is which viruses remain to be treated and against which viral infections drug discovery is focusing in 2021. Due to climate change, frequent international travel, profound demographic shift, increasing air pollution, overpopulation in certain regions, changes in the population age distribution, virus infections are in a continuous flux. Global warming might lead to a continuous re-emergence of hemorrhagic fever viruses in the future, bringing these pathogens closer to the northern countries. Globally prevalent respiratory tract infections, such as influenza, respiratory syncytial virus (RSV) and rhinoviruses, have been declining thanks to current containment measures for SARS-CoV-2; however, increase in air pollution is expected to increase mortality and morbidity due to these viral respiratory infections.

Virus infections of zoonotic origin (e.g., rabies, coronaviruses) will never be eradicated, and new viruses might emerge using wild or domesticated animals as vectors. Other infections that were almost eradicated due to vaccination (e.g., measles, polio) may come back given irregular immunization coverage caused by inequitable global access to vaccines and growing vaccine hesitancy across the world. Bioterrorism could, in theory, also make use of deadly viruses as weapons. For those viruses known to be life-threatening, prophylactic treatment can be considered. Likewise, the population may benefit from the development of antivirals against viruses that may cause cancer or neurological diseases; in this way antivirals may contribute to the general fitness of the society. In this review, we describe the efforts that have been conducted during the last 3 years to discover antivirals against several of these treats. This review highlights some important new discoveries, and it is not meant to give a complete survey of every ongoing effort.

2. Severe acute respiratory syndrome coronavirus type 2 (SARS-CoV-2)

The ongoing coronavirus disease 2019 (COVID-19) pandemic caused by the highly transmissible and pathogenic severe acute respiratory syndrome coronavirus 2 (SARS-CoV-2) can be anticipated to be resolved by vaccination; however, it might be important for future prevention and control to develop a pan-coronavirus vaccine (see also discussion in Section 13), as initiated by VBI vaccines Inc. with a trivalent eVLP Vaccine candidate encoding the antigenic spike proteins of SARS-CoV-2, SARS-CoV, and Middle East respiratory syndrome coronavirus (MERS-CoV) in a single particle.[1] This expectation is obviously contingent on a number of factors including equitable access to vaccines, widespread acceptance, high uptake, and sustained coverage specially against rapidly emerging variants. Even in the event of SARS-CoV-2 becoming an endemic pathogen, the future emergence of related epidemic or pandemic outbreaks cannot be ruled out, making critical the availability of effective therapeutic strategies for those who are infected.

The diverse repertoire of structural, non-structural, and accessory proteins encoded by the SARS-CoV-2 RNA genome[2] reveals multiple opportunities for antiviral therapeutic targeting, which are all being relentlessly pursued as it can be expected that a cocktail of compounds acting at different steps of the viral life cycle will be needed to efficiently suppress viral replication in vivo. These may include inhibitors of viral entry, polyprotein proteolytic processing, and viral RNA synthesis, which can be additionally complemented by other classes of antivirals interfering with host-mediated regulatory functions. More detailed information about this quickly evolving drug discovery panorama can be retrieved from recent literature reviews.[3–5]

At present, the only investigational drug to have received emergency use authorization for the treatment of adults and children from 12 year old with severe COVID-19 is the pyrrolo[2,1-f][triazin-4-amino]adenine 1'-cyano substituted C-nucleoside ProTide remdesivir **1** (GS-5734, Veklury®, Gilead Sciences, Fig. 1),[6] an RNA-dependent RNA polymerase (RdRp) inhibitor.[7] In a placebo-controlled trial, remdesivir was found to speed the recovery time in hospitalized adult patients with confirmed SARS-CoV-2 lower respiratory tract infection.[8] However, the results of a recently published meta-analysis of randomized trials suggested little or no mortality benefit associated with the intravenous administration of remdesivir in COVID-19 patients.[9] Being a phosphoramidate prodrug, remdesivir is converted on cell entry to its corresponding modified nucleoside monophosphate by the sequential action of esterases and phosphoramidases; such metabolic step occurs primarily in liver cells where the expression of these hydrolytic enzymes is more pronounced. As this prodrug design was originally not intended for lung delivery, debate is ongoing on whether respiratory

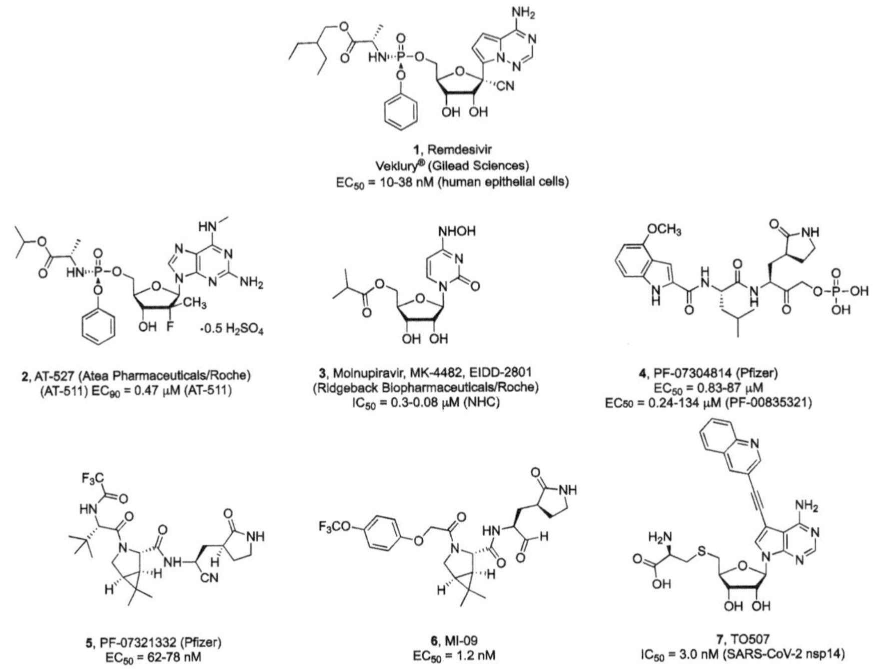

Fig. 1 Selected nucleoside and protease inhibitors against SARS-CoV-2.

viruses such as SARS-CoV-2 that predominantly infect alveolar epithelial cells might be better inhibited by nucleoside prodrugs that could be selectively converted to the active drug in the infected tissue. In a recent study, the nucleoside GS-441524, which turned out to be a major metabolite present in the bloodstream after administration of remdesivir, was demonstrated to significantly reduce viral titers and improve inflammation in the lungs of SARS-CoV-2-infected mice,[10] suggesting that a more favorable lung tissue distribution could be achieved with this compound.[11] Regardless of this consideration, remdesivir remains an intriguing molecule that may inspire future structural optimization programs for the development of potent RdRp inhibitors against SARS-CoV-2 infection. One possibility could be the synthesis of more lipophilic prodrugs of remdesivir or its analogues for the delivery of 5'-monophosphates in the cells.[12]

The potential impact of the prodrug moiety on the inhibitory effect of nucleoside analogues towards SARS-CoV-2 replication can be further surmised by the in vitro inactivity of sofosbuvir against this virus in Vero cells versus an EC_{50} of 6.3 μM displayed by its cyclic phosphate prodrug congener.[13] Two moderately active nucleoside lead compounds are

2'-methyl-cytidine and 2'-methyl-7-fluoro-7-deazaadenosine that displayed antiviral activity against SARS-CoV-2 in Vero cells at 9.2 and 7.6 μM, respectively.[13]

AT-527 (Atea Pharmaceuticals/Roche), compound **2** (Fig. 1), is an oral double prodrug of a 2'-fluoro-2'-methyl guanosine analogue that recently entered clinical trials for the treatment of mild or moderate COVID-19 disease in non-hospitalized patients [NCT04396106 and NCT04709835 (Phase II); NCT04889040 (Phase III)]. Its free base, designated as AT-511, was described to exert sub-micromolar activity (EC_{90} = 0.47 μM) in human airway epithelial cells (HAE) infected with SARS-CoV-2[14]; however, this activity profile was not supported by an independent study using similar cellular assays,[15] leaving the in vivo efficacy of this molecule to be ascertained.

Another broad-spectrum polymerase inhibitor currently in Phase II clinical trials for the treatment of COVID-19 is the 5'-isobutyryl ester prodrug of N^4-hydroxycytidine (NHC) known as molnupiravir (**3**, EIDD-2801; Ridgeback Biotherapeutic in partnership with Merck), which was recently shown to be well tolerated and effectively reduce nasopharyngeal SARS-CoV-2 infectious virus and viral RNA.[16] This orally administered antiviral agent was shown to potently inhibit SARS-CoV-2 replication as well as suppress viral transmission in vivo in different animal models.[17,18] Therapeutic and prophylactic administration of **3** in immunodeficient mice implanted with authentic human lung tissue (LoM) revealed a >4 log reduction of infectious virus titers and effective prevention of SARS-CoV-2 infection, respectively.[19] However, the biochemical mechanism mediating the mutagenic effect of N^4-hydroxycytidine has yet to be elucidated and may be a concern for long term treatment.[20]

At present, there are 36 studies listed in the ClinicalTrials.gov registry to assess the efficacy and safety of the purine nucleoside prodrug analogue favipiravir (T705) as oral treatment against COVID-19. A meta-analysis of all completed clinical trials till December 2020 revealed a positive clinical improvement in patients receiving favipiravir during seven days after hospitalization compared to control groups, but no significant beneficial effect in reducing mortality for patients with mild to moderate COVID-19.[21] Electron cryomicroscopy (cryoEM) studies revealed that favipiravir ribonucleoside triphosphate (favipiravir-RTP) interacts with the catalytic site of SARS-CoV-2 RdRp in a nonproductive binding mode that may account for the inefficient incorporation into a RNA primer strand observed in polymerase activity assays.[22]

At the moment, it is unclear which specific combination of sugar, nucleobase, and prodrug form would be ideal for enabling potent and tissue-specific antiviral activity against SARS-CoV-2 in vivo.

However, an issue to be faced in developing nucleoside-based inhibitors is associated with an increased risk of toxicity compared to other drug classes, as their antiviral activity is exerted by competing with natural substrates for RdRp-catalyzed incorporation.[23] Modifications introduced at the 4'-position of the potent antitumor compound gemcitabine to decrease toxic effects have been attempted without success.[24]

The virally encoded chymotrypsin-like or main protease ($3CL^{pro}$ or M^{pro}) is a critical component of the SARS-CoV-2 life cycle. Along with another cysteine protease, the papain-like protease (PL^{pro}), the SARS-CoV-2 M^{pro} processes the virus p1a/p1ab polyproteins translated from the viral RNA into a series of individual non-structural proteins (nsps), including the M^{pro} itself, which are essential for viral replication and transcription. Of note, no close human analogues of the coronavirus M^{pro} are known. Thereby, selective SARS-CoV-2 M^{pro} inhibition represents an attractive approach to COVID-19 treatment that could avoid unintended drug-target interactions. Furthermore, the antiviral efficacy of small molecule M^{pro} inhibitors is not expected to be affected by spike protein variants, as the SARS-CoV-2 M^{pro} and the viral spike protein are distinct protein entities within the viral proteome.

Impressive drug discovery efforts from both academia and industry have provided clear evidence of the ability of various M^{pro} small molecule inhibitors to efficiently suppress SARS-CoV-2 replication in vitro, as recently reviewed.[25] Nonetheless, this research is still at an early discovery stage and only two viable clinical candidate have been identified so far.

The M^{pro} protease inhibitor PF-07304814 (**4**) is a phosphate prodrug that was shown to undergo in vivo metabolic conversion to its parent derivative designated as PF-00835321.[26] PF-00835231 inhibited recombinant SARS-CoV-2 M^{pro} activity with nanomolar potency (Ki = 0.271 nM) with excellent off-target selectivity, while exhibiting potent and highly selective antiviral activity in both cellular assays and in vivo murine SARS-CoV-2 model either as a single agent or in combination with remdesivir. A Phase I clinical trial in patients hospitalized with SARS-CoV-2 virus infection to assess the safety, tolerability, and pharmacokinetics of intravenously dosed **4** has been completed and is awaiting publication of results (NCT04535167).

PF-07321332 (compound **5**) is the latest orally active inhibitor of SARS-CoV-2 main protease to have reportedly entered phase III clinical trials in 2021 and is currently under investigation for its safety in combination with the antiretroviral protease inhibitor ritonavir (RTV) in non-hospitalized COVID-19 patients with increased risk of developing severe illness (NCT04960202).[27] Compound **5** strongly inhibited SARS-CoV-2 M^{pro} biochemical activity with an inhibitory constant (Ki) value of 3.11 nM, while no inhibitory effects were observed against a range of mammalian proteases. In the SARS-CoV-2 infection Vero E6 cell assay, PF-07321332 displayed potent inhibition of viral replication (EC_{50} = 74.5 nM), which was further confirmed using physiologically relevant in vitro assays in human adenocarcinoma-derived alveolar basal epithelial cells constitutively expressing ACE2 (A549-ACE2 (EC_{50} = 77.9 nM, EC_{90} = 215 nM) and differentiated normal human bronchial epithelial (dNHBE) cells (EC_{50} = 61.8 nM, EC_{90} = 181 nM). The in vivo oral antiviral potency, safety, and pharmacokinetics profile of PF-07321332 were demonstrated in a mouse-adapted SARS-CoV-2 model and an ongoing Phase I clinical trial in healthy human participants (NCT04756531).

M^{pro} inhibitors containing a bicyclic proline similar to **5** were also disclosed by Yang et al. Among a set of 32 compounds, two compounds indicated as MI-09 (compound **6**) and MI-30 displayed potent SARS-CoV-2 M^{pro} inhibitory activity (IC_{50} = 15.2 and 17.2 nM, respectively) combined with nanomolar antiviral activity in SARS-CoV-2 infected human alveolar epithelial cells (HPAEpiC). These compounds were additionally evaluated in a transgenic mouse model of SARS-CoV-2 infection, where their oral or intraperitoneal administration could significantly reduce lung viral load and ameliorate lung pathological damage.[28] These latest discoveries are particularly noteworthy as oral formulations of highly efficacious antiviral agents hold significant potential for more convenient and widespread therapeutic intervention.

More recently, the two methyltransferases (MTases), nsp14 and nsp16, employed by SARS-CoV-2 to protect the 5'-end of its newly synthesized mRNA are also receiving attention as potential therapeutic targets.[29,30] RNA capping is crucial to ensure efficient viral translation and evade clearance by host innate immune responses. The nsp14 MTase catalyzes the transfer of a methyl group from *S*-adenosyl-L-methionine (SAM) to the 7-position of the terminal guanosine residue. The rational modification of the *S*-adenosyl-L-homocysteine (SAH), which is formed as main product during this process, led to 7-substituted deaza analogues such as **7** with strong binding affinities and substantial inhibitory activity against this enzyme.[30]

All discussed results are preliminary and in this quickly evolving scenario it is hard to tell which treatment approach(es) may deliver antiviral drugs with substantial clinical efficacy for the treatment of COVID-19.

3. Influenza viruses

Influenza viruses (IVs, types A, B, C, and D) are commonly circulating respiratory pathogens of the *Orthomyxoviridae* family, whose genome is organized in a number of (-)-ss RNA segments.[31] Clinically, influenza A (IAV) strains are the most significant due to evolutionary advantages conferred by the propensity of their genome to undergo reassortment.[32] This may contribute to the emergence of highly virulent pandemic subtypes with potential for cross-species transmission and ability to escape vaccine or drug pressure.[33] Early classes of antiviral drugs licensed for flu infections, which interfere with the activity of either the viral M2 ion channel protein (e.g., amantadine and rimantadine) or the surface enzyme neuraminidase (NA) [e.g., oseltamivir (Tamiflu), zanamivir (Relenza), peramivir (Rapivab), laninamivir octanoate (Inavir, approved only in Japan)],[34] have been associated with various levels of resistance[35,36]; thereby, they can be anticipated to be slowly superseded in the mid term by other small-molecules inhibitors with similar or distinct inhibition mechanisms.

Significant efforts are ongoing to develop second-generation inhibitors targeting highly prevalent M2 proton channel mutants of adamantane-resistant influenza A strains that may raise the genetic barrier to resistance.[37,38] A recently disclosed chemical modification at the 5-position of oseltamivir allowed for an additional interaction within the NA active site that produced new analogues with restored affinity for drug resistant strains.[39,40] Among those, 4-phenylthiobenzyl substituted compound **8** (Fig. 2) exhibited potent inhibitory activity against various influenza virus subtypes comparable to oseltamivir in cell based assays, while being 86-fold more active against the predominant H5N1-H274Y mutant strain in a neuraminidase inhibitory assay ($IC_{50} = 32.8$ nM).[39]

Alternative approaches for treating the IAV infection sought to abolish the influenza RNA-dependent RNA polymerase (RdRp) activity preventing transcription and replication of the viral genome.[41] High-resolution structural analysis revealed this enzyme to be a highly conserved

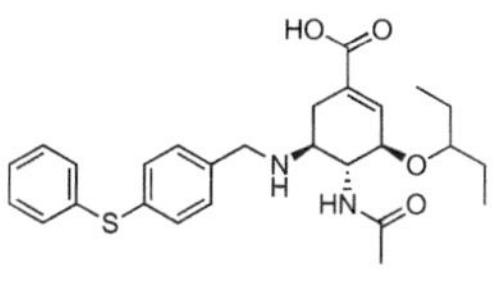

8
EC_{50} = 0.056-0.39 μM
IC_{50} (NA) = 32.81 nM (H5N1-H274Y)

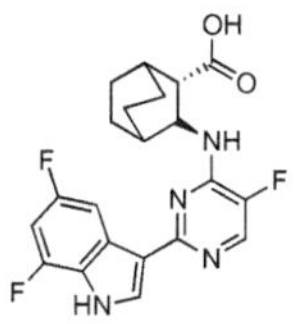

9
IC_{50} = 11-22 nM

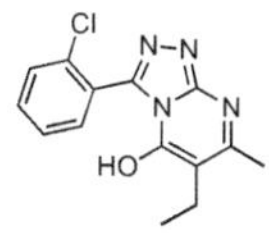

10, ANA-1
EC_{50} = 0.09-1.23 μM

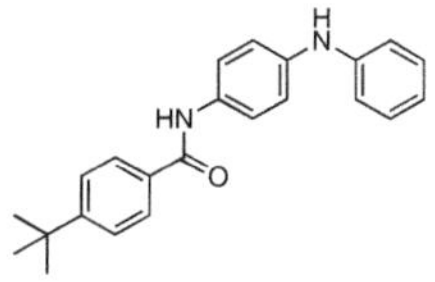

11, S119-8
EC_{50} = 1.43-15.1 μM

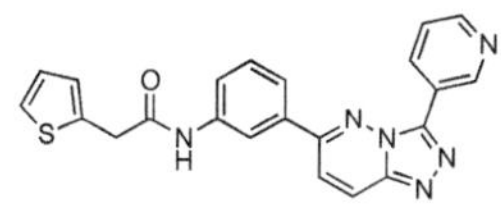

12, M85
IC_{50} = 0.35-2.05 μM

Fig. 2 Examples of mechanistically distinct influenza virus inhibitors.

heterotrimeric complex organized in different closely interacting subunits decorated by flexible domains,[42] which provided distinctly exploitable targeting sites for antiviral drug discovery. Among selected clinical candidates, only the oral polyheterocyclic ester prodrug baloxavir marboxil (BMX, Xofluza®, Roche and Shionogi) received global approval for the treatment of acute influenza.[43] In 2020, this indication was extended to post-exposure prophylaxis for patients 12 years of age or older as the first single-dose preventive treatment of influenza.[44] This drug is known to inhibit the cap-dependent endonuclease (CEN) activity of the polymerase acidic (PA) subunit of influenza A and B viruses, thereby blocking initiation of mRNA synthesis.[45] However, mutations in the PA subunit associated with BMX resistance were detected during clinical trials,[46] prompting further medicinal chemistry efforts towards the exploration of alternative scaffolds.[47–49]

Pimodivir (JNJ63623872 or VX787, Janssen Pharmaceutica), an oral azaindole inhibitor of the influenza A polymerase basic 2 (PB2) cap-binding domain,[50] was found to significantly decrease viral load in a Phase II study[51]; however, its development was discontinued after Phase III results in hospitalized patients with influenza A showed no added benefit compared with the standard of care (SOC) (NCT03376321). Notwithstanding, PB2 remains an attractive therapeutic target for its key role in viral transcription, and further lead optimization efforts yielded a 7-fluoroindole bioisostere of pimodivir (compound **9**, Fig. 2) with potent anti-IAV activity (EC_{50} = 11–22 nM, A H1N1 and H3N2 strains), improved metabolic stability, good pharmacokinetics, and in vivo efficacy.[52]

The purine analogue favipiravir (Avigan®, Toyama Chemical), which blocks influenza virus proliferation either by inducing chain termination during RNA synthesis or through lethal mutagenesis, entered clinical practice in Japan in 2014.[53] However, due to teratogenic concerns, approval was granted upon condition of limiting its use against novel and re-emerging influenza virus infections unresponsive to SOC. Results from USA phase III clinical trials in adults with uncomplicated influenza are pending (NCT02026349; NCT02008344).

A parallel approach for inhibiting the RdRP function, which may lead to a next-generation of anti-influenza therapeutics, relies on the disruption of the polymerase assembly process, during which protein-protein interactions (PPIs) occur between the different subunits concurring to the formation of a functional complex.[54] From data-driven and biochemical assay-based high throughput screenings of structurally unrelated heterocyclic libraries, numerous hit compounds emerged as potent antiviral agents and strong inhibitors of the viral polymerase activity, which were shown to primarily disrupt the PA-PB1 interaction. Compound **10** (ANA-1, Fig. 2) was effective against multiple IAV subtypes in cell culture, reduced lung virus titers in infected mice, and protected animals from lethal challenge upon intranasal administration.[55]

The discovery that the small molecule nucleozin can induce loss of viral viability by causing abnormal aggregation of viral nucleoprotein (NP) in the cytoplasm with consequent blockage of its nuclear accumulation[56] provided an opportunity for the development of a novel class of antiviral drugs for the treatment of IAV infections.[57–59] Compound S119-8 (**11**, Fig. 2) was found to exert broad-spectrum micromolar activity against numerous influenza A and B strains and its combination with oseltamivir led to a strong synergism.[59]

An increasing number of studies have examined direct inhibitors of the viral fusion protein hemagglutinin (HA), a surface glycoprotein that mediates viral attachment and entry, as a mean to complement other classes of anti-influenza agents for achieving pan-subtype efficacy.[60–62] However, doubts remain about the utility of targeting HA for single agent antiviral therapy due to its high susceptibility to mutations.

On the other hand, given the low probability of changes occurring to the host genome during the viral infection cycle, interfering with cellular processes involved in the entry process of influenza is seen as a promising strategy to combat treatment failures due to drug resistance. Two host directed antiviral agents are currently being evaluated in clinical trials.

The antiprotozoal agent nitazoxanide (Alinia) is a repurposed thiazolide prodrug of tizoxanide that prevents correct assembly of new viral particles by interfering with cellular factors involved in the maturation of viral hemagglutinin.[63] After successfully completing a Phase IIb/III study in adults and adolescents with acute uncomplicated influenza,[64] it now moved into Phase 3 (NCT02612922). Fludase (DAS-181, Ansun BioPharma) is a recombinant fusion protein that adheres to the host epithelial cells and removes the sialic acid receptor from the cell-surface preventing viral entry.[65] Results from a Phase 2 trial suggested that the drug was well tolerated and significantly reduced viral load and viral shedding.[66]

Recently, M85 (compound **12**, Fig. 2) was reported as an inhibitor of class II PI3K host kinase activity required for IAV internalization into target cells via endocytosis.[67] The compound had low cytotoxicity, inhibited viral replication of IAV, IVB, HCV, and human rhinoviruses in various cell types, while exerting a synergistic antiviral action in combination with oseltamivir that protected mice against a lethal challenge of influenza A H1N1 virus.

4. Respiratory syncytial virus (RSV)

Human respiratory syncytial virus (RSV, major antigenic subtypes A and B) is a *Orthopneumovirus* of the *Pneumoviridae* family containing a negative sense, single stranded linear 15 kb RNA genome made of 10 genes encoding 11 proteins.[68] RSV is responsible for lower respiratory tract infections that can be life-threatening among patients with impaired immune function like premature infants and the frail elderly.[69,70] Narrow, suboptimal preventive and therapeutic options for RSV infection, confined to monoclonal antibody palivizumab (Synagis) and broad-spectrum nucleoside analogue ribavirin (aerosol formulation),[71] along with the potential for nosocomial outbreaks inflict a heavy clinical and economic burden. Significant global efforts are underway by several major pharmaceutical companies to develop RSV fusion (F) protein inhibitors to prevent viral entry into host cells, as recently reviewed in detail elsewhere,[72] which seems to suggest that the first marketed drug against RSV infections might be within close reach. Among the latest candidates to advance to late-stage development are 5-chloroindole and spirocyclopropyl azaindole-based analogues rilematovir (**13**, JNJ-53718678, Janssen Pharmaceutica, Fig. 3)[73] and sisunatovir (**14**, RV521; ReViral, Fig. 3),[74] respectively, which are currently being evaluated for oral administration in hospitalized infants and vulnerable

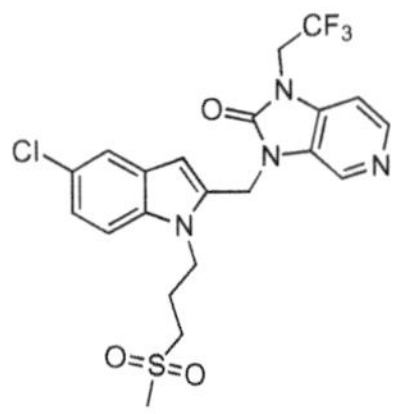

13, rilematovir (Janssen Pharmaceutica)
EC_{50} = 0.5 nM

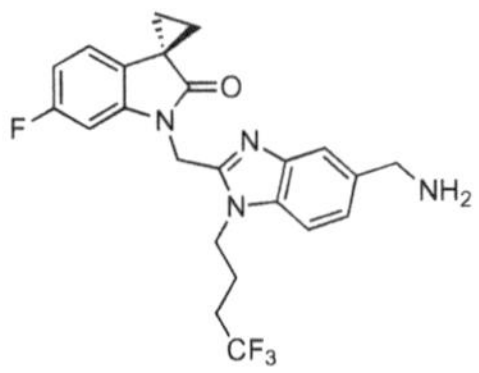

14, sisunatovir (ReViral)
EC_{50} = 1.3 nM

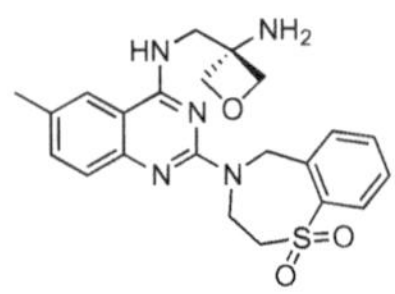

15, ziresovir (Ark Biosciences/Roche)
EC_{50} = 5 nM

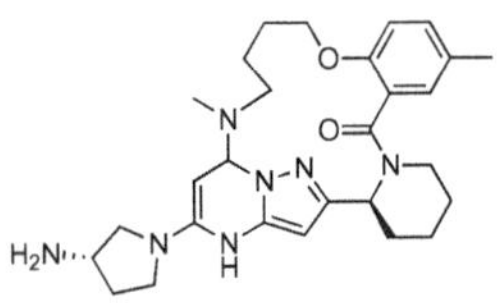

16
EC_{50} = 2.0 nM

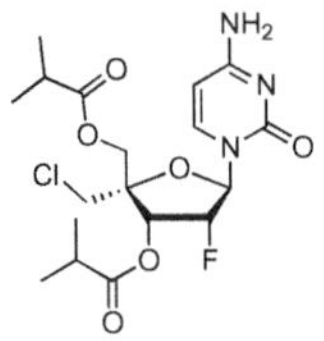

17, lumicitabine (Alios BioPharma)
EC_{50} = 0.26 μM

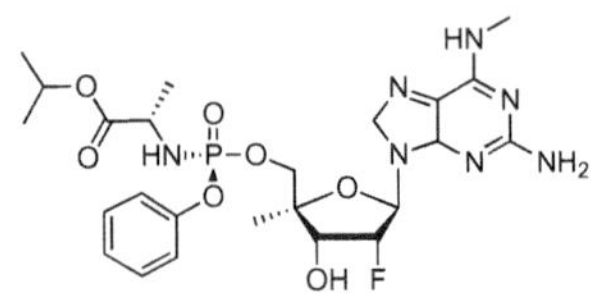

18 (Atea Pharmaceuticals)
EC_{50} = 0.052 μM

Fig. 3 Selected RSV fusion and replication inhibitors.

adult populations. Phase II studies are also in progress for ziresovir (**15**, AK0529; in-licensed by Ark Biosciences from Roche, Fig. 3), which was evolved from a benzoazepine quinoline chemical series.[75] On the other hand, despite promising data from preclinical and preliminary clinical studies,[76] results from a recently completed Phase II clinical trial examining the efficacy of the pyrazolo[1,5-a]pyrimidines candidate presatovir (GS-5806; Gilead Sciences) revealed a lack of therapeutic benefit in adult hematopoietic cell transplant recipients[77,78] and low barrier to resistance.[79] While the clinical utility of RSV fusion inhibitors remains to be fully validated, the identification of new heterocyclic leads binding the viral surface F glycoprotein continues to be of interest to the medicinal chemistry community. A lead optimization strategy based on a machine learning algorithm led to novel compounds integrating the popular benzimidazole scaffold.[80] The most active analogue in this series demonstrated subnanomolar *in vitro* potency comparable to JNJ-53718678 (EC_{50} = 0.5 vs 0.9 nM) against wild-type RSV with a CC_{50} > 20 μM; however, it was significantly less potent against the clinically reported viral escape mutant D489E, emphasizing the need for RSV combination therapy rather than monotherapy.

Macrocyclic pyrazolo[1,5-a]pyrimidine derivatives, as exemplified by a 15-membered ring containing compound X, displaying activities in the low nanomolar to subnanomolar range against wild-type RSV A2 were recently disclosed by means of molecular docking studies. Compound **16** combined excellent potency against both wild-type and drug-resistant mutant RSV F proteins owing to productive interactions of the rigid structure, suggesting the potential for developing inhibitors with anti-RSV activities.[81]

The latest compound to enter the anti-RSV drug discovery pipeline is the post-entry replication inhibitor EDP-938 (Enanta Pharmaceuticals), whose mode of action has been linked to the inhibition of the viral nucleoprotein (N), which serves as a polymerase co-factor in the RSV ribonucleoprotein (RNP) complex responsible for carrying out RNA synthesis. EDP-938 exhibited EC_{50} values in the 21–64 nM range against a panel of RSV A and B laboratory strains and clinical isolates when using primary human bronchial epithelial cells (HBECs) as in vitro model. Further in vitro studies revealed a high barrier to resistance as well as no cross-resistance between EDP-938 and other classes of RSV inhibitors, but rather a potentially synergistic in vitro antiviral effect. The antiviral efficacy was further confirmed in vivo using an non-human African Green monkey model of RSV infection.[82] In current Phase II clinical trials, hospitalized and non-hospitalized infants as well as adult allogeneic hematopoietic cell transplant (HCT) recipients are undergoing treatment with EDP-938.

Several nucleoside polymerase inhibitors have been described with potent *in vitro* anti-RSV activity[72,83]; however, this discovery approach has been delayed by the unavailability of sufficient structural data on the large (L) protein and poor understanding of its interactions (e.g., with the tetrameric phosphoprotein P to form an active RNA polymerase complex) and enzymatic activities.[84,85] The most active nucleoside analogues are prodrugs sharing a 2'-deoxy-2'-fluoro-ribose sugar moiety bearing different 4'-substituents in combination with various nucleobases (e.g., cytosine, amino-pyrrolotriazine, and N^6-methylated diaminopurine). Lumicitabine (**17**, ALS-8176; Alios BioPharma, Fig. 3), the 3',5'-di-*O*-isobutyryl prodrug of 2′-fluoro-4′-chloromethyl-cytidine, emerged as a potent ($EC_{50} = 0.26$ μM) and selective inhibitor of RSV replication with no concomitant inhibition of human polymerases.[86] After achieving clinical proof of concept in human RSV challenge studies, however, the development of this compound was ultimately discontinued due to unclear reasons following the early termination of a study in hospitalized infants (NCT03333317). The potent in vitro anti-RSV activity ($EC_{50} = 0.052$ μM) was described for a phosphoramidate prodrug of 2'-deoxy-2'-fluoro-4'-methyl nucleoside analogue bearing a modified purine base (**18**, Fig. 3) in HAE cultures with no cytotoxicity found in bone marrow stem cell assay.[87] Preliminary oral bioavailability data were reported in hamsters along with the demonstration of the conversion to the active triphosphate form in lung tissue. Remdesivir was shown to effectively inhibit RSV A2 replication *in vitro* with an EC_{50} of 530 nM in primary human lung cells.[88] An in vivo efficacy study was conducted by subjecting RSV A2-infected African green monkeys to intravenous administration of 10 mg/kg delivered once daily, which resulted in a >2-$\log_{10}$ reduction in the peak lung viral load.

In parallel, non-nucleosides are also being investigated mainly as allosteric inhibitors of the L polymerase along with molecules targeting other viral proteins involved in the virus life cycle such as the attachment G protein, another viral protein involved in the initiation of the RSV infection, as exhaustively discussed by Cockerill et al.[72]

5. Rhinoviruses

The *Picornaviridae* is a large and diverse family of (+)-ssRNA viruses whose genome is packed within an icosahedral capsid made of four viral proteins (VP1-4). Notable members include human rhinoviruses (HRVs, type A, B, C),[89] which are ubiquitous and persistent pathogens most frequently associated with infections of the upper respiratory tract (i.e., common colds) that can occasionally cause an aggravation of preexisting pulmonary and middle ear inflammatory conditions in vulnerable patient populations.[90,91] Their (+)-ssRNA genome (~7.21 kb) is translated into a single polyprotein that is cleaved by virally encoded proteases 2A and 3C to provide 11 structural and non-structural proteins. At present, no antiviral drug has been licensed for therapeutic use in rhinovirus or other enterovirus infections. Numerous active small molecules have been discovered through extensive in vitro screening campaigns that fall mainly into two mechanistic classes, i.e., capsid and 3C protease ($3C^{pro}$) inhibitors, with the former being the most advanced avenue for rhinovirus treatment.[92,93] However, lead compounds were either not developed or failed to meet the clinical criteria for their successful use because of their toxicity, poor pharmacology, or rapid resistance development [e.g., pleconaril, pirodavir, vapendavir, and rupintrivir). Moreover, a major challenge will be identifying anti-HRV compounds with a broad activity spectrum due to the circulation of approximately 160 different viral serotypes characterized by high mutation rates.

Discovery of novel capsid binders continues through the exploration of structurally diverse heterocyclic scaffolds, some of which showed potent in vitro activities.[94,95] Although their precise uncoating inhibition mechanism has not been fully validated, it is believed that these compounds occupy the hydrophobic pleconaril-binding pocket of VP1, resembling its mode of action. It is of interest to note that pleconaril-resistant viruses feature amino acid substitutions with bulkier residues in this pocket that sterically preclude binding.[96] Besides reducing the size of the molecules to fit the narrower pocket of HRV mutants,[97,98] a better strategy to raise the barrier to resistance is identifying inhibitors that stabilize the viral capsid dynamics inducing uncoating by targeting alternative binding sites. Recent cryo-EM studies demonstrated that the pyrazolopyrimidine derivative OB-5-340 (compound **19**, Fig. 4) binds to a less conserved region of VP1 that only partly overlaps with the attachment site of other capsid inhibitors,[99] which could explain its potent inhibitory activity against HRV-B5 (average $IC_{50} = 0.08\ \mu M$) and other high-level pleconaril-resistant HRV strains.[100]

The RdRp of rhinoviruses could be a viable target for the development of antiviral agents; however, investigations in this direction have not yet produced any concrete results, except for the report of the inhibitory activity of the nucleoside analogue 7-deaza-2'-methyladenosine (7DMA, MK0608) in HAE cultures infected with several HRV-C strains ($EC_{50} = 1.8–12\ nM$).[101]

Despite having distinct cellular tropisms and clinical manifestations, rhinoviruses are structurally and genetically related to other human pathogens of the same genus *Enterovirus* in the *Picornaviridae* family. Prior to the advent of genetic sequencing, enterovirus D68 (EV-D68) was misclassified as rhinovirus 87 due to its almost exclusive association with respiratory illnesses, unlike most enteroviruses that infect a variety of cell types.[102] EV-D68 is an emerging pathogen that in recent years has been increasingly associated with episodes of neurological disease outbreaks.[103]

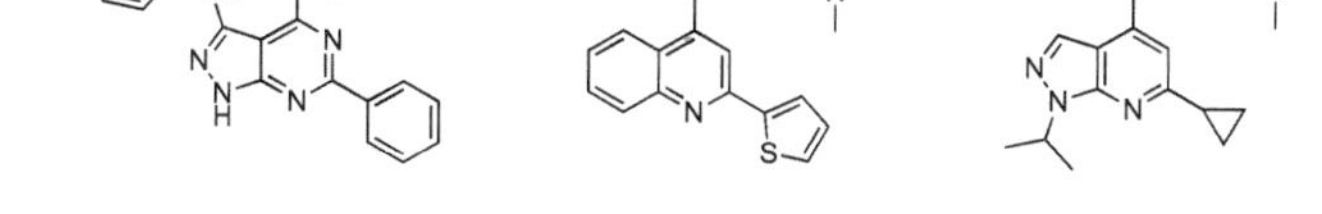

19, OBR-5-340
$IC_{50} = 0.08\ \mu M$

20
$EC_{50} = 0.01\text{-}0.06\ \mu M$

21, UAWJ172
$EC_{50} = 0.025\text{-}0.1\ \mu M$

Fig. 4 Examples of small molecule inhibitors active against HRVs and EV-D68.

The viral proteins that are being exploited as targets to develop anti-EV-D68 agents are those for which inhibitors were selected in the anti-HRV drug discovery process, since their high degree of sequence conservation among different enteroviruses might allow for broad-spectrum antiviral activity.[104] As an example, fluoxetine-modified inhibitors with broad spectrum activity against enteroviruses and rhinoviruses have been reported that target the non-structural 2C protein.[105] This protein has been proposed to play multifunctional roles in the viral life cycle of nonpolio enteroviruses, and significant efforts are being made in identifying antivirals against this target.

Structure–activity relationship studies designed to improve the activity of EV-D68 inhibitor dibucaine yielded a series of analogues containing a quinoline scaffold, as represented by compound **20** (Fig. 3), which exhibited EC_{50} values in the nanomolar range and a high selectivity against different EV-D68 strains.[106] Recently, a systematic modification of the pyrazolopyridine scaffold led to several lead compounds including **21** (Fig. 3) endowed with excellent broad-spectrum antiviral potency against several viruses (e.g., EV-D68, EV-A71, and CVB3) coupled and minimal cytotoxicity.[107]

Apart from targeting specific steps in the viral life cycle via direct-acting antivirals, host factors that are essential for viral replication are equally attractive drug targets for enterovirus inhibition and have been covered in a recent review.[93] At the moment, the only small molecule in clinical development (Phase III) for the treatment of common colds caused by enterovirus/rhinovirus infections is the broad-spectrum antiviral nitazoxanide.

6. Ebolavirus (EBOV)

Within the *Filoviridae* family of negative-sense, single-stranded RNA (ssRNA) viruses, Ebola virus (EBOV) and Marburg virus (MARV) cause clinically similar acute illnesses that can evolve into severe systemic infections culminating in multiorgan failure.[108] With reported case-fatality rates as high as 90% (e.g., for the highly pathogenic Zaire EBOV), EBOV stands out as one of the deadliest viruses known to mankind and a potential bioterrorism threat.[109] Since 1976, when filoviruses were discovered, more than 25 major Ebola virus disease (EVD) outbreaks have been recorded in several West and Central African countries that claimed thousands of lives. The 2020 regulatory approval of two vaccines [single-dose Ervebo (Merck) and two-component Zabdeno-and-Mvabea (Janssen Pharmaceutica)] validated adaptive immunity as a viable mean to control EBOV transmission,[110] however optimal impact on outbreak response remains to be defined mainly due to challenging settings to routine immunization.[111]

At present, immune therapy is very much on the front line of EVD management following the approval in 2020 of two human monoclonal antibody (mAb) products (i.e., Inmazeb and Ebanga) for intravenous administration[112–114]; however, orally small molecule therapeutics are seen as more convenient options in the event of high risk exposure and chronic infections to reach immunologically privileged sites that might harbor the virus. Nonetheless, screening for antifiloviral lead compound identification has been severely hampered by the requirement for biosafety level-4 (BSL-4) containment facilities combined with the limitations associated with surrogate BSL-2 cell-based systems to faithfully model the viral life cycle.[115]

The 19 kb EBOV genome encodes seven structural proteins and several non-structural proteins, potentially offering multiple druggable targets.[115] Nonetheless, the most promising routes towards antiviral treatment have primarily focused either on the inhibition of the viral entry or genome transcription/replication steps, as recently reviewed.[114–116]

Various small molecules targeting all key aspects of the viral entry process have been actively discovered, however, most have only been studied in preclinical models. Some of the latest examples with the largest potential for further profiling as filovirus entry inhibitors are illustrated in Fig. 5. Selected adamantane carboxamide derivatives, exemplified by **22**, exhibited nanomolar antiviral potency against both BSL2 vesicular stomatitis virus (VSV) pseudotyped EBOV (pEBOV) and wild type EBOV combined with in vitro favorable liver metabolic stability.[117] Using various filoviral

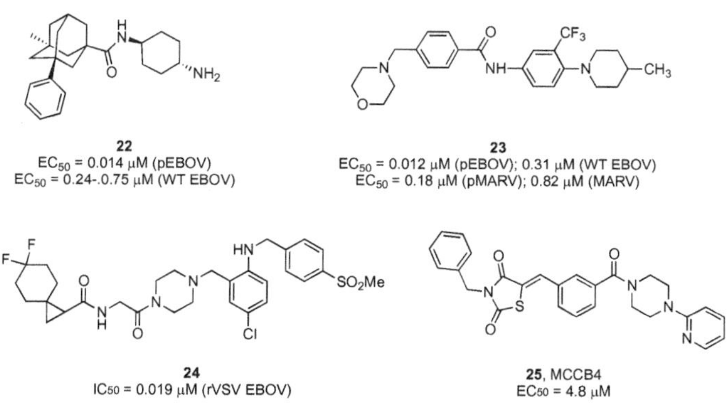

Fig. 5 Selected compounds with antiviral activity against EBOV and MARV.

replication-defective and replicative infectivity assays, Gaisina et al. discovered compounds sharing a common 4-(aminomethyl)benzamide core with broad spectrum activity against both EBOV (Mayinga) and MARV (Angola).[118] In addition to low micromolar potency ($EC_{50} < 1$ μM), these compounds (exemplified by **23**) exhibited favorable pharmacokinetic and pharmacological properties without detected cytotoxicity. Preliminary mechanistic studies suggested that the inhibitory activity of these compounds resulted from the direct binding and inhibition of the EBOV envelope glycoprotein (GP) protein, clearly indicating GP as an attractive target for therapeutic intervention other than the key immunogen for eliciting antibody responses.

Besides interfering with the processes controlled by the virus, other viable means of blocking viral entry rely on host-directed inhibitors interfering with cellular regulatory networks. Diamide piperazine derivatives were shown to inhibit EBOV by targeting the endosomal host receptor Niemann-Pick C1 (NPC1) and preventing its subsequent interaction with GP, which is considered an essential step for infection.[119] Systematic optimization of the original scaffold facilitated modulation of the inhibitory potency and hydrophobicity of this series of analogues. The most promising antiviral profile against EBOV was observed for compound **24** ($EC_{50} = 0.019$ μM in a replication competent VSV vector),[120] as supported by in vivo data on oral administration of **24** in combination with ritonavir.

A series of epoxysuccinyl-based derivatives were found to inhibit lysosomal cysteine cathepsins, which are proteases required for cleavage of highly glycosylated sequences from the GP and unmasking of its binding site for further interactions with NPC1.[121] The most potent compounds exhibited antiviral activity against EBOV and MARV with EC_{50} values in the low nanomolar range.

Several nucleoside and nucleotide polymerase inhibitors showed promising antiviral activities in cell-based assays often combined with therapeutic efficacy in animal models; however, none of these agents proved to be clinically useful. Prior to repurposing for SARS-CoV-2, Remdesivir entered clinical trials for EVD treatment following evidence of effective post-exposure protection in EBOV-infected rhesus monkeys.[122,123] However, its experimental use as part of an emergency response to the 2018 EVD outbreak in the Democratic Republic of the Congo (Phase III PALM trial) was associated with inferior survival rates compared to 2 mAb therapies, resulting in treatment discontinuation.[124] Favipiravir displayed in vivo efficacy as postexposure prophylaxis (PEP) and early treatment in a therapeutic mouse model; however, survival was reduced at a later stage of the infection.[53] The Phase II evaluation of its therapeutic efficacy in patients with Guinea and West African EVD was not conclusive. Another early-stage development candidate is imino-C-nucleoside galidesivir (BXC4430),[125,126] for which a Phase I clinical trial has been recently completed to evaluate its intramuscular administration in healthy subjects (NCT03800173). Based on promising in vitro data,[127] the lipid prodrug brincidofovir (CMX001), which was originally developed to treat DNA viral infections and is now approved for treatment of smallpox,[128] was authorized for compassionate use during the West African EVD epidemic but the trial was later discontinued due to lack of efficiency in preclinical animal models.[129]

The potential utility of *S*-adenosyl-L-homocysteine hydrolase (SAHase) inhibitors to provide protection against a lethal Ebola virus challenge has been demonstrated in vivo for known carbocyclic nucleosides 3-deazaadenosine and 3-deazaneplanocin A.[130,131]

MCCB4 (compound **25**, Fig. 5) is a non-nucleoside polymerase inhibitor identified by in silico driven design that displayed selective anti-EBOV activity in a minigenome assay. This compound was suggested to interfere with viral RNA synthesis by binding to a hydrophobic pocket on the nucleoprotein (NP), which is an essential component of the ribonucleoprotein complex.[132]

7. Human immunodeficiency virus (HIV)

In developed countries, access to antiretroviral therapy (ART) has drastically improved the prognosis and life expectancy of human immunodeficiency virus (HIV)-positive individuals. However, HIV transmission remains unabated with an estimated worldwide average of 1.5 million people infected in 2020.[133] In addition, an increasing trend in HIV infection rates can be expected as a result of reduced access to HIV testing services and medical care during the COVID-19 pandemic. In the absence of an effective vaccine for HIV, the use of small molecule antiretroviral medications as a prevention strategy represents a potent alternative for preventing transmission among HIV-negative individuals at risk of contracting the infection and a major endeavor in antiviral drug discovery.

In 2012, the HIV medication Truvada® (Gilead Sciences), a fixed dose combination product containing the two nucleoside reverse transcriptase inhibitors (NRTIs) emtricitabine (FTC, 200 mg) and tenofovir disoproxil fumarate (TDF, 300 mg), became the first oral regimen to be approved worldwide for daily use in HIV pre-exposure prophylaxis (PrEP). The safety and efficacy (>90%) of this HIV prevention regimen has been demonstrated, while resistance was found to be rare (<0.1%).[134] However, one issue is that it is only used use to <20% of the people in the USA that could benefit from it.[134]

A second tenofovir-based formulation Descovy® given once-daily (Gilead Sciences, tenofovir alafenamide (TAF)/FTC 25 mg/200 mg) was associated with comparable efficiency in reducing HIV acquisition compared with Truvada® but a higher bone and renal safety in a Phase III study (DISCOVER trial), which led to its FDA approval in 2019 for at-risk uninfected adults and adolescents.[135]

However, the medication adherence must be high for sustained efficacy, and the potential to induce resistance occurs when PrEP is used during undiagnosed acute HIV infections.[136] Obviously, PrEP does not prevent viral infections with strains that are resistant against the selected drugs. Many challenges still remain for prevalence in the African region. The PrEP drug must be cheap, easy to access, easy to administer, allow optimal adherence, while the person should be regularly diagnosed to monitor the effect. The benefits, however, clearly outweigh the disadvantages. A variety of long-acting (LA) antiviral compounds and different extended-release formulations are now under investigation to overcome shortcomings associated with once daily self-administered oral therapy, particularly adherence, non-compliance, and cumulative toxicity.[137]

The beginning of 2021 saw the approval of the first long-acting intramuscular injectable anti-HIV treatment for monthly use (Cabenuva®), which was obtained from the coformulation of the integrase inhibitor cabotegravir (200 mg, CAB, ViiV Healthcare, **26**, Fig. 6) and the non-nucleoside reverse transcriptase inhibitor (NNRTI) rilpivirine (300 mg, RPV, Janssen Pharmaceutica).[138] At the same time, a CAB long-acting (CAB-LA) nanosuspension is being clinically evaluated as HIV PrEP

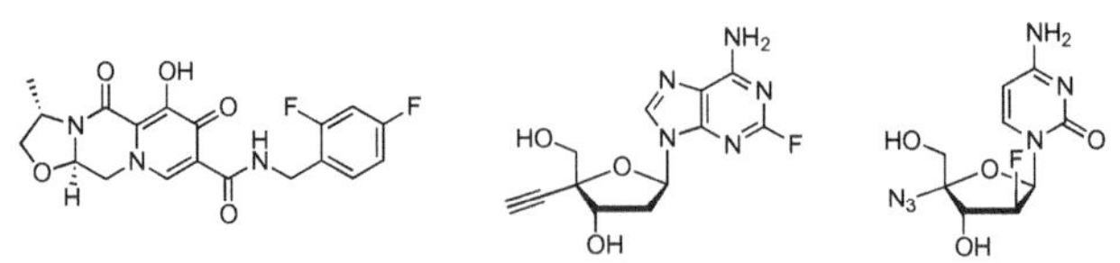

26, Cabotegravir (ViiV Healthcare)
IC_{50} = 0.2 nM

27, Islatravir (Merck)
EC_{50} = 0.3 nM

28, FNC
EC_{50} = 0.06 nM

Fig. 6 Long-acting antivirals under investigation for HIV pre-exposure prophylaxis.

monotherapy.[139] Two phase III studies (HPTN083 and HPTN084) enrolling different HIV-uninfected risk populations were prematurely unblinded after establishing the superior efficacy of injectable CAB administered every month or every two months versus once-daily oral Truvada.[140,141] However, during the HPTN 083 trial, infections occurred in four cases in spite of preventive CAB-LA exposure but were detected only at a later stage in the collected blood samples most likely due to long-term viral suppression slowing antibody production. This delay in HIV diagnosis led to the emergence of drug resistance, highlighting the need for more sensitive viral load testing.[140,142]

Islatavir (4′-ethynyl-2-fluoro-2′-deoxyadenosine, EFdA, MK-8591, Merck, **27**, Fig. 6) is a highly potent investigational nucleoside analogue that inhibits HIV reverse transcriptase with an unique mode of action by blocking primer translocation without causing direct chain termination. Its active triphosphate metabolite exhibits an exceptionally long intracellular half-life and is responsible for prolonged virological effects, which enables decreased dosing frequencies making this compound suitable for PrEP. Phase III clinical trials are currently undergoing to compare the efficacy of islatravir administered once monthly for lowering the risk of HIV infection to current daily PrEP. Recent clinical studies demonstrated the feasibility of formulating islatavir as an implantable drug for PrEP purposes along with the potential for a once-yearly administration.[143]

The low oral absorption of the NNRTI dapivirine led to its formulation for topical delivery as an intravaginal silicone ring (IVR), which is currently in the WHO list of prequalified medicines for HIV-1 prevention. In phase II trials, the slow release of dapivirine over one month was shown to reduce the risk of HIV acquisition by 30% compared with placebo, and high acceptance and adherence were observed among the study population.[144]

Lenacapavir is a capsid inhibitor that appears to disrupt multiple stages of the HIV life cycle. A subcutaneous injectable formulation of this drug candidate displayed a long half-life and is being investigated as a twice-yearly injectable formulation for HIV PrEP.[145]

An additional agent that could potentially be suitable for PrEP in the future owing to its high potency and slow systemic clearance is 2′-deoxy-2′-β-fluoro-4′-azidocytidine (FNC, **28**, Fig. 6),[146] which was recently shown to combine the direct inhibition in HIV-1 reverse transcription with an indirect antiviral effect on the host antiviral factor A3G, which may explain its long-lasting action.

8. Yellow fever virus (YFV)

Yellow fever virus (YFV) is a vector-borne (+)-ssRNA virus that shares the same genomic structure and geographical distribution as other pathogens of the genus Flavivirus (*Flaviviridae* family) responsible for causing severe dengue, Zika virus disease, Japanese encephalitis, and West Nile fever.[147] Most individuals contracting YFV infection do not experience symptoms, however in symptomatic patients a wide spectrum of clinical manifestations has been documented ranging from a mild acute febrile illness to a severe viral hemorrhagic fever associated with hepatic dysfunction and renal failure, which is fatal in approximately 50% of cases.[148]

YFV transmission occurs through two main coexisting cycles involving non-human primates (NHPs) or humans and infected mosquito species of the *Aedes* and *Haemagogus* genus.[149] In the jungle cycle, infected NHPs serve as sylvatic reservoir of YFV, virtually dismissing the possibility of eradicating this virus. Viremic humans who encroached on NHP forest habitats can introduce the virus into densely populated urbanized areas, where transmission can be perpetuated by competent local vector populations (e.g., *Aedes aegypti*).

The disease is preventable; a single subcutaneous injection of YF 17D live-attenuated vaccine administered from the age of 9 months appears to confer lifelong immunity.[150] While vaccination campaigns and vector-control efforts have concurred to reduce the incidence of YFV infections in many high-risk areas, endemic sylvatic circulation continues in tropical and subtropical regions of Africa and South America where immunization coverage is insufficient.[151,152] Recent outbreaks of YFV-related hemorrhagic disease in rural areas of Angola, Democratic Republic of Congo, and Brazil have renewed fears of a resurgence of urban transmission of this old viral threat.[153] Although YFV is absent in the Asia-Pacific region, more frequent exchanges between Africa and Asia caused imported yellow fever cases, which are seen as a growing public health concern since suitable ecological and climatic conditions are present in these non-endemic areas that could support local transmission.[154] Notably, the increasing geographic distribution of *Aedes aegypti* due to climate change is expected to exacerbate the risk of global spread and transmission of YFV across previously unaffected countries such as North America, China, and southern Europe by 2050.[155] By 2100, even parts of southern Canada and northern Europe may offer suitable ecological niches to sustain its survival.[156]

In this scenario, antiviral-based treatments are an important complement to the YF vaccine in building a healthcare response to counteract this potential threat, especially for individuals unwilling to be vaccinated or for whom vaccination is contraindicated, such as those experiencing YF vaccine-associated viscerotropic (YEL-AVD) and neutropic (YEL-AND) diseases similar to those caused by the wild-type virus.[157]

Repurposing compounds previously investigated or marketed for other therapeutic purposes has been largely pursued for yellow fever disease treatment. Through in vitro screening of a small library of nucleoside analogues, 2'-methyl-7-deazaadenosine, 2'-methyl-7-fluoro-7-deazaadenosine, and a ProTide of 2',2"-dichlorouridine were identified as effective inhibitors of YFV replication with EC_{50} values in the 0.25–2.1 μM range and no cytotoxicity up to 100 μM.[158]

Recent in vitro and preclinical studies in a murine model revealed a potential therapeutic use of the HCV approved nucleoside phosphoramidate prodrug sofosbuvir in the treatment of YFV infections. This prodrug has the advantage of being metabolically activated primarily in the liver, which is a major site of YFV infection. Sofosbuvir was shown to bind to a conserved region of the YFV RNA polymerase (NS5) and exert its antiviral activity (EC_{50} = 4.2–4.8 μM) against both vaccine and wild-type strains of YFV when tested in human hepatoma cell-based assays. In addition, prophylactic therapeutic treatment with sofosbuvir 1 day prior to infection enhanced survival of YFV-infected animals and reduced liver injury.[159] The off-label, compassionate use of Sofosbuvir to treat two YFV-infected patients with acute liver failure led to a reduction in viremia and disease improvement.[160]

Another RdRp nucleoside inhibitor to be tested as potential anti-YFV candidate is the C-nucleoside galidesivir (BCX4430). Despite displaying a moderate in vitro anti-YFV activity (EC_{90} = 27.6 μM), galidesivir exhibited potent preclinical efficacy in a Syrian golden hamster model of yellow fever infection. Complete survival of infected animals and improved disease parameters were observed upon intraperitoneal treatment with this nucleoside for 4 or 7 days, which also primed an immune response resulting in protection against a secondary YFV challenge. A phase Ib study to determine the safety and antiviral effects of intravenous galidesivir in YF patients was prematurely terminated (NCT03891420).[161]

The active 5'-triphosphate metabolite of remdesivir, which accumulates intracellularly and serves as substrate for the RdRp of several RNA viruses, was shown to be an efficient in vitro inhibitor of YFV polymerase with an IC_{50} of 0.26 μM.[162] The YFV polymerase emerged as a superior target for remdesivir among other flaviviral polymerases, while being less efficiently inhibited than coronavirus polymerases.

In addition, some broad-spectrum compounds built on different heterocyclic scaffolds have shown inhibition efficacy against YFV in vitro, however their targets remain undefined.[163–165]

9. Poliovirus (PV)

At the launch of the Global Polio Eradication Initiative in 1988, paralytic poliomyelitis was an endemic disease in more than 125 countries.[166] Today, the global incidence of wild poliovirus infections has plunged by over 99% thanks to effective vaccination programs. However, while wild poliovirus type 2 (WPV2) and type 3 (WPV3) were declared eradicated in 2015 and 2019, respectively, sporadic cases associated with the circulation of wild type poliovirus type 1 are still being recorded in Pakistan and Afghanistan, despite sustained efforts to eliminate this virus.[167]

Besides potential breaches of poliovirus containment in research laboratories and vaccine manufacturing facilities, a major menace to global eradication is posed by circulating vaccine-derived polioviruses (cVDPVs) evolved from the use of live attenuated oral poliovirus vaccine (OPV).[168] These variants are genetically distinct from wild polioviruses but cannot be distinguished in terms of pathogenicity and transmissibility.[169] At present, cVDPVs, particularly strains of serotype 2, are responsible for the majority of paralytic polio outbreaks worldwide.[170]

In addition, individuals with a primary immunodeficiency disorder (PID) exposed to OPV may excrete immunodeficiency-related vaccine-derived polioviruses (iVDPVs) for prolonged periods of time, which are likely to constitute a reservoir of virus after eradication.[171] The development of effective antiviral drugs seems to be the best option to halt chronic excretion of iVDPVs and minimize the likelihood of potential polio outbreaks post eradication.[172] The viral proteins considered as targets for anti-polio drug development are those exploited for the identification of anti-rhinovirus agents, due to their high degree of conservation among enteroviruses. In the past, many small molecules capable of inhibiting poliovirus replication have been described (e.g., ribavirin, pleconaril, vapendavir, and pirodavir); however, none progressed further than Phase II evaluation.

Since a potential poliovirus antiviral drug is considered to have little or no commercial value for its development to be pursued by pharmaceutical

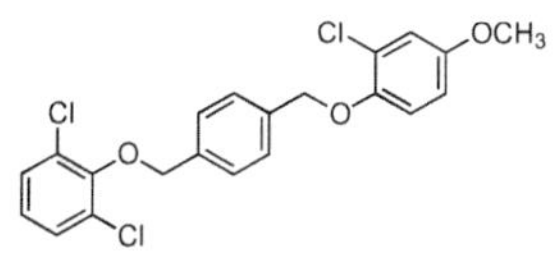

29, pocapavir, V-073 (ViroDefense Inc.)
EC_{50} = 0.003-0.126 μM

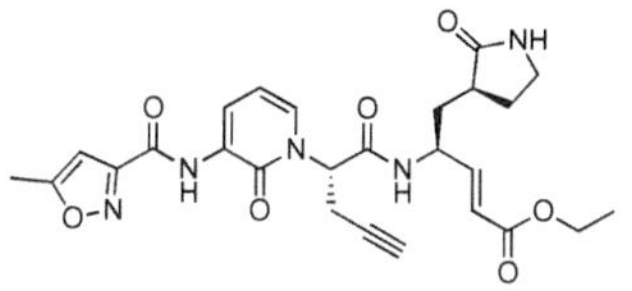

30, V-7404 (AG-7404) (ViroDefense Inc.)
EC_{50} = 0.080-0.674 μM

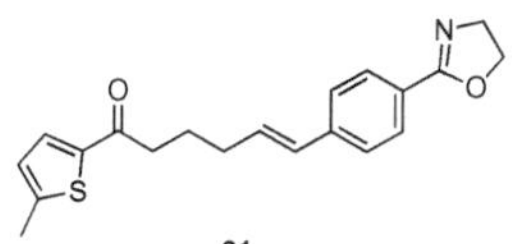

31
EC_{50} = 0.027-0.33 μM

Fig. 7 Examples of PV small molecule inhibitors.

companies, a partnership known as the Poliovirus Antiviral Initiative (PAI) was established to advance the discovery of effective poliovirus antiviral drugs. Two antiviral agents with different modes of action, the capsid binder pocapavir (V-073, compound **29**, Fig. 8)[173,174] and the peptidomimetic inhibitor of 3C protease V-7404 (compound **30**, Fig. 7),[175] are now under clinical characterization (ViroDefense Inc. & the Bill & Melinda Gates Foundation), for potential use as combination therapy.

Recently, some pyrazolopyridine-based compounds were reported to suppress WPV-1 replication in cell culture with EC_{50} values as low as 5 μM by targeting the viral 2C protein.[176] Another report indicated the anticancer nucleoside gemcitabine as an active anti-polio compound in vitro (IC_{50} = 0.3 μM) able to completely protect HeLa cells against the cytopathic effects of WPV1 at 25 μM.[177] Suppression of poliovirus replication was shown to be correlated with the direct inhibition of RNA chain elongation by gemcitabine triphosphate.

Screening of a compound library through a series of in vitro assays against wild-type and vaccine-derived PV strains identified a lead compound (**31**, Fig. 7) with broad spectrum antiviral activity in the nanomolar range. On the basis of time-of-addition studies, it was hypothesized that **31** could be targeting multiple stages of poliovirus replication; however, other related structural analogues appeared to interfere with the viral entry step.[178]

10. Hepatitis C virus (HCV)

Hepatis C virus (HCV) is a bloodborne, hepatotropic RNA virus and the only member of the genus Hepacivirus within the *Flaviviridae* family causing both acute and chronic infections in humans.[179] Its genome consists of a 9.6 kb (+)-ss-RNA that is translated into a single large polyprotein precursor. Subsequent polyprotein processing requires two viral proteases (NS2 and NS3/4A), which are autocatalytically excised from the polyprotein, and at least two host cellular peptidases to yield three HCV structural (S) and seven nonstructural (NS) proteins. Several direct-acting antiviral agents (DAAs) have been licensed for clinical use that target three of these NS proteins implicated in different steps of the HCV infectious cycle, i.e., the RdRp (NS5B), NS3/4A protease, and NS5A, a multifunctional RNA-binding protein involved in viral replication, assembly, and regulation of cellular pathways.[180] Various fixed-dose combination regimens comprising DAAs with synergistic modes of action have further been approved to offset compensatory evolutionary mechanisms leading to resistance development and viral rebound. In addition, this strategy can reduce treatment cost and duration, improve adherence, and limit dose-limiting toxicity.

The pharmacokinetics and pharmacodynamics of DAA regimens has been recently reviewed.[181] Most of them consist of two drugs, either an NS5B and NS5A inhibitor [e.g., sofosbuvir + ledipasvir (Harvoni®), sofosbuvir + velpatasvir (Epclusa®)] or an NS5A and NS3/4A inhibitor [e.g., elbasvir + grazoprevir (Zepatier®), glecaprevir + pibrentasvir (Maviret®, Mavyret®)]. Also a triple combination targeting NS5B, NS5A, and NS3/4A [e.g., Sofosbuvir + Velpatasvir + Voxilaprevir, Vosevi®][182] as well as a multiple regimen (Paritaprevir + Ritonavir + Ombitasvir and Dasabuvir)[183] have been approved for the treatment of chronic HCV genotype (GT)1-6 and GT1 infection in adults, respectively. With the availability of these oral combination therapies, the standard of care for chronic HCV infection has dramatically improved. Today, most patients, including those previously refractory to treatment such as individuals with HCV and HIV co-infection, decompensated liver disease, and renal impairment, can be successfully cured, as indicated by sustained virologic response (SVR) rates (defined as aviremia 24 weeks after therapy completion) up to 95%.[184] Recently, an 8-week regimen of glecaprevir and pibrentasevir given to untreated patients with chronic HCV genotype 1–6 infection and compensated cirrhosis was shown to afford a virological cure rate as high as 99.7%, irrespective of any pre-treatment patient or viral characteristics.[185] While a short treatment of acute HCV infection with sofosbuvir and ribavirin was demonstrated to be suboptimal,[186] high efficacy was instead obtained with a 6-week regimen of sofosbuvir and ledipasvir in patients with HCV GT1 monoinfection[187] as well as GT1 or 4 HCV infection and HIV-1 coinfection.[188] DAA treatment may also lead to partial restoration of HCV-specific immune responses in chronically infected patients, an effect that might be improved with early administration.[189] DAA re-treatment of chronic HCV infections has also been shown to be highly effective.[190] Re-treatment regimens based on a combination of sofosbuvir, velpatasvir, and voxilaprevir[191] or glecaprevir and pibrentasvir[192] for DAA-experienced patients have been approved. In addition, the beneficial clinical effects of DAAs has led to a significant reduction in the waitlist for liver transplant due to chronic HCV infection, while increasing the survival of patients with HCV-associated hepatocellular carcinoma.[193]

Overall, this scenario offers good prospects that the WHO program to eliminate viral hepatitis C as a public health problem can meet its goals by 2030.[194] Nonetheless, at present, there are still approximately 58 million people worldwide living with chronic HCV infection and 1.5 million new infections annually,[195] reflecting the fact that significant challenges remain to be addressed including accurate diagnosis of HCV infected patients, treatment of patients with decompensated cirrhosis or renal impairment, serious complications due to drug-drug interactions, high cost of treatment for underdeveloped countries, and occurrence of resistance associated variants (RAVs) to DAAs.[181] In particular, the development of new curative HCV drugs that may be effective against DAA-resistant mutants and broadly active on multiple HCV genotypes remains an important to optimize current therapeutic regimens and facilitate the global eradication of HCV, especially in the absence of a prophylactic vaccine.

Drug discovery efforts are actively ongoing to discover new pan-genotypic inhibitors of NS5A that are expected to decrease the incidence of treatment failure associated with low barrier to resistance of most marketed NS5A inhibitors for genotype 1 subtype a (GT1a)[196–198] as well as improve the pharmacokinetic profiles of existing drugs by prodrug synthesis to enhance solubility properties and simplify drug dosage formulations (e.g., pibrentasvir).[199]

The macrocyclic scaffold structure shared by approved HCV NS3/4A protease inhibitors has been modified to modulate both potency and drug-like properties of the inhibitors as well as prevent resistance due to pre-existing resistance-associated substitutions (RASs) and/or baseline polymorphisms among diverse genotypes.[200,201]

Aiming at identifying alternative uridine-containing sugar-modified NS5B inhibitors, chemical modifications were applied to the sugar moiety

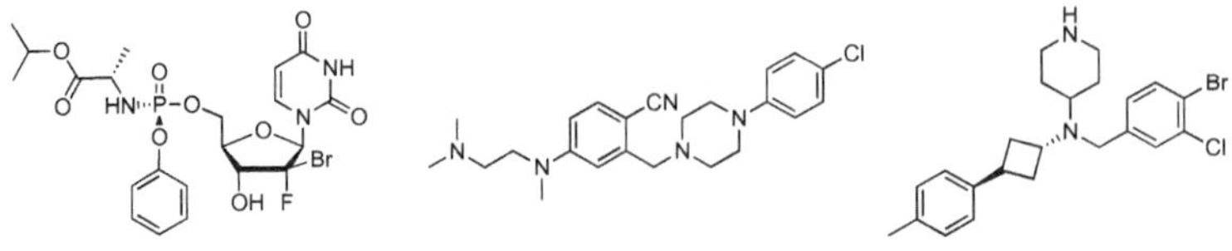

32
EC_{50} = 21-180 nM

33, L0909
EC_{50} = 0.022 μM

34
EC_{50} = 0.047 μM

Fig. 8 Structures of anti-HCV lead candidates with different modes of action.

of sofosbuvir, such as the introduction of a fluorine atom at the 4′ position of a 2′-methyl-uridine phosphoramidate prodrug (AL-335)[202] or the replacement of the 2’-methyl-2’-fluoro moiety with a dihalogenated functional group, such as in compound **32** (Fig. 8).[203] This Sp phosphoramidate diastereomer of 2′-bromo,2′-fluoro-uridine exhibited potent inhibition of HCV in cell culture (EC_{50} =21–180 nM) combined with pan-genotypic anti-HCV replicon activity. Its corresponding 5′-triphosphate was selective for HCV NS5B polymerase of GT1-6 with no concomitant inhibition of human or cellular mitochondrial RNA polymerases, while exhibiting favorable liver accumulation according to a preliminary liver pharmacokinetic study in beagle dogs. Alternatively, guanosine containing nucleoside phosphoramidate prodrug analogues also displayed potent activity against HCV infection and in some cases potential value as new clinical anti-HCV agents. This include a 2’,4’-fluoro substituted guanosine nucleotide analogue (AL-611)[204] and AT-527 (compound **2** in Fig. 1), which entered clinical development owing to the attractive broad spectrum coverage of major HCV genotypes.[205] Second-generation non-nucleoside inhibitors of NS5B, such as *N*-benzoxaborole benzofuran analogues, are also undergoing structural optimization programs.[206]

Recently, there has also been a surge in research efforts to find anti-HCV compounds with different modes of action, i.e., targeting stages of HCV infection either preceding or following viral replication. Although the full understanding of these complex multifactorial processes remains elusive, several related factors and processes have been investigated as potential therapeutic targets. These compounds could have an important synergist effect with DAAs towards HCV elimination or serve as preventive measures against HCV infection and reinfection.

The design of preclinical candidate L0909 (compound **33**, Fig. 8), featuring a 2-((4-arylpiperazin-1-yl)methyl)benzonitrile scaffold, was the result of a structure−activity relationship (SAR) study that optimized a modestly potent anti-HCV 2-(arylamino)methylbenzonitrile lead inhibitor by introducing a piperazine moiety. L0909, which displayed nanomolar antiviral activity in a HCVcc system ($EC_{50} = 0.022$ μM), appeared to act on the HCV entry stage. A synergistic effect against HCV was observed for L0909 in combination with DAAs together with a high sensitivity to clinical resistant HCV mutants. Further in vivo pharmacological and preclinical results supported the favorable anti-HCV profile of this compound.[207]

A series of polyheterocyclic derivatives featuring a 4-aminopiperidine scaffold as common structural element have been shown to be involved with the modulation of the HCV assembly process.[208] Antiviral synergy was observed for the combination of a starting screening hit with various marketed HCV therapeutics in cell-based assays of HCV, and an advantageous hepatotropic distribution was confirmed by in vivo pharmacokinetic studies. Lead optimization studies defined compound **34** (Fig. 8) as a promising candidate with nanomolar anti-HCV activity for further in vivo and and preclinical efficacy profiling.

In principle, broad spectrum antiviral activity against HCV can only be achieved with compounds acting by multiple modes of action (e.g., ribavirin) or through host-cell directed strategies, such as those targeting the human enzyme dihydroorotate dehydrogenase (DHODH).[209]

11. Rabies virus (RABV)

Rabies virus (RABV) is the archetypal species of the genus Lyssavirus (*Rhabdoviridae* family), a group of neurotropic viruses that carry a (-)-ss RNA genome and can infect both humans and animals.[210] Rabies disease is the deadliest among all viral infections with a 100% fatality rate after the onset of clinical symptoms.[211] Upon entering the peripheral nerves, RABV reaches the central nervous system (CNS) by means of axonal transport, where it preferentially replicates evading the host immune system and causing a severe and incurable neurological disease.[212]

Effective vaccines are available for pre-exposure immunization as well as human rabies immune globulin (HRIG) for post-exposure prophylaxis (PEP) intervention early after infection, which can successfully prevent the disease from progressing towards symptoms.[213,214] However, the risk of acquiring rabies infections is perpetuated through the low level of routine immunization coverage in high risk countries, manifesting in an estimated 59,000 human deaths claimed by RABV each year worldwide. RABV can be expected to persist globally as a pervasive public health threat because of its zoonotic transmission, which occurs in approximately 99% of cases through the transfer of infectious saliva from bites of infected dogs.

Antiviral molecules that may be able to suppress viral replication in an infected individual during the symptomatic phase of encephalitic rabies will be the next major step in fighting human RABV infections, however direct dosing in the cerebrospinal fluid is still considered an unmet challenge.[215] The development of antiviral compounds is also desirable as cost-effective and easily storable alternatives to rabies immunoglobulins currently used for PEP, especially considering the heavy economic burden associated with RABV prophylaxis and the fact that most infections (95%) occur in Africa and Asia where the supply of these biologics is limited.[216]

The promising in vitro inhibitory activity of the broad-spectrum RNA polymerase inhibitor ribavirin against RABV did not translate into in vivo antiviral efficacy in preclinical animal models and clinical trials in humans.[217] Some ribavirin analogues named EICAR and EICNR displayed increased in vitro activity (EC_{50} = 0.9–3.8 μM) as inhibitors of viral replication and transcription, most likely via IMPDH inhibition, compared to ribavirin (EC_{50} = 18.6 μM); however, a more systematic evaluation is needed to define their anti-RABV potential.[218]

Independent studies demonstrated the capability of favipiravir to effectively reduce RABV replication in vitro[219,220]; however, little or no effect on survival in a mouse model of RABV infection was observed. Recently, in vivo imaging studies demonstrated that favipiravir treatment 1 h or 2 days after intramuscular inoculation of RABV in mice could suppress viral replication in the periphery, but double or triple doses were needed to suppress replication in the CNS.[221]

Two adamantane derivatives, 1-adamantyl-(5-bromo-2-methoxybenzyl) amine (ABMA) and its 1-dimethyl analogue DABMA, were described to prevent RABV viral gene expression and genome replication by interfering with different steps of the late endosomal pathway.[222] When delivered 2 h pre-infection, these early-stage post-entry inhibitors exhibited in vitro micromolar activities against RABV.

More recently, the implementation of a high-throughput screening (HTS) protocol based on a single-cycle RABV reporter strain allowed to identify a first-in-class, highly potent (EC_{50} = 2–52 nM) and selective small molecule inhibitor of RABV, designated as GRP-60367 (compound **35**, Fig. 9). Mechanistic characterization demonstrated that GRP-60367 blocks RABV entry by specifically targeting a previously unknown druggable site on the G protein, representing a suitable lead compound for the design of next-generation antivirals for RABV treatment and prevention.[223]

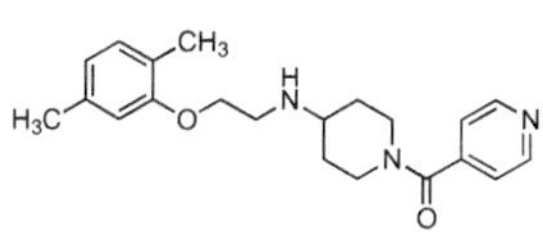

35, GRP-60367
EC_{50} = 2-52 nM

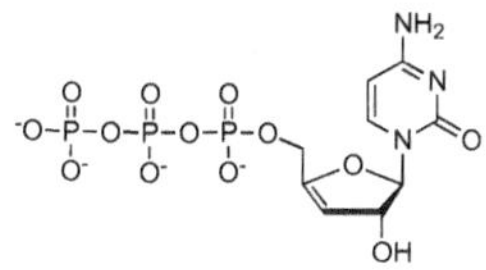

36, ddhCTP

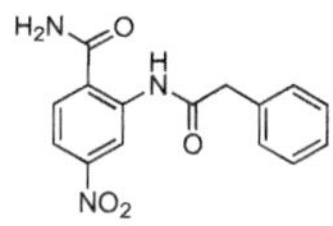

37, AS-48
IC_{50} = 0.6-3.0 μM

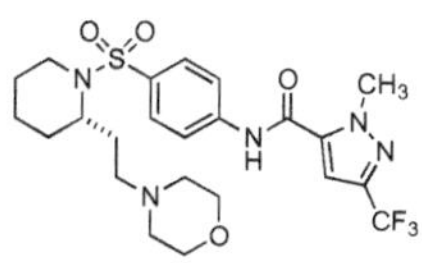

38, ERDRP-0519
EC_{50} = 60 nM

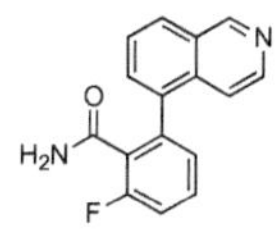

39, GHP-88309
EC_{50} = 0.6 μM

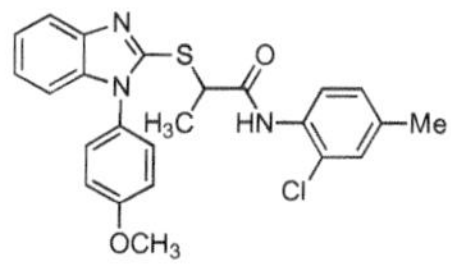

40, JMN3-003
EC_{50} = 0.03 μM

Fig. 9 Small molecules active against RABV and MeV infection.

Notably, a broadly conserved radical *S*-adenosyl-L-methionine (SAM) enzyme known as viperin appeared to be implicated in the inhibition of the replication of many RNA viruses, including RABV.[224] Recently, it was demonstrated that this enzyme synthesizes a broad spectrum antiviral nucleotide, i.e., 3′-deoxy-3′,4′-didehydro-cytidine triphosphate (ddhCTP, **36** in Fig. 9), by catalyzing the dehydration of CTP via a radical mechanism. This study established ddhCTP as a naturally occurring chain terminator encoded by the human genome.[225] Additional studies suggested that viperin exerts its antiviral effect by activating a component of the protein ubiquitination machinery with subsequent degradation of cellular and viral proteins essential for viral replication.[226]

Host-targeted therapies to combat RABV infections have not been very successful to date.[216]

12. Measles virus (MeV)

Measles virus (MeV) is a (-)-ssRNA virus classified in the genus Morbillivirus within the Paramyxoviridae family that is recognized as the causative agent of one of the most contagious respiratory acute disease affecting humans. Its basic reproduction number (i.e., average cases caused by an infected individual during the infectious period) is estimated to vary between 12 and 18, surpassing by far that of SARS-CoV-2 (i.e., 2.5–3.5).[227] A safe, affordable, and effective life-attenuated vaccine is available to prevent outbreaks and ensure immunity against MeV; however, this virus is still prevalent in Third World countries where vaccine access and distribution are inefficient. Recently, developed countries where endemic MeV transmission was considered eliminated experienced a re-emergence of this virus following an alarming decline in immunization coverage due to vaccination hesitancy. In the period between 2017 and 2019, an approximate 130% increase of deaths directly attributed to measles was documented worldwide (207,500 in 2019), mostly among children under 5 years of age.[228] The risk of globally recurring measles outbreaks is anticipated to raise due to the COVID-19 pandemic as many countries suspended childhood immunization measles campaigns leaving dangerous immunity gaps.[228,229] Cost-effective and readily distributable small molecule antivirals are urgently needed for outbreak control especially among under-vaccinated populations in vulnerable countries where the clinical need is highest and to aid in the long term measles eradication by synergizing with vaccination to close gaps in herd immunity due to vaccine refusal.

In the absence of a specific antiviral agent approved for treating MeV infections, various small molecule inhibitors targeting viral components as well as cellular factors have been evaluated in vitro and in animal models; however, further clinical development has been delayed by the low commercial value of a potential measles drug, ethical challenges associated with the design of clinical trials among a prevailing pediatric patient population, and the limited opportunities for intervention associated with the predominantly acute nature of the disease.[230] In rare cases, MeV may invade the central nervous system (CNS) during or after infection leading to severe neurological manifestations that complicate potential antiviral treatments.[231]

A homology model of the MeV fusion (F) protein trimer guided the rational design of a class of nonpeptidic fusion protein inhibitors featuring an anilide scaffold that appeared to exert a stabilizing effect on the prefusion state of F trimers.[232,233] These compounds, as exemplified by AS-48 (compound **37**, Fig. 9), exhibited broad spectrum activity across various MeV strains as well as closely related pathogens of the *Morbillivirus* genus such as the canine distemper virus (CDV), however their moderate antiviral potency (EC_{50} values in the low-micromolar range) observed in cellular assays hampered their further evaluation in animal models.

Among nucleoside polymerase inhibitors, remdesivir was evaluated against various paramyxoviruses by employing reporter virus assays and emerged as a potentially viable candidate to treat MeV infections with an EC_{50} of 0.037 μM.[234] The introduction of a fluorine substituent at the 2'-position of 4'-azido-cytidine (R1479) yielded two additional analogues, namely 2'-fluoro and 2'-difluoro-4'-azido-cytidine, which displayed low micromolar activities (EC_{50} = 0.37 and 0.34 μM, respectively) superior to the parent compound (1.9 μM) when tested in a recombinant measle virus model).[235]

The investigation of non-nucleoside inhibitors of MeV RdRp has delivered molecules with promising antiviral profiles that could advance the path towards the development of effective measles therapeutics. ERDRP-0519 (compound **38**, Fig. 9) is an orally bioavailable pan-morbillivirus inhibitor derived from an optimization process of hits containing the pyrazole carboxamide scaffold that were identified by a cell-based HTS campaign.[236] Compound **38** showed potent in vitro inhibitory activity against MeV and CDV strains. Initial animal in vivo characterization focused on testing the efficacy of **38** in a lethal CDV-ferret surrogate model of human MeV. CDV is known to cause lethal measles-like disease in ferrets. This study showed that oral treatment of CDV-infected ferrets 3 days after exposure completely suppressed disease clinical signs. All treated animals survived and developed a robust immune response against the virus providing protection against reinfection when challenged with a later lethal virus dose.[237]

Recently, the in vivo oral efficacy of this compound was further validated in non-human primates known to develop a human measles-like disease.[238] Prophylactic and post-exposure therapeutic treatment of squirrel monkeys (*Saimiri sciureus*) reduced virus shedding and prevented measles disease. Mechanistically, ERDRP-0519 was demonstrated to target the MeV polymerase and specifically inhibit all phosphodiester bond formation during both the *de novo* initiation of RNA synthesis and RNA elongation steps.[239]

A high-throughput screen for inhibitors of human parainfluenza virus type-3 (HPIV3) led to the identification of another broad-spectrum paramyxovirus allosteric inhibitor, i.e., GHP-88309 (compound **39**, Fig. 9), which was also active against MeV in well-differentiated human airway organoid cultures.[240] This compound was also shown to act as an allosteric inhibitor of viral polymerase activity by interfering with the initiation phase of viral synthesis. In vivo proof-of-concept efficacy was established in a lethal Sendai virus mouse surrogate model of human HPIV3 disease. Oral administration of **39** 48 h after infection provided complete protection against infection and led to immunoprotection acquisition against reinfection.

A series of hit compounds emerging from a large scale HTS was shown to constitute a mechanistically novel class of MeV inhibitors able to interfere with host factors required for viral RdRp activity. Among these, JMN3-003 (compound **40**, Fig. 9) showed potent and broad activity against MeV as well as a selection of clinically significant members of the para- and orthomyxovirus families. A high barrier against rapid viral escape from inhibition for this class of compounds was demonstrated by in vitro adaptation studies.[241] Other efforts to target host factors as a strategy to develop anti-measle compounds include among others repurposing kinase inhibitors.[242]

13. New coronaviruses (CoVs)

Coronaviruses (CoVs) are diverse (+)-ssRNA viruses classified in the *Coronaviridae* family and *Orthocoronavirinae* subfamily, where they are further taxonomically clustered into four genera designated as α, β, γ, and δ.[243] Seasonal coronavirus infections are caused in humans by a total of four α-CoVs (HCoV-NL63 and HCoV-229E) and β-CoVs (HCoV-HKU1 and HCoV-OC43), which circulate endemically and are typically responsible for mild to moderate gastrointestinal and upper-tract respiratory illnesses.[244] However, the last two decades have witnessed the alarming epidemic and pandemic impact of this class of human pathogens with the emergence of three viral pneumonia outbreaks caused by previously unknown β-CoVs (SARS-CoV in 2003, MERS-CoV in 2012, and SARS-CoV-2 in 2019).[245,246] While new cases of SARS have not been reported since 2004, MERS continues to circulate in the Middle East with the latest confirmed cases and deaths dating between March and July 2021.[247] As of the 8 september 2021, the underway SARS-CoV-2 pandemic has affected more than 221 million people and claimed 4.5 million lives.[248] The pathogenesis of these infections may vary from mild symptoms to severe respiratory distress and in some cases multiple organ failure.[249] Upon binding specific receptors on the surface of respiratory epithelial cells [i.e., angiotensin-converting enzyme 2 (ACE2) for SARS-CoV and SARS-CoV-2, dipeptidyl peptidase 4 (DPP4) for MERS-CoV], these viruses enter the airways and start replicating primarily in the lower respiratory tract where they can trigger an aberrant host immune response.

These three highly pathogenic CoVs are alleged to have originated in bats, which are known to be primary natural hosts for human β-CoVs.[250] While adaptation from bats in an intermediary host was shown to have facilitated transfer to humans in the case of SARS-CoV and MERS-CoV (i.e., via infected palm civets and dromedary camels, respectively),[246] no clear evidence has been found to support this transmission mode for the new virus.[245] Phylogenetic analysis suggested that the SARS-CoV-2 lineage persisted for decades in bats where it evolved as to be capable of direct transmission and replication in human cells.[251]

Given the zoonotic origin of these viruses as well as their rapid mutation frequency often resulting in variants of higher infectivity, new CoVs related to SARS, MERS, and SARS-CoV-2 may not be excluded in the future and demand urgent action. The development and stockpiling of antiviral therapeutics with broad anti-coronavirus coverage against the currently known pathogenic CoVs must be prioritized for pandemic preparedness and prevention.[252]

The most logic strategy towards the discovery of drugs that would inhibit a wide range of closely and distantly related viral pathogens is targeting viral proteins that feature the highest degree of sequence conservation among different species and thus may be less likely to mutate. A recent patent literature search indicates that several potential targets can be considered for pan-coronavirus drug design (e.g., RdRp, M^{pro}, PL^{pro}, RNA helicase, spike glycoprotein).[253] A map of conserved regions among the main human coronavirus proteins has been recently reported based on the analysis of a large set of viral genomic sequences.[254]

The viral main protease M^{pro} is an ideal target for broad spectrum antiviral inhibition as M^{pro} structural and functional similarities across different coronaviruses are known. For instance, SARS-CoV and SARS-CoV-2 M^{pro} sequences share 96% amino acid identity and a 100% sequence homology occur at the level of their substrate binding active site.[255]

In an initial study, a structure-based approach led to the identification of a series of α-ketoamide-based peptidomimetic M^{pro} inhibitors with equipotent activity against α-CoVs (HCoV-229E) and β-CoVs (SARS-CoV and MERS-CoV) in cell cultures.[256] Scaffold optimization yielded compound **41** (Fig. 10) featuring a pyridone ring between the P3 and P2 amide bonds that inhibited SARS-CoV-2 M^{pro} with an IC_{50} of 0.67 μM, while displaying IC_{50} values corresponding to 0.90 and 0.58 μM for inhibiting the SARS-CoV M^{pro} and the MERS-CoV M^{pro}, respectively.[255] Compound **41** was further tested in a SARS-CoV replicon where it showed the ability to inhibit RNA replication at $EC_{50} = 1.75$ μM. In human Calu-3 cells infected with SARS-CoV-2, an EC_{50} of 4 to 5 μM was observed.

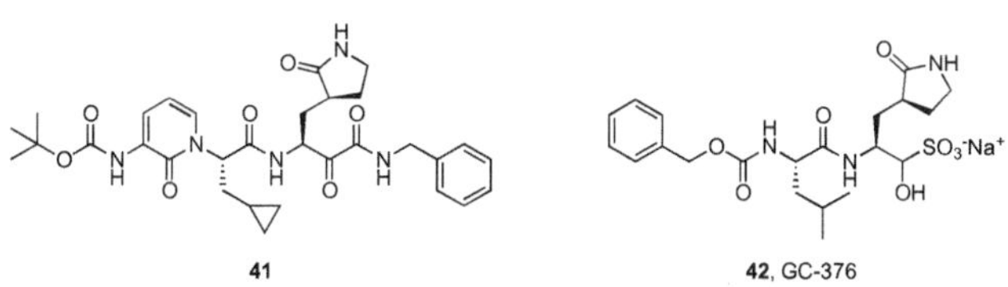

Fig. 10 Structures of some potential pan-coronavirus antivirals.

Boceprevir and the peptidomimetic GC376 (**42**, Fig. 10) were also found to possess a broad spectrum multitargeting profile against SARS-CoV-2, SARS-CoV, MERS-CoV, and HCoV.[257] Deuterated variants of GC376 were recently reported to have improved potency compared to the parent compound in vitro, while in vivo efficacy was demonstrated by the observation of enhanced survival rates in infected mice following treatment at 24 h postinfection.[258]

PF-00835231 (the parent of clinical candidate PF-07304814 **4**, Fig. 1) was originally developed as a SARS-CoVM^{pro} inhibitor in response to the 2002-2003 SARS outbreak.[259] Pan-coronavirus antiviral activity was also observed with PF-00835231 against a variety of α-CoVs (NL63-CoV, PEDV, FIPV), β-CoVs (HKU4-CoV, HKU5-CoV, HKU9-CoV, MHV-CoV, OC43-CoV, HKU1- CoV), and γ-CoVs (IBV-CoV) with Ki values ranging from 30 pM to 4 nM in in fluorescence resonance energy transfer (FRET) protease activity assays, suggesting a potential therapeutic use against emerging coronaviruses.[26]

The orally bioavailable M^{pro} inhibitor PF-07321332 **5** (Fig. 1) currently in clinical trials for COVID-19 could inhibit the proteolytic activity of M^{Pro} of various β-CoVs (SARS-CoV, HCoV-HKU1, HCoV-OC43, MERS-CoV) as well as endemic α-CoVs (HCoV-229E, HCoV-NL63) in FRET assays with Ki values ranging from 4.94 to 226 nM.[27] The potent pan-human coronavirus antiviral activity of PF07321332 was further validated in cytopathic effect (CPE) cellular assays against SARS-CoV (EC_{90} = 317 nM), MERS-CoV (EC_{90} = 351 nM), and HCoV-229E (EC_{90} = 620 nM).

Since the RdRp (also named nsp 12) and its mode of substrate recognition during viral RNA synthesis are highly conserved among diverse coronaviruses (e.g., RdRp of SARS-CoV-2 shares 99.1% similarity and 96% amino acid identity with that of SARS-CoV),[260] a broad antiviral activity spectrum can likewise be expected from nucleoside analogues that are incorporated into the growing RNA chain after intracellular phosphorylation and either induce delayed chain termination or lead to base-pair mismatches resulting in lethal mutation rates, as epitomized by remdesivir and molnupiravir, respectively. Besides SARS-CoV-2, the in vitro antiviral activity spectrum of remdesivir (**1**, Fig. 1) was revealed to further cover SARS-CoV (EC_{50} = 0.069 μM) and MERS-CoV (EC_{50} = 0.025–0.12 μM) as well as endemic HCoV-OC43 and HCoV-229E and porcine CoVs.[261] Further in vivo efficacy studies conducted in a rhesus macaque and mouse model of MERS-CoV and SARS-CoV infection, respectively, demonstrated the prophylactic and therapeutic value of this drug.

The parent nucleoside of mutagenic SARS-CoV-2 antiviral molnupiravir **3** (Fig. 1) was shown to be highly active in vitro against SARS-CoV-2, while also conferring a potent inhibition of MERS-CoV (IC_{50} = 0.024 μM) and SARS-CoV (IC_{50} = 0.14 μM) in primary human epithelial cell cultures. Interestingly, the antiviral activity of NHC was extended to zoonotic SARS- and MERS-like CoVs (SHC014, HKU3, and HKU5), suggesting that an increase (up to 20%) in the natural amino acid variation of the RdRp did not affect the activity of this nucleoside prodrug.[17]

AT-511, the free base congener of clinical candidate AT-527 (**2**, Fig. 1), was also shown to have a broad antiviral activity spectrum among α-CoVs (HCoV-229E) and β-CoVs (HCoV-OC43, SARS-CoV) in different cell lines with EC_{50} values in the micromolar and sub-micromolar range. Surprisingly, only moderate activity was observed against MERS-CoV (EC_{50} = 26-27 μM), suggesting that the mode of action of this nucleoside analogue might not be limited to RdRp activity inhibition and subsequent non-obligate chain termination but implicate the additional interaction with another functional domain of the polymerase complex [i.e., nidovirus RdRp-associated nucleotidyltransferase (NiRAN)] crucial for viral replication.[14]

Further potential drug candidates for the treatment of multiple coronavirus infections could target the viral entry step, and specifically the spike (S) protein-mediated membrane fusion. Among a series of lipopeptide analogues, EK1C4 exhibited the most potent inhibitory activity against membrane fusion and pseudovirus infection of SARS-CoV-2 with IC_{50} values of 1.3 and 15.8 nM, respectively, while also effectively inhibiting the replication of SARS-CoV, MERS-CoV, and other HCov.[262] A protective effect against HCoV-OC43 infection was observed upon intranasal application of EK1C4 in mice before or after viral challenge, suggesting that EK1C4 could be useful for both prevention and treatment of infection by the currently circulating SARS-CoV-2 and other potentially emerging CoVs.

Other broad spectrum antivirals targeting coronaviruses with an unknown mode of actions have been identified by screening programs[263] and drug repurposing campaign.[264] Targeting host factors such as kinase pathways,[265] cysteine protease like cathepsin L,[266] or compounds that deplete cellular polyamines by inhibiting their biosynthesis (e.g., fluoromethylornithine) and in this way reduce viral attachment and entry[267] maybe also represent cellular approaches towards broad spectrum antivirals.

14. Conclusion

The fight against virus infections is based upon the prophylactic (or preventive) administration of vaccines or therapeutic use of antiviral drugs, although, as testified by PrEP, antivirals (e.g., Truvada® and Descovy®) have also been formally approved for the prevention of HIV infections. Generally, when a virus has economic implications but is mostly harmless, like in the case of rhinovirus infections, a drug that reduces symptoms and infectivity may suffice. By contrast, for viral infections with an enormous impact on social (e.g., HCV-related cirrhosis and hepatocellular carcinoma) or healthcare (e.g., SARS-CoV-2) systems, it is vital to develop drugs that can eradicate the causative pathogen. The development of drugs for antiviral prophylaxis is particularly significant for viral infections leading to severe illnesses and high case fatality rates, and especially those for which the production of a vaccine is problematic. As corroborated in the past by the development of beta-lactam antibiotics, an infection can most effectively be treated using compounds whose mode of action either triggers the formation of covalent bonds or avoids that these bonds are broken. Accordingly, several nucleoside analogues and protease inhibitors have proven their value as antiviral agents so far, but we can envisage more broad-spectrum therapeutic applications for these classes of drugs in the future. It should also be appreciated that heterocyclic-based antiviral drugs may unforeseeably interact with proteins of the host, leading to pharmacological toxicity that can only be discovered during clinical Phase III and IV studies. These compounds need to bind to a viral target protein with high affinity, thus easily leading to the selection of resistant viruses. Targeting host factors remains a challenging approach as the virus may hijack other pathways once the cellular receptors or enzymes that are normally exploited for its in vivo infection, replication, and transmission are blocked.

Overall, the availability of a vaccine for a given viral infection may also dictate the priorities for antiviral drug discovery research. Smallpox (variola virus) was the first virus infection to be eradicated owing to vaccination. Poliovirus maybe the next example, however the final steps towards its global eradication appear to pose a few unique challenges. In the foreseeable future, we may witness the elimination of mumps and measles viruses providing that a continued effort in vaccination coverage is ensured, while rubella virus is virtually eradicated. With the availability of DAAs, HCV can be expected to be eradicated within a few years, while for hepatitis B virus (HBV) both vaccines and antivirals are applicable. The continuously emerging new influenza virus mutants may also be counteracted by vaccines and antivirals, whereas for RSV antiviral drug treatment seems to become increasingly imminent. For several other virus infections, such as rhino or hemorrhagic fever infections, vaccines and antivirals are anxiously awaited. Finally, several successful vaccines have been launched to prevent SARS-CoV-2 infections, and effective antivirals will likely follow.

Depending on their nature, viral infectious diseases may be amenable to either prophylaxis, drug treatment or both. As amply demonstrated in this review, these are important times for the field and many approaches are being pursued in parallel to combat viral infections with a continuing impact on human health. The pace of antiviral drug discovery needs to move well beyond a single target, and it can be anticipated that the next decade will deliver antiviral cocktails against most of these infections.

References

1. VBI Vaccines. *VBI-2901: Pan-Coronavirus Vaccine Candidate Overview, VBI Vaccines*; 2020; pp. 6–9.
2. Wu, F.; Zhao, S.; Yu, B.; Chen, Y.-M.; Wang, W.; Song, Z.-G.; Hu, Y.; Tao, Z.-W.; Tian, J.-H.; Pei, Y.-Y.; Yuan, M.-L.; Zhang, Y.-L.; Dai, F.-H.; Liu, Y.; Wang, Q.-M.; Zheng, J.-J.; Xu, L.; Holmes, E. C.; Zhang, Y.-Z. A New Coronavirus Associated With Human Respiratory Disease in China. *Nature* **2020**, *579*, 265–269.
3. Su, H.; Zhou, F.; Huang, Z.; Ma, X.; Natarajan, K.; Zhang, M.; Huang, Y.; Su, H. Molecular Insights into Small-Molecule Drug Discovery for SARS-CoV-2. *Angew. Chem. Int. Ed.* **2021**, *60*, 9789–9802.
4. Namchuk, M. N. Early Returns on Small Molecule Therapeutics for SARS-CoV-2. *ACS Infect. Dis.* **2021**, 7, 1298–1302.
5. Shang, Z.; Chan, S. Y.; Liu, W. J.; Li, P.; Huang, W. Recent Insights Into Emerging Coronavirus: SARS-CoV-2. *ACS Infect. Dis.* **2021**, 7, 1369–1388.
6. U.S. Food & Drug Administration. *FDA Approves First Treatment for COVID-19*; 2020. https://www.fda.gov/news-events/press-announcements/fda-approves-first-treatment-covid-19.
7. Kokic, G.; Hillen, H. S.; Tegunov, D.; Dienemann, C.; Seitz, F.; Schmitzova, J.; Farnung, L.; Siewert, A.; Höbartner, C.; Cramer, P. Mechanism of SARS-CoV-2 Polymerase Stalling by Remdesivir. *Nat. Commun.* **2021**, *12*, 279.
8. Beigel, J. H.; Tomashek, K. M.; Dodd, L. E.; Mehta, A. K.; Zingman, B. S.; Kalil, A. C.; Hohmann, E.; Chu, H. Y.; Luetkemeyer, A.; Kline, S.; Lopez de Castilla, D.; Finberg, R. W.; Dierberg, K.; Tapson, V.; Hsieh, L.; Patterson, T. F.; Paredes, R.; Sweeney, D. A.; Short, W. R.; Touloumi, G.; Lye, D. C.; Ohmagari, N.; Oh, M.-D.; Ruiz-Palacios, G. M.; Benfield, T.; Fätkenheuer, G.; Kortepeter, M. G.; Atmar, R. L.; Creech, C. B.; Lundgren, J.; Babiker, A. G.; Pett, S.; Neaton, J. D.; Burgess, T. H.; Bonnett, T.; Green, M.; Makowski, M.; Osinusi, A.; Nayak, S.; Lane, H. C. Remdesivir for the Treatment of Covid-19—Final Report. *New Engl. J. Med.* **2020**, *383*, 1813–1826.
9. Pan, H.; Peto, R.; Henao-Restrepo, A. M.; Preziosi, M. P.; Sathiyamoorthy, V.; Abdool Karim, Q.; Alejandria, M. M.; Hernández García, C.; Kieny, M. P.;

Malekzadeh, R.; Murthy, S.; Reddy, K. S.; Roses Periago, M.; Abi Hanna, P.; Ader, F.; Al-Bader, A. M.; Alhasawi, A.; Allum, E.; Alotaibi, A.; Alvarez-Moreno, C. A.; Appadoo, S.; Asiri, A.; Aukrust, P.; Barratt-Due, A.; Bellani, S.; Branca, M.; Cappel-Porter, H. B. C.; Cerrato, N.; Chow, T. S.; Como, N.; Eustace, J.; García, P. J.; Godbole, S.; Gotuzzo, E.; Griskevicius, L.; Hamra, R.; Hassan, M.; Hassany, M.; Hutton, D.; Irmansyah, I.; Jancoriene, L.; Kirwan, J.; Kumar, S.; Lennon, P.; Lopardo, G.; Lydon, P.; Magrini, N.; Maguire, T.; Manevska, S.; Manuel, O.; McGinty, S.; Medina, M. T.; Mesa Rubio, M. L.; Miranda-Montoya, M. C.; Nel, J.; Nunes, E. P.; Perola, M.; Portolés, A.; Rasmin, M. R.; Raza, A.; Rees, H.; Reges, P. P. S.; Rogers, C. A.; Salami, K.; Salvadori, M. I.; Sinani, N.; Sterne, J. A. C.; Stevanovikj, M.; Tacconelli, E.; Tikkinen, K. A. O.; Trelle, S.; Zaid, H.; Røttingen, J. A.; Swaminathan, S. Repurposed Antiviral Drugs for Covid-19—Interim WHO Solidarity Trial Results. *N. Engl. J. Med.* **2021**, *384*, 497–511.

10. Li, Y.; Cao, L.; Li, G.; Cong, F.; Li, Y.; Sun, J.; Luo, Y.; Chen, G.; Li, G.; Wang, P.; Xing, F.; Ji, Y.; Zhao, J.; Zhang, Y.; Guo, D.; Zhang, X. Remdesivir Metabolite GS-441524 Effectively Inhibits SARS-CoV-2 Infection in Mouse Models. *J. Med. Chem.* **2021**.
11. Yan, V. C.; Muller, F. L. Advantages of the Parent Nucleoside GS-441524 Over Remdesivir for Covid-19 Treatment. *ACS Med. Chem. Lett.* **2020**, *11*, 1361–1366.
12. Schooley, R. T.; Carlin, A. F.; Beadle, J. R.; Valiaeva, N.; Zhang, X.-Q.; Clark, A. E.; McMillan, R. E.; Leibel, S. L.; McVicar, R. N.; Xie, J.; Garretson, A. F.; Smith, V. I.; Murphy, J.; Hostetler, K. Y. Rethinking Remdesivir: Synthesis, Antiviral Activity and Pharmacokinetics of Oral Lipid Prodrugs. *bioRxiv* **2021**.
13. Zandi, K.; Amblard, F.; Musall, K.; Downs-Bowen, J.; Kleinbard, R.; Oo, A.; Cao, D. D.; Liang, B.; Russell, O. O.; McBrayer, T.; Bassit, L.; Kim, B.; Schinazi, R. F. Repurposing Nucleoside Analogs for Human Coronaviruses. *Antimicrob. Agents Chemother.* **2021**, *65*.
14. Good, S. S.; Westover, J.; Jung, K. H.; Zhou, X.-J.; Moussa, A.; Colla, P. L.; Collu, G.; Canard, B.; Sommadossi, J.-P. AT-527, a Double Prodrug of a Guanosine Nucleotide Analog, is a Potent Inhibitor of SARS-CoV-2 *In Vitro* and a Promising Oral Antiviral for Treatment of COVID-19. *Antimicrob. Agents Chemother.* **2021**, *65*, e02420–e02479.
15. Do, T. N. D.; Donckers, K.; Vangeel, L.; Chatterjee, A. K.; Gallay, P. A.; Bobardt, M. D.; Bilello, J. P.; Cihlar, T.; De Jonghe, S.; Neyts, J.; Jochmans, D. A Robust SARS-CoV-2 Replication Model in Primary Human Epithelial Cells at the Air Liquid Interface to Assess Antiviral Agents. *Antiviral Res.* **2021**, *192*.
16. Fischer, W.; Eron, J. J.; Holman, W.; Cohen, M. S.; Fang, L.; Szewczyk, L. J.; Sheahan, T. P.; Baric, R.; Mollan, K. R.; Wolfe, C. R.; Duke, E. R.; Azizad, M. M.; Borroto-Esoda, K.; Wohl, D. A.; Loftis, A. J.; Alabanza, P.; Lipansky, F.; Painter, W. P. Molnupiravir, an Oral Antiviral Treatment for COVID-19. *medRxiv* **2021**.
17. Sheahan, T. P.; Sims, A. C.; Zhou, S.; Graham, R. L.; Pruijssers, A. J.; Agostini, M. L.; Leist, S. R.; Schäfer, A.; Dinnon, K. H.; Stevens, L. J.; Chappell, J. D.; Lu, X.; Hughes, T. M.; George, A. S.; Hill, C. S.; Montgomery, S. A.; Brown, A. J.; Bluemling, G. R.; Natchus, M. G.; Saindane, M.; Kolykhalov, A. A.; Painter, G.; Harcourt, J.; Tamin, A.; Thornburg, N. J.; Swanstrom, R.; Denison, M. R.; Baric, R. S. An orally Bioavailable Broad-Spectrum Antiviral Inhibits SARS-CoV-2 in Human Airway Epithelial Cell Cultures and Multiple Coronaviruses in Mice. *Sci. Transl. Med.* **2020**, 12.
18. Cox, R. M.; Wolf, J. D.; Plemper, R. K. Therapeutically Administered Ribonucleoside Analogue MK-4482/EIDD-2801 Blocks SARS-CoV-2 Transmission in Ferrets. *Nat. Microbiol.* **2021**, *6*, 11–18.

19. Wahl, A.; Gralinski, L. E.; Johnson, C. E.; Yao, W.; Kovarova, M.; Dinnon, K. H.; Liu, H.; Madden, V. J.; Krzystek, H. M.; De, C.; White, K. K.; Gully, K.; Schäfer, A.; Zaman, T.; Leist, S. R.; Grant, P. O.; Bluemling, G. R.; Kolykhalov, A. A.; Natchus, M. G.; Askin, F. B.; Painter, G.; Browne, E. P.; Jones, C. D.; Pickles, R. J.; Baric, R. S.; Garcia, J. V. SARS-CoV-2 Infection is Effectively Treated and Prevented by EIDD-2801. *Nature* **2021**, *591*, 451–457.
20. Menéndez-Arias, L. Decoding Molnupiravir-Induced Mutagenesis in SARS-CoV-2. *J. Biol. Chem.* **2021**, *297*, 100867.
21. Hassanipour, S.; Arab-Zozani, M.; Amani, B.; Heidarzad, F.; Fathalipour, M.; Martinez-de-Hoyo, R. The efficacy and Safety of Favipiravir in Treatment of COVID-19: A Systematic Review and Meta-Analysis of Clinical Trials. *Sci. Rep.* **2021**, *11*, 11022.
22. Naydenova, K.; Muir, K. W.; Wu, L.-F.; Zhang, Z.; Coscia, F.; Peet, M. J.; Castro-Hartmann, P.; Qian, P.; Sader, K.; Dent, K.; Kimanius, D.; Sutherland, J. D.; Löwe, J.; Barford, D.; Russo, C. J. Structure of the SARS-CoV-2 RNA-Dependent RNA Polymerase in the Presence of Favipiravir-RTP. *Proc. Natl. Acad. Sci. U. S. A.* **2021**, *118*.
23. Tian, L.; Qiang, T.; Liang, C.; Ren, X.; Jia, M.; Zhang, J.; Li, J.; Wan, M.; YuWen, X.; Li, H.; Cao, W.; Liu, H. RNA-Dependent RNA Polymerase (RdRp) Inhibitors: The Current Landscape and Repurposing for the COVID-19 Pandemic. *Eur. J. Med. Chem.* **2021**, *213*, 113201.
24. Zheng, Z.; Groaz, E.; Snoeck, R.; De Jonghe, S.; Herdewijn, P.; Andrei, G. Influence of 4′-Substitution on the Activity of Gemcitabine and Its ProTide Against VZV and SARS-CoV-2. *ACS Med. Chem. Lett.* **2021**, *12*, 88–92.
25. Yang, H.; Yang, J. A Review of the Latest Research on Mpro Targeting SARS-COV Inhibitors. *RSC Med. Chem.* **2021**, *12*, 1026–1036.
26. Boras, B.; Jones, R. M.; Anson, B. J.; Arenson, D.; Aschenbrenner, L.; Bakowski, M. A.; Beutler, N.; Binder, J.; Chen, E.; Eng, H.; Hammond, H.; Hammond, J.; Haupt, R. E.; Hoffman, R.; Kadar, E. P.; Kania, R.; Kimoto, E.; Kirkpatrick, M. G.; Lanyon, L.; Lendy, E. K.; Lillis, J. R.; Logue, J.; Luthra, S. A.; Ma, C.; Mason, S. W.; McGrath, M. E.; Noell, S.; Obach, R. S.; O'Brien, M. N.; O'Connor, R.; Ogilvie, K.; Owen, D.; Pettersson, M.; Reese, M. R.; Rogers, T. F.; Rossulek, M. I.; Sathish, J. G.; Shirai, N.; Steppan, C.; Ticehurst, M.; Updyke, L. W.; Weston, S.; Zhu, Y.; Wang, J.; Chatterjee, A. K.; Mesecar, A. D.; Frieman, M. B.; Anderson, A. S.; Allerton, C. Discovery of a Novel Inhibitor of Coronavirus 3CL Protease for the Potential Treatment of COVID-19. *bioRxiv* **2021**.
27. Owen, D. R.; Allerton, C. M. N.; Anderson, A. S.; Aschenbrenner, L.; Avery, M.; Berritt, S.; Boras, B.; Cardin, R. D.; Carlo, A.; Coffman, K. J.; Dantonio, A.; Di, L.; Eng, H.; Ferre, R.; Gajiwala, K. S.; Gibson, S. A.; Greasley, S. E.; Hurst, B. L.; Kadar, E. P.; Kalgutkar, A. S.; Lee, J. C.; Lee, J.; Liu, W.; Mason, S. W.; Noell, S.; Novak, J. J.; Obach, R. S.; Ogilvie, K.; Patel, N. C.; Pettersson, M.; Rai, D. K.; Reese, M. R.; Sammons, M. F.; Sathish, J. G.; Singh, R. S. P.; Steppan, C. M.; Stewart, A. E.; Tuttle, J. B.; Updyke, L.; Verhoest, P. R.; Wei, L.; Yang, Q.; Zhu, Y. An Oral SARS-CoV-2 Mpro Inhibitor Clinical Candidate for the Treatment of COVID-19. *medRxiv* **2021**.
28. Qiao, J.; Li, Y.-S.; Zeng, R.; Liu, F.-L.; Luo, R.-H.; Huang, C.; Wang, Y.-F.; Zhang, J.; Quan, B.; Shen, C.; Mao, X.; Liu, X.; Sun, W.; Yang, W.; Ni, X.; Wang, K.; Xu, L.; Duan, Z.-L.; Zou, Q.-C.; Zhang, H.-L.; Qu, W.; Long, Y.-H.-P.; Li, M.-H.; Yang, R.-C.; Liu, X.; You, J.; Zhou, Y.; Yao, R.; Li, W.-P.; Liu, J.-M.; Chen, P.; Liu, Y.; Lin, G.-F.; Yang, X.; Zou, J.; Li, L.; Hu, Y.; Lu, G.-W.; Li, W.-M.; Wei, Y.-Q.; Zheng, Y.-T.; Lei, J.;

Yang, S. SARS-CoV-2 Mpro Inhibitors With Antiviral Activity in a Transgenic Mouse Model. *Science* **2021**, *371*, 1374–1378.
29. Bobileva, O.; Bobrovs, R.; Kanepe, I.; Patetko, L.; Kalnins, G.; Sisovs, M.; Bula, A. L.; Grinberga, S.; Boroduskis, M.; Ramata-Stunda, A.; Rostoks, N.; Jirgensons, A.; Tars, K.; Jaudzems, K. Potent SARS-CoV-2 mRNA Cap Methyltransferase Inhibitors by Bioisosteric Replacement of Methionine in SAM Cosubstrate. *ACS Med. Chem. Lett.* **2021**, *12*, 1102–1107.
30. Otava, T.; Sala, M.; Li, F. L.; Fanfrlik, J.; Devkota, K.; Perveen, S.; Chau, I.; Pakarian, P.; Hobza, P.; Vedadi, M.; Boura, E.; Nencka, R. The Structure-Based Design of SARS-CoV-2 nsp14 Methyltransferase Ligands Yields Nanomolar Inhibitors. *ACS Infect. Dis.* **2021**, 7, 2214–2220.
31. Krammer, F.; Smith, G. J. D.; Fouchier, R. A. M.; Peiris, M.; Kedzierska, K.; Doherty, P. C.; Palese, P.; Shaw, M. L.; Treanor, J.; Webster, R. G.; García-Sastre, A. Influenza. *Nat. Rev. Dis. Primers.* **2018**, *4*, 3.
32. Kim, H.; Webster, R. G.; Webby, R. J. Influenza Virus: Dealing with a Drifting and Shifting Pathogen. *Viral Immunol.* **2018**, *31*, 174–183.
33. Harrington, W. N.; Kackos, C. M.; Webby, R. J. The Evolution and Future of Influenza Pandemic Preparedness. *Exp. Mol. Med.* **2021**, *53*, 737–749.
34. Mtambo, S. E.; Amoako, D. G.; Somboro, A. M.; Agoni, C.; Lawal, M. M.; Gumede, N. S.; Khan, R. B.; Kumalo, H. M. Influenza Viruses: Harnessing the Crucial Role of the M2 Ion-Channel and Neuraminidase Toward Inhibitor Design. *Molecules* **2021**, *26*, 880.
35. He, W.; Zhang, W.; Yan, H.; Xu, H.; Xie, Y.; Wu, Q.; Wang, C.; Dong, G. Distribution and Evolution of H1N1 Influenza A Viruses With Adamantanes-Resistant Mutations Worldwide From 1918 to 2019. *J. Med. Virol.* **2021**, *93*, 3473–3483.
36. Bloom, J. D.; Gong, L. I.; Baltimore, D. Permissive Secondary Mutations Enable the Evolution of Influenza Oseltamivir Resistance. *Science* **2010**, *328*, 1272–1275.
37. Hu, Y.; Hau, R. K.; Wang, Y.; Tuohy, P.; Zhang, Y.; Xu, S.; Ma, C.; Wang, J. Structure–Property Relationship Studies of Influenza A Virus AM2-S31N Proton Channel Blockers. *ACS Med. Chem. Lett.* **2018**, *9*, 1111–1116.
38. Wang, Y.; Hu, Y.; Xu, S.; Zhang, Y.; Musharrafieh, R.; Hau, R. K.; Ma, C.; Wang, J. In Vitro Pharmacokinetic Optimizations of AM2-S31N Channel Blockers Led to the Discovery of Slow-Binding Inhibitors With Potent Antiviral Activity Against Drug-Resistant Influenza A Viruses. *J. Med. Chem.* **2018**, *61*, 1074–1085.
39. Zhang, J.; Murugan, N. A.; Tian, Y.; Bertagnin, C.; Fang, Z.; Kang, D.; Kong, X.; Jia, H.; Sun, Z.; Jia, R.; Gao, P.; Poongavanam, V.; Loregian, A.; Xu, W.; Ma, X.; Ding, X.; Huang, B.; Zhan, P.; Liu, X. Structure-Based Optimization of N-Substituted Oseltamivir Derivatives as Potent Anti-Influenza A Virus Agents with Significantly Improved Potency Against Oseltamivir-Resistant N1-H274Y Variant. *J. Med. Chem.* **2018**, *61*, 9976–9999.
40. Zhang, J.; Poongavanam, V.; Kang, D.; Bertagnin, C.; Lu, H.; Kong, X.; Ju, H.; Lu, X.; Gao, P.; Tian, Y.; Jia, H.; Desta, S.; Ding, X.; Sun, L.; Fang, Z.; Huang, B.; Liang, X.; Jia, R.; Ma, X.; Xu, W.; Murugan, N. A.; Loregian, A.; Huang, B.; Zhan, P.; Liu, X. Optimization of N-Substituted Oseltamivir Derivatives as Potent Inhibitors of Group-1 and -2 Influenza A Neuraminidases, Including a Drug-Resistant Variant. *J. Med. Chem.* **2018**, *61*, 6379–6397.
41. Yuan, S.; Wen, L.; Zhou, J. Inhibitors of Influenza A Virus Polymerase. *ACS Infect. Dis.* **2018**, *4*, 218–223.
42. te Velthuis, A. J. W.; Fodor, E. Influenza Virus RNA Polymerase: Insights Into the Mechanisms of Viral RNA Synthesis. *Nat. Rev. Microbiol.* **2016**, *14*, 479–493.

43. Hayden, F. G.; Sugaya, N.; Hirotsu, N.; Lee, N.; de Jong, M. D.; Hurt, A. C.; Ishida, T.; Sekino, H.; Yamada, K.; Portsmouth, S.; Kawaguchi, K.; Shishido, T.; Arai, M.; Tsuchiya, K.; Uehara, T.; Watanabe, A. Baloxavir Marboxil for Uncomplicated Influenza in Adults and Adolescents. *N. Engl. J. Med.* **2018**, *379*, 913–923.
44. Ikematsu, H.; Hayden, F. G.; Kawaguchi, K.; Kinoshita, M.; de Jong, M. D.; Lee, N.; Takashima, S.; Noshi, T.; Tsuchiya, K.; Uehara, T. Baloxavir Marboxil for Prophylaxis against Influenza in Household Contacts. *N. Engl. J. Med.* **2020**, *383*, 309–320.
45. Yang, T. Baloxavir Marboxil: The First Cap-Dependent Endonuclease Inhibitor for the Treatment of Influenza. *Ann. Pharmacother.* **2019**, *53*, 754–759.
46. Hashimoto, T.; Baba, K.; Inoue, K.; Okane, M.; Hata, S.; Shishido, T.; Naito, A.; Wildum, S.; Omoto, S. Comprehensive Assessment of Amino Acid Substitutions in the Trimeric RNA Polymerase Complex of Influenza A Virus Detected in Clinical Trials of Baloxavir Marboxil. *Influenza Other Respir. Viruses* **2021**, *15*, 389–395.
47. Ivashchenko, A. A.; Mitkin, O. D.; Jones, J. C.; Nikitin, A. V.; Koryakova, A. G.; Ryakhovskiy, A.; Karapetian, R. N.; Kravchenko, D. V.; Aladinskiy, V.; Leneva, I. A.; Falynskova, I. N.; Glubokova, E. A.; Govorkova, E. A.; Ivachtchenko, A. V. Non-Rigid Diarylmethyl Analogs of Baloxavir as Cap-Dependent Endonuclease Inhibitors of Influenza Viruses. *J. Med. Chem.* **2020**, *63*, 9403–9420.
48. Credille, C. V.; Morrison, C. N.; Stokes, R. W.; Dick, B. L.; Feng, Y.; Sun, J.; Chen, Y.; Cohen, S. M. SAR Exploration of Tight-Binding Inhibitors of Influenza Virus PA Endonuclease. *J. Med. Chem.* **2019**, *62*, 9438–9449.
49. Miyagawa, M.; Akiyama, T.; Taoda, Y.; Takaya, K.; Takahashi-Kageyama, C.; Tomita, K.; Yasuo, K.; Hattori, K.; Shano, S.; Yoshida, R.; Shishido, T.; Yoshinaga, T.; Sato, A.; Kawai, M. Synthesis and SAR Study of Carbamoyl Pyridone Bicycle Derivatives as Potent Inhibitors of Influenza Cap-dependent Endonuclease. *J. Med. Chem.* **2019**, *62*, 8101–8114.
50. Clark, M. P.; Ledeboer, M. W.; Davies, I.; Byrn, R. A.; Jones, S. M.; Perola, E.; Tsai, A.; Jacobs, M.; Nti-Addae, K.; Bandarage, U. K.; Boyd, M. J.; Bethiel, R. S.; Court, J. J.; Deng, H. B.; Duffy, J. P.; Dorsch, W. A.; Farmer, L. J.; Gao, H.; Gu, W. X.; Jackson, K.; Jacobs, D. H.; Kennedy, J. M.; Ledford, B.; Liang, J. L.; Maltais, F.; Murcko, M.; Wang, T. S.; Wannamaker, M. W.; Bennett, H. B.; Leeman, J. R.; McNeil, C.; Taylor, W. P.; Memmott, C.; Jiang, M.; Rijnbrand, R.; Bral, C.; Germann, U.; Nezami, A.; Zhang, Y. G.; Salituro, F. G.; Bennani, Y. L.; Charifson, P. S. Discovery of a Novel, First-in-Class, Orally Bioavailable Azaindole Inhibitor (VX-787) of Influenza PB2. *J. Med. Chem.* **2014**, *57*, 6668–6678.
51. Finberg, R. W.; Lanno, R.; Anderson, D.; Fleischhackl, R.; van Duijnhoven, W.; Kauffman, R. S.; Kosoglou, T.; Vingerhoets, J.; Leopold, L. Phase 2b Study of Pimodivir (JNJ-63623872) as Monotherapy or in Combination With Oseltamivir for Treatment of Acute Uncomplicated Seasonal Influenza A: TOPAZ Trial. *J. Infect. Dis.* **2019**, *219*, 1026–1034.
52. McGowan, D. C.; Balemans, W.; Embrechts, W.; Motte, M.; Keown, J. R.; Buyck, C.; Corbera, J.; Funes, M.; Moreno, L.; Cooymans, L.; Tahri, A.; Eymard, J.; Stoops, B.; Strijbos, R.; Van den Berg, J.; Fodor, E.; Grimes, J. M.; Koul, A.; Jonckers, T. H. M.; Raboisson, P.; Guillemont, J. Design, Synthesis, and Biological Evaluation of Novel Indoles Targeting the Influenza PB2 Cap Binding Region. *J. Med. Chem.* **2019**, *62*, 9680–9690.
53. Łagocka, R.; Dziedziejko, V.; Kłos, P.; Pawlik, A. Favipiravir in Therapy of Viral Infections. *J. Clin. Med.* **2021**, *10*.

54. Massari, S.; Desantis, J.; Nizi, M. G.; Cecchetti, V.; Tabarrini, O. Inhibition of Influenza Virus Polymerase by Interfering With Its Protein–Protein Interactions. *ACS Infect. Dis.* **2021**, 7, 1332–1350.
55. Yuan, S.; Chu, H.; Zhao, H.; Zhang, K.; Singh, K.; Chow, B. K. C.; Kao, R. Y. T.; Zhou, J.; Zheng, B.-J. Identification of a Small-Molecule Inhibitor of Influenza Virus via Disrupting the Subunits Interaction of the Viral Polymerase. *Antiviral Res.* **2016**, *125*, 34–42.
56. Kao, R. Y.; Yang, D.; Lau, L. S.; Tsui, W. H.; Hu, L.; Dai, J.; Chan, M. P.; Chan, C. M.; Wang, P.; Zheng, B. J.; Sun, J.; Huang, J. D.; Madar, J.; Chen, G.; Chen, H.; Guan, Y.; Yuen, K. Y. Identification of Influenza A Nucleoprotein as an Antiviral Target. *Nat. Biotechnol.* **2010**, *28*, 600–605.
57. Woodring, J. L.; Lu, S.-H.; Krasnova, L.; Wang, S.-C.; Chen, J.-B.; Chou, C.-C.; Huang, Y.-C.; Cheng, T.-J. R.; Wu, Y.-T.; Chen, Y.-H.; Fang, J.-M.; Tsai, M.-D.; Wong, C.-H. Disrupting the Conserved Salt Bridge in the Trimerization of Influenza A Nucleoprotein. *J. Med. Chem.* **2020**, *63*, 205–215.
58. Sethy, B.; Hsieh, C.-F.; Lin, T.-J.; Hu, P.-Y.; Chen, Y.-L.; Lin, C.-Y.; Tseng, S.-N.; Horng, J.-T.; Hsieh, P.-W. Design, Synthesis, and Biological Evaluation of Itaconic Acid Derivatives as Potential Anti-Influenza Agents. *J. Med. Chem.* **2019**, *62*, 2390–2403.
59. White, K. M.; Abreu, P., Jr.; Wang, H.; De Jesus, P. D.; Manicassamy, B.; García-Sastre, A.; Chanda, S. K.; DeVita, R. J.; Shaw, M. L. Broad Spectrum Inhibitor of Influenza A and B Viruses Targeting the Viral Nucleoprotein. *ACS Infect. Dis.* **2018**, *4*, 146–157.
60. White, K.; Esparza, M.; Liang, J.; Bhat, P.; Naidoo, J.; McGovern, B. L.; Williams, M. A. P.; Alabi, B. R.; Shay, J.; Niederstrasser, H.; Posner, B.; García-Sastre, A.; Ready, J.; Fontoura, B. M. A. Aryl Sulfonamide Inhibits Entry and Replication of Diverse Influenza Viruses via the Hemagglutinin Protein. *J. Med. Chem.* **2021**, *64*, 10951–10966.
61. Wu, G.; Yu, G.; Yu, Y.; Yang, S.; Duan, Z.; Wang, W.; Liu, Y.; Yu, R.; Li, J.; Zhu, T.; Gu, Q.; Li, D. Chemoreactive-Inspired Discovery of Influenza A Virus Dual Inhibitor to Block Hemagglutinin-Mediated Adsorption and Membrane Fusion. *J. Med. Chem.* **2020**, *63*, 6924–6940.
62. Gaisina, I. N.; Peet, N. P.; Cheng, H.; Li, P.; Du, R.; Cui, Q.; Furlong, K.; Manicassamy, B.; Caffrey, M.; Thatcher, G. R. J.; Rong, L. Optimization of 4-Aminopiperidines as Inhibitors of Influenza A Viral Entry That Are Synergistic With Oseltamivir. *J. Med. Chem.* **2020**, *63*, 3120–3130.
63. Rossignol, J. F.; La Frazia, S.; Chiappa, L.; Ciucci, A.; Santoro, M. G. Thiazolides, a New Class of Anti-Influenza Molecules Targeting Viral Hemagglutinin at the Post-Translational Level. *J. Biol. Chem.* **2009**, *284*, 29798–29808.
64. Haffizulla, J.; Hartman, A.; Hoppers, M.; Resnick, H.; Samudrala, S.; Ginocchio, C.; Bardin, M.; Rossignol, J. F. Effect of Nitazoxanide in Adults and Adolescents With Acute Uncomplicated Influenza: A Double-Blind, Randomised, Placebo-Controlled, Phase 2b/3 Trial. *Lancet Infect. Dis.* **2014**, *14*, 609–618.
65. Malakhov, M. P.; Aschenbrenner, L. M.; Smee, D. F.; Wandersee, M. K.; Sidwell, R. W.; Gubareva, L. V.; Mishin, V. P.; Hayden, F. G.; Kim, D. H.; Ing, A.; Campbell, E. R.; Yu, M.; Fang, F. Sialidase Fusion Protein as a Novel Broad-Spectrum Inhibitor of Influenza Virus Infection. *Antimicrob. Agents Chemother.* **2006**, *50*, 1470–1479.
66. Moss, R. B.; Hansen, C.; Sanders, R. L.; Hawley, S.; Li, T.; Steigbigel, R. T. A Phase II Study of DAS181, a Novel Host Directed Antiviral for the Treatment of Influenza Infection. *J. Infect. Dis.* **2012**, *206*, 1844–1851.

67. O'Hanlon, R.; Leyva-Grado, V. H.; Sourisseau, M.; Evans, M. J.; Shaw, M. L. An Influenza Virus Entry Inhibitor Targets Class II PI3 Kinase and Synergizes With Oseltamivir. *ACS Infect. Dis.* **2019**, *5*, 1779–1793.
68. Collins, P. L.; Fearns, R.; Graham, B. S. Respiratory Syncytial Virus: Virology, Reverse Genetics, and Pathogenesis of Disease. *Curr. Top. Microbiol. Immunol.* **2013**, *372*, 3–38.
69. Falsey, A. R.; Hennessey, P. A.; Formica, M. A.; Cox, C.; Walsh, E. E. Respiratory Syncytial Virus Infection in Elderly and High-Risk Adults. *N. Engl. J. Med.* **2005**, *352*, 1749–1759.
70. Nair, H.; Nokes, D. J.; Gessner, B. D.; Dherani, M.; Madhi, S. A.; Singleton, R. J.; O'Brien, K. L.; Roca, A.; Wright, P. F.; Bruce, N.; Chandran, A.; Theodoratou, E.; Sutanto, A.; Sedyaningsih, E. R.; Ngama, M.; Munywoki, P. K.; Kartasasmita, C.; Simões, E. A.; Rudan, I.; Weber, M. W.; Campbell, H. Global Burden of Acute Lower Respiratory Infections Due to Respiratory Syncytial Virus in Young Children: A Systematic Review and Meta-Analysis. *Lancet* **2010**, *375*, 1545–1555.
71. Domachowske, J. B.; Anderson, E. J.; Goldstein, M. The Future of Respiratory Syncytial Virus Disease Prevention and Treatment. *Infect. Dis. Ther.* **2021**, *10*, 47–60.
72. Cockerill, G. S.; Good, J. A. D.; Mathews, N. State of the Art in Respiratory Syncytial Virus Drug Discovery and Development. *J. Med. Chem.* **2019**, *62*, 3206–3227.
73. Vendeville, S.; Tahri, A.; Hu, L. L.; Demin, S.; Cooymans, L.; Vos, A.; Kwanten, L.; Van den Berg, J.; Battles, M. B.; McLellan, J. S.; Koul, A.; Raboisson, P.; Roymans, D.; Jonckers, T. H. M. Discovery of 3-({5-Chloro-1- 3-(methylsulfonyl)propyl -1H-indol-2-yl}methyl)-1-(2,2,2 -trifluoroethyl)-1,3-dihydro-2H-imidazo 4,5-c pyridin-2-one (JNJ-53718678), a Potent and Orally Bioavailable Fusion Inhibitor of Respiratory Syncytial Virus. *J. Med. Chem.* **2020**, *63*, 8046–8058.
74. Cockerill, G. S.; Angell, R. M.; Bedernjak, A.; Chuckowree, I.; Fraser, I.; Gascon-Simorte, J.; Gilman, M. S. A.; Good, J. A. D.; Harland, R.; Johnson, S. M.; Ludes-Meyers, J. H.; Littler, E.; Lumley, J.; Lunn, G.; Mathews, N.; McLellan, J. S.; Paradowski, M.; Peeples, M. E.; Scott, C.; Tait, D.; Taylor, G.; Thom, M.; Thomas, E.; Barber, C. V.; Ward, S. E.; Watterson, D.; Williams, G.; Young, P.; Powell, K. Discovery of Sisunatovir (RV521), an Inhibitor of Respiratory Syncytial Virus Fusion. *J. Med. Chem.* **2021**, *64*, 3658–3676.
75. Zheng, X. F.; Gao, L.; Wang, L. S.; Liang, C. G.; Wang, B. X.; Liu, Y. F.; Feng, S.; Zhang, B.; Zhou, M. W.; Yu, X.; Xiang, K. L.; Chen, L.; Guo, T.; Shen, H. C.; Zou, G.; Wu, J. Z.; Yun, H. Y. Discovery of Ziresovir as a Potent, Selective, and Orally Bioavailable Respiratory Syncytial Virus Fusion Protein Inhibitor. *J. Med. Chem.* **2019**, *62*, 6003–6014.
76. Mackman, R. L.; Sangi, M.; Sperandio, D.; Parrish, J. P.; Eisenberg, E.; Perron, M.; Hui, H.; Zhang, L.; Siegel, D.; Yang, H.; Saunders, O.; Boojamra, C.; Lee, G.; Samuel, D.; Babaoglu, K.; Carey, A.; Gilbert, B. E.; Piedra, P. A.; Strickley, R.; Iwata, Q.; Hayes, J.; Stray, K.; Kinkade, A.; Theodore, D.; Jordan, R.; Desai, M.; Cihlar, T. Discovery of an Oral Respiratory Syncytial Virus (RSV) Fusion Inhibitor (GS-5806) and Clinical Proof of Concept in a Human RSV Challenge Study. *J. Med. Chem.* **2015**, *58*, 1630–1643.
77. Chemaly, R. F.; Dadwal, S. S.; Bergeron, A.; Ljungman, P.; Kim, Y. J.; Cheng, G. S.; Pipavath, S. N.; Limaye, A. P.; Blanchard, E.; Winston, D. J.; Stiff, P. J.; Zuckerman, T.; Lachance, S.; Rahav, G.; Small, C. B.; Mullane, K. M.; Patron, R. L.; Lee, D. G.; Hirsch, H. H.; Waghmare, A.; McKevitt, M.; Jordan, R.; Guo, Y.; German, P.; Porter, D. P.; Gossage, D. L.; Watkins, T. R.; Marty, F. M.; Chien, J. W.; Boeckh, M. A Phase 2, Randomized, Double-blind,

Placebo-Controlled Trial of Presatovir for the Treatment of Respiratory Syncytial Virus Upper Respiratory Tract Infection in Hematopoietic-Cell Transplant Recipients. *Clin. Infect Dis.* **2020**, *71*, 2777–2786.

78. Marty, F. M.; Chemaly, R. F.; Mullane, K. M.; Lee, D. G.; Hirsch, H. H.; Small, C. B.; Bergeron, A.; Shoham, S.; Ljungman, P.; Waghmare, A.; Blanchard, E.; Kim, Y. J.; McKevitt, M.; Porter, D. P.; Jordan, R.; Guo, Y.; German, P.; Boeckh, M.; Watkins, T. R.; Chien, J. W.; Dadwal, S. S. A Phase 2b, Randomized, Double-Blind, Placebo-Controlled Multicenter Study Evaluating Antiviral Effects, Pharmacokinetics, Safety, and Tolerability of Presatovir in Hematopoietic Cell Transplant Recipients With Respiratory Syncytial Virus Infection of the Lower Respiratory Tract. *Clin. Infect Dis.* **2020**, *71*, 2787–2795.
79. Porter, D. P.; Guo, Y.; Perry, J.; Gossage, D. L.; Watkins, T. R.; Chien, J. W.; Jordan, R. Assessment of Drug Resistance during Phase 2b Clinical Trials of Presatovir in Adults Naturally Infected With Respiratory Syncytial Virus. *Antimicrob. Agents Chemother.* **2020**, *64*.
80. Pribut, N.; Kaiser, T. M.; Wilson, R. J.; Jecs, E.; Dentmon, Z. W.; Pelly, S. C.; Sharma, S.; Bartsch, P. W.; Burger, P. B.; Hwang, S. S.; Le, T.; Sourimant, J.; Yoon, J.-J.; Plemper, R. K.; Liotta, D. C. Accelerated Discovery of Potent Fusion Inhibitors for Respiratory Syncytial Virus. *ACS Infect. Dis.* **2020**, *6*, 922–929.
81. Yamaguchi-Sasaki, T.; Tokura, S.; Ogata, Y.; Kawaguchi, T.; Sugaya, Y.; Takahashi, R.; Iwakiri, K.; Abe-Kumasaka, T.; Yoshida, I.; Arikawa, K.; Sugiyama, H.; Kanuma, K. Discovery of a Potent Dual Inhibitor of Wild-Type and Mutant Respiratory Syncytial Virus Fusion Proteins. *ACS Med. Chem. Lett.* **2020**, *11*, 1145–1151.
82. Rhodin, M. H. J.; McAllister, N. V.; Castillo, J.; Noton, S. L.; Fearns, R.; Kim, I. J.; Yu, J.; Blaisdell, T. P.; Panarese, J.; Shook, B. C.; Or, Y. S.; Goodwin, B.; Lin, K. EDP-938, a Novel Nucleoprotein Inhibitor of Respiratory Syncytial Virus, Demonstrates Potent Antiviral Activities In Vitro and in a Non-Human Primate Model. *PLoS Pathog.* **2021**, *17*, e1009428.
83. Jordan, P. C.; Stevens, S. K.; Deval, J. Nucleosides for the Treatment of Respiratory RNA Virus Infections. *Antivir. Chem. Chemother.* **2018**, *26*. https://doi.org/10.1177/2040206618764483.
84. Cao, D.; Gao, Y.; Roesler, C.; Rice, S.; D'Cunha, P.; Zhuang, L.; Slack, J.; Domke, M.; Antonova, A.; Romanelli, S.; Keating, S.; Forero, G.; Juneja, P.; Liang, B. Cryo-EM Structure of the Respiratory Syncytial Virus RNA Polymerase. *Nat. Commun.* **2020**, *11*, 368.
85. Gilman, M. S. A.; Liu, C.; Fung, A.; Behera, I.; Jordan, P.; Rigaux, P.; Ysebaert, N.; Tcherniuk, S.; Sourimant, J.; Eléouët, J. F.; Sutto-Ortiz, P.; Decroly, E.; Roymans, D.; Jin, Z.; McLellan, J. S. Structure of the Respiratory Syncytial Virus Polymerase Complex. *Cell* **2019**, *179*, 193–204.e114.
86. Wang, G. Y.; Deval, J.; Hong, J.; Dyatkina, N.; Prhavc, M.; Taylor, J.; Fung, A.; Jin, Z. N.; Stevens, S. K.; Serebryany, V.; Liu, J. W.; Zhang, Q. L.; Tam, Y. E.; Chanda, S. M.; Smith, D. B.; Symons, J. A.; Blatt, L. M.; Beigelman, L. Discovery of 4'-Chloromethyl-2'-deoxy-3',5'-di-O-isobutyryl-2'-fluorocytidine (ALS-8176), A First-in-Class RSV Polymerase Inhibitor for Treatment of Human Respiratory Syncytial Virus Infection. *J. Med. Chem.* **2015**, *58*, 1862–1878.
87. Sommadossi, J.-P.; Moussa, A. *β-D-2′-Deoxy-2′-Substituted-4′-Substituted-2-Substituted-N6-Substituted-6-Aminopurine Nucleotides for the Treatment of Paramyxovirus and Orthomyxovirus Infections*; WO/2018/009623, January 11, 2018, 2018.
88. Mackman, R. L.; Hui, H. C.; Perron, M.; Murakami, E.; Palmiotti, C.; Lee, G.; Stray, K.; Zhang, L.; Goyal, B.; Chun, K.; Byun, D.; Siegel, D.; Simonovich, S.; Du Pont, V.; Pitts, J.; Babusis, D.; Vijjapurapu, A.; Lu, X.; Kim, C.; Zhao, X.;

Chan, J.; Ma, B.; Lye, D.; Vandersteen, A.; Wortman, S.; Barrett, K. T.; Toteva, M.; Jordan, R.; Subramanian, R.; Bilello, J. P.; Cihlar, T. Prodrugs of a 1'-CN-4-Aza-7,9-dideazaadenosine C-Nucleoside Leading to the Discovery of Remdesivir (GS-5734) as a Potent Inhibitor of Respiratory Syncytial Virus With Efficacy in the African Green Monkey Model of RSV. *J. Med. Chem.* **2021**, *64*, 5001–5017.

89. Jacobs, S. E.; Lamson, D. M.; St George, K.; Walsh, T. J. Human Rhinoviruses. *Clin. Microbiol. Rev.* **2013**, *26*, 135–162.
90. Heikkinen, T.; Chonmaitree, T. Importance of Respiratory Viruses in Acute Otitis Media. *Clin. Microbiol. Rev.* **2003**, *16*, 230–241.
91. Gern, J. E. How rhinovirus Infections Cause Exacerbations of Asthma. *Clin. Exp. Allergy* **2015**, *45*, 32–42.
92. Coultas, J. A.; Cafferkey, J.; Mallia, P.; Johnston, S. L. Experimental Antiviral Therapeutic Studies for Human Rhinovirus Infections. *J. Exp. Pharmacol.* **2021**, *13*, 645–659.
93. Bauer, L.; Lyoo, H.; van der Schaar, H. M.; Strating, J. R. P. M.; van Kuppeveld, F. J. M. Direct-Acting Antivirals and Host-Targeting Strategies to Combat Enterovirus Infections. *Curr. Opin. Virol.* **2017**, *24*, 1–8.
94. Kim, J.; Shin, J. S.; Ahn, S.; Han, S. B.; Jung, Y.-S. 3-Aryl-1,2,4-Oxadiazole Derivatives Active Against Human Rhinovirus. *ACS Med. Chem. Lett.* **2018**, *9*, 667–672.
95. Da Costa, L.; Scheers, E.; Coluccia, A.; Casulli, A.; Roche, M.; Di Giorgio, C.; Neyts, J.; Terme, T.; Cirilli, R.; La Regina, G.; Silvestri, R.; Mirabelli, C.; Vanelle, P. Structure-Based Drug Design of Potent Pyrazole Derivatives Against Rhinovirus Replication. *J. Med. Chem.* **2018**, *61*, 8402–8416.
96. Ledford, R. M.; Patel, N. R.; Demenczuk, T. M.; Watanyar, A.; Herbertz, T.; Collett, M. S.; Pevear, D. C. VP1 Sequencing of All Human Rhinovirus Serotypes: Insights Into Genus Phylogeny and Susceptibility to Antiviral Capsid-Binding Compounds. *J. Virol.* **2004**, *78*, 3663–3674.
97. Lacroix, C.; Laconi, S.; Angius, F.; Coluccia, A.; Silvestri, R.; Pompei, R.; Neyts, J.; Leyssen, P. In Vitro Characterisation of a Pleconaril/Pirodavir-Like Compound With Potent Activity Against Rhinoviruses. *Virol. J.* **2015**, *12*, 106.
98. Egorova, A.; Kazakova, E.; Jahn, B.; Ekins, S.; Makarov, V.; Schmidtke, M. Novel pleconaril Derivatives: Influence of Substituents in the Isoxazole and Phenyl Rings on the Antiviral Activity Against Enteroviruses. *Eur. J. Med. Chem.* **2020**, *188*, 112007.
99. Wald, J.; Pasin, M.; Richter, M.; Walther, C.; Mathai, N.; Kirchmair, J.; Makarov, V. A.; Goessweiner-Mohr, N.; Marlovits, T. C.; Zanella, I.; Real-Hohn, A.; Verdaguer, N.; Blaas, D.; Schmidtke, M. Cryo-EM Structure of Pleconaril-Resistant Rhinovirus-B5 Complexed to the Antiviral OBR-5-340 Reveals Unexpected Binding Site. *Proc. Natl. Acad. Sci. U.S.A.* **2019**, *116*, 19109–19115.
100. Makarov, V. A.; Braun, H.; Richter, M.; Riabova, O. B.; Kirchmair, J.; Kazakova, E. S.; Seidel, N.; Wutzler, P.; Schmidtke, M. Pyrazolopyrimidines: Potent Inhibitors Targeting the Capsid of Rhino- and Enteroviruses. *ChemMedChem* **2015**, *10*, 1629–1634.
101. Mello, C.; Aguayo, E.; Rodriguez, M.; Lee, G.; Jordan, R.; Cihlar, T.; Birkus, G. Multiple Classes of Antiviral Agents Exhibit *In Vitro* Activity Against Human Rhinovirus Type C. *Antimicrob. Agents Chemother.* **2014**, *58*, 1546–1555.
102. Blomqvist, S.; Savolainen, C.; Råman, L.; Roivainen, M.; Hovi, T. Human Rhinovirus 87 and Enterovirus 68 Represent a Unique Serotype With Rhinovirus and Enterovirus Features. *J. Clin. Microbiol.* **2002**, *40*, 4218–4223.
103. Holm-Hansen, C. C.; Midgley, S. E.; Fischer, T. K. Global Emergence of Enterovirus D68: A Systematic Review. *Lancet Infect. Dis.* **2016**, *16*, E64–E75.

104. Hu, Y.; Musharrafieh, R.; Zheng, M.; Wang, J. Enterovirus D68 Antivirals: Past, Present, and Future. *ACS Infect. Dis.* **2020**, *6*, 1572–1586.
105. Bauer, L.; Manganaro, R.; Zonsics, B.; Hurdiss, D. L.; Zwaagstra, M.; Donselaar, T.; Welter, N. G. E.; van Kleef, R.; Lopez, M. L.; Bevilacqua, F.; Raman, T.; Ferla, S.; Bassetto, M.; Neyts, J.; Strating, J.; Westerink, R. H. S.; Brancale, A.; van Kuppeveld, F. J. M. Rational Design of Highly Potent Broad-Spectrum Enterovirus Inhibitors Targeting the Nonstructural Protein 2C. *PLoS Biol.* **2020**, *18*.
106. Musharrafieh, R.; Zhang, J.; Tuohy, P.; Kitamura, N.; Bellampalli, S. S.; Hu, Y.; Khanna, R.; Wang, J. Discovery of Quinoline Analogues as Potent Antivirals Against Enterovirus D68 (EV-D68). *J. Med. Chem.* **2019**, *62*, 4074–4090.
107. Hu, Y.; Kitamura, N.; Musharrafieh, R.; Wang, J. Discovery of Potent and Broad-Spectrum Pyrazolopyridine-Containing Antivirals Against Enteroviruses D68, A71, and Coxsackievirus B3 by Targeting the Viral 2C Protein. *J. Med. Chem.* **2021**, *64*, 8755–8774.
108. Jacob, S. T.; Crozier, I.; Fischer, W. A.; Hewlett, A.; Kraft, C. S.; Vega, M.-A. D. L.; Soka, M. J.; Wahl, V.; Griffiths, A.; Bollinger, L.; Kuhn, J. H. Ebola Virus Disease. *Nat. Rev. Dis. Primers* **2020**, *6*, 13.
109. Bray, M. Defense Against Filoviruses Used as Biological Weapons. *Antiviral Res.* **2003**, *57*, 53–60.
110. Satyanarayana, M. Vaccines Could Make Big Ebola Outbreaks a Thing of the Past. *C&EN Global Enterprise* **2021**, *99*, 16–19.
111. Bausch, D. G. The Need for a New Strategy for Ebola Vaccination. *Nat. Med.* **2021**, *27*, 580–581.
112. Satyanarayana, M. First Ebola Treatment Approved by FDA. *C&EN Global Enterprise* **2020**, *98*, 13.
113. https://www.fda.gov/drugs/drug-safety-and-availability/fda-approves-treatment-ebola-virus.
114. Hansen, F.; Feldmann, H.; Jarvis, M. A. Targeting Ebola Virus Replication Through Pharmaceutical Intervention. *Expert Opin. Investig. Drugs* **2021**, *30*, 201–226.
115. Hoenen, T.; Groseth, A.; Feldmann, H. Therapeutic Strategies to Target the Ebola Virus Life Cycle. *Nat. Rev. Microbiol.* **2019**, *17*, 593–606.
116. Edwards, M. R.; Basler, C. F. Current Status of Small Molecule Drug Development for Ebola Virus and Other Filoviruses. *Curr. Opin. Virol.* **2019**, *35*, 42–56.
117. Plewe, M. B.; Sokolova, N. V.; Gantla, V. R.; Brown, E. R.; Naik, S.; Fetsko, A.; Lorimer, D. D.; Dranow, D. M.; Smutney, H.; Bullen, J.; Sidhu, R.; Master, A.; Wang, J. R.; Kallel, E. A.; Zhang, L. H.; Kalveram, B.; Freiberg, A. N.; Henkel, G.; McCormack, K. Discovery of Adamantane Carboxamides as Ebola Virus Cell Entry and Glycoprotein Inhibitors. *ACS Med. Chem. Lett.* **2020**, *11*, 1160–1167.
118. Gaisina, I. N.; Peet, N. P.; Wong, L.; Schafer, A. M.; Cheng, H.; Anantpadma, M.; Davey, R. A.; Thatcher, G. R. J.; Rong, L. Discovery and Structural Optimization of 4-(Aminomethyl)benzamides as Potent Entry Inhibitors of Ebola and Marburg Virus Infections. *J. Med. Chem.* **2020**, *63*, 7211–7225.
119. Cote, M.; Misasi, J.; Ren, T.; Bruchez, A.; Lee, K.; Filone, C. M.; Hensley, L.; Li, Q.; Ory, D.; Chandran, K.; Cunningham, J. Small Molecule Inhibitors Reveal Niemann-Pick C1 is Essential for Ebola Virus Infection. *Nature* **2011**, *477*, 344–U122.
120. Liu, H.; Tian, Y.; Lee, K.; Krishnan, P.; Wang, M. K.; Whelan, S.; Mevers, E.; Soloveva, V.; Dedic, B.; Liu, X.; Cunningham, J. M. Identification of Potent Ebola Virus Entry Inhibitors With Suitable Properties for in Vivo Studies. *J. Med. Chem.* **2018**, *61*, 6293–6307.
121. van der Linden, W. A.; Schulze, C. J.; Herbert, A. S.; Krause, T. B.; Wirchnianski, A. A.; Dye, J. M.; Chandran, K.; Bogyo, M. Cysteine Cathepsin Inhibitors as Anti-Ebola Agents. *ACS Infect. Dis.* **2016**, *2*, 173–179.

122. Warren, T. K.; Jordan, R.; Lo, M. K.; Ray, A. S.; Mackman, R. L.; Soloveva, V.; Siegel, D.; Perron, M.; Bannister, R.; Hui, H. C.; Larson, N.; Strickley, R.; Wells, J.; Stuthman, K. S.; Van Tongeren, S. A.; Garza, N. L.; Donnelly, G.; Shurtleff, A. C.; Retterer, C. J.; Gharaibeh, D.; Zamani, R.; Kenny, T.; Eaton, B. P.; Grimes, E.; Welch, L. S.; Gomba, L.; Wilhelmsen, C. L.; Nichols, D. K.; Nuss, J. E.; Nagle, E. R.; Kugelman, J. R.; Palacios, G.; Doerffler, E.; Neville, S.; Carra, E.; Clarke, M. O.; Zhang, L.; Lew, W.; Ross, B.; Wang, Q.; Chun, K.; Wolfe, L.; Babusis, D.; Park, Y.; Stray, K. M.; Trancheva, I.; Feng, J. Y.; Barauskas, O.; Xu, Y.; Wong, P.; Braun, M. R.; Flint, M.; McMullan, L. K.; Chen, S.-S.; Fearns, R.; Swaminathan, S.; Mayers, D. L.; Spiropoulou, C. F.; Lee, W. A.; Nichol, S. T.; Cihlar, T.; Bavari, S. Therapeutic Efficacy of the Small Molecule GS-5734 Against Ebola Virus in Rhesus Monkeys. *Nature* **2016**, *531*, 381–385.
123. Siegel, D.; Hui, H. C.; Doerffler, E.; Clarke, M. O.; Chun, K.; Zhang, L.; Neville, S.; Carra, E.; Lew, W.; Ross, B.; Wang, Q.; Wolfe, L.; Jordan, R.; Soloveva, V.; Knox, J.; Perry, J.; Perron, M.; Stray, K. M.; Barauskas, O.; Feng, J. Y.; Xu, Y.; Lee, G.; Rheingold, A. L.; Ray, A. S.; Bannister, R.; Strickley, R.; Swaminathan, S.; Lee, W. A.; Bavari, S.; Cihlar, T.; Lo, M. K.; Warren, T. K.; Mackman, R. L. Discovery and Synthesis of a Phosphoramidate Prodrug of a Pyrrolo[2,1-f][triazin-4-amino] Adenine C-Nucleoside (GS-5734) for the Treatment of Ebola and Emerging Viruses. *J. Med. Chem.* **2017**, *60*, 1648–1661.
124. Mulangu, S.; Dodd, L. E.; Davey, R. T., Jr.; Tshiani Mbaya, O.; Proschan, M.; Mukadi, D.; Lusakibanza Manzo, M.; Nzolo, D.; Tshomba Oloma, A.; Ibanda, A.; Ali, R.; Coulibaly, S.; Levine, A. C.; Grais, R.; Diaz, J.; Lane, H. C.; Muyembe-Tamfum, J. J.; Sivahera, B.; Camara, M.; Kojan, R.; Walker, R.; Dighero-Kemp, B.; Cao, H.; Mukumbayi, P.; Mbala-Kingebeni, P.; Ahuka, S.; Albert, S.; Bonnett, T.; Crozier, I.; Duvenhage, M.; Proffitt, C.; Teitelbaum, M.; Moench, T.; Aboulhab, J.; Barrett, K.; Cahill, K.; Cone, K.; Eckes, R.; Hensley, L.; Herpin, B.; Higgs, E.; Ledgerwood, J.; Pierson, J.; Smolskis, M.; Sow, Y.; Tierney, J.; Sivapalasingam, S.; Holman, W.; Gettinger, N.; Vallée, D.; Nordwall, J. A Randomized, Controlled Trial of Ebola Virus Disease Therapeutics. *N. Engl. J. Med.* **2019**, *381*, 2293–2303.
125. Warren, T. K.; Wells, J.; Panchal, R. G.; Stuthman, K. S.; Garza, N. L.; Van Tongeren, S. A.; Dong, L.; Retterer, C. J.; Eaton, B. P.; Pegoraro, G.; Honnold, S.; Bantia, S.; Kotian, P.; Chen, X.; Taubenheim, B. R.; Welch, L. S.; Minning, D. M.; Babu, Y. S.; Sheridan, W. P.; Bavari, S. Protection Against Filovirus Diseases by a Novel Broad-Spectrum Nucleoside Analogue BCX4430. *Nature* **2014**, *508*, 402–405.
126. Taylor, R.; Kotian, P.; Warren, T.; Panchal, R.; Bavari, S.; Julander, J.; Dobo, S.; Rose, A.; El-Kattan, Y.; Taubenheim, B.; Babu, Y.; Sheridan, W. P. BCX4430—A Broad-Spectrum Antiviral Adenosine Nucleoside Analog Under Development for the Treatment of Ebola Virus Disease. *J. Infect. Public Health* **2016**, *9*, 220–226.
127. McMullan, L. K.; Flint, M.; Dyall, J.; Albariño, C.; Olinger, G. G.; Foster, S.; Sethna, P.; Hensley, L. E.; Nichol, S. T.; Lanier, E. R.; Spiropoulou, C. F. The Lipid Moiety of Brincidofovir is Required for In Vitro Antiviral Activity Against Ebola Virus. *Antiviral Res.* **2016**, *125*, 71–78.
128. https://www.fda.gov/drugs/drug-safety-and-availability/fda-approves-drug-treat-smallpox.
129. Dunning, J.; Kennedy, S. B.; Antierens, A.; Whitehead, J.; Ciglenecki, I.; Carson, G.; Kanapathipillai, R.; Castle, L.; Howell-Jones, R.; Pardinaz-Solis, R.; Grove, J.; Scott, J.; Lang, T.; Olliaro, P.; Horby, P. W.; Team, R.-B. T. Experimental Treatment of Ebola Virus Disease With Brincidofovir. *PLoS One* **2016**, *11*, e0162199.

130. Bray, M.; Driscoll, J.; Huggins, J. W. Treatment of lethal Ebola Virus Infection in Mice with a Single Dose of an S-Adenosyl-L-Homocysteine Hydrolase Inhibitor. *Antiviral Res.* **2000**, *45*, 135–147.
131. Huggins, J.; Zhang, Z.-X.; Bray, M. Antiviral Drug Therapy of Filovirus Infections: S-Adenosylhomocysteine Hydrolase Inhibitors Inhibit Ebola Virus In Vitro and in a Lethal Mouse Model. *J. Infect. Dis.* **1999**, *179*, S240–S247.
132. Easton, V.; McPhillie, M.; Garcia-Dorival, I.; Barr, J. N.; Edwards, T. A.; Foster, R.; Fishwick, C.; Harris, M. Identification of a Small Molecule Inhibitor of Ebola Virus Genome Replication and Transcription Using In Silico Screening. *Antiviral Res.* **2018**, *156*, 46–54.
133. https://www.who.int/news-room/fact-sheets/detail/hiv-aids.
134. Riddell, J.t.; Amico, K. R.; Mayer, K. H. HIV Preexposure Prophylaxis: A Review. *JAMA* **2018**, *319*, 1261–1268.
135. Mayer, K. H.; Molina, J. M.; Thompson, M. A.; Anderson, P. L.; Mounzer, K. C.; De Wet, J. J.; DeJesus, E.; Jessen, H.; Grant, R. M.; Ruane, P. J.; Wong, P.; Ebrahimi, R.; Zhong, L.; Mathias, A.; Callebaut, C.; Collins, S. E.; Das, M.; McCallister, S.; Brainard, D. M.; Brinson, C.; Clarke, A.; Coll, P.; Post, F. A.; Hare, C. B. Emtricitabine and Tenofovir Alafenamide vs Emtricitabine and Tenofovir Disoproxil Fumarate for HIV pre-exposure Prophylaxis (DISCOVER): Primary Results From a Randomised, Double-blind, Multicentre, Active-controlled, Phase 3, Non-Inferiority Trial. *Lancet* **2020**, *396*, 239–254.
136. Gibas, K. M.; van den Berg, P.; Powell, V. E.; Krakower, D. S. Drug Resistance During HIV Pre-Exposure Prophylaxis. *Drugs* **2019**, *79*, 609–619.
137. Straubinger, T.; Kay, K.; Bies, R. Modeling HIV Pre-Exposure Prophylaxis. *Front. Pharmacol.* **2020**, *10*.
138. https://www.fda.gov/news-events/press-announcements/fda-approves-first-extended-release-injectable-drug-regimen-adults-living-hiv.
139. Engelman, K. D.; Engelman, A. N. Long-Acting Cabotegravir for HIV/AIDS Prophylaxis. *Biochemistry* **2021**, *60*, 1731–1740.
140. Marzinke, M. A.; Grinsztejn, B.; Fogel, J. M.; Piwowar-Manning, E.; Li, M.; Weng, L.; McCauley, M.; Cummings, V.; Ahmed, S.; Haines, C. D.; Bushman, L. R.; Petropoulos, C.; Persaud, D.; Adeyeye, A.; Kofron, R.; Rinehart, A.; St Clair, M.; Rooney, J. F.; Pryluka, D.; Coelho, L.; Gaur, A.; Middelkoop, K.; Phanuphak, N.; Cohen, M. S.; Hendrix, C. W.; Anderson, P.; Hanscom, B.; Donnell, D.; Landovitz, R. J.; Eshleman, S. H. Characterization of HIV Infection in Cisgender Men and Transgender Women Who Have Sex With Men Receiving Injectable Cabotegravir for HIV Prevention: HPTN 083. *J. Infect. Dis.* **2021**.
141. https://www.who.int/news/item/09-11-2020-trial-results-reveal-that-long-acting-injectable-cabotegravir-as-prep-is-highly-effective-in-preventing-hiv-acquisition-in-women.
142. Maxmen, A. Achilles Heel Spotted for Promising HIV-Prevention Drug. *Nature* **2021**, *591*, 357–358.
143. Markowitz, M.; Grobler, J. A. Islatravir for the Treatment and Prevention of Infection With the Human Immunodeficiency Virus Type 1. *Curr. Opin. HIV AIDS* **2020**, *15*, 27–32.
144. Baeten, J. M.; Palanee-Phillips, T.; Mgodi, N. M.; Mayo, A. J.; Szydlo, D. W.; Ramjee, G.; Gati Mirembe, B.; Mhlanga, F.; Hunidzarira, P.; Mansoor, L. E.; Siva, S.; Govender, V.; Makanani, B.; Naidoo, L.; Singh, N.; Nair, G.; Chinula, L.; Parikh, U. M.; Mellors, J. W.; Balán, I. C.; Ngure, K.; van der Straten, A.; Scheckter, R.; Garcia, M.; Peda, M.; Patterson, K.; Livant, E.; Bunge, K.; Singh, D.; Jacobson, C.; Jiao, Y.; Hendrix, C. W.; Chirenje, Z. M.; Nakabiito, C.; Taha, T. E.; Jones, J.; Torjesen, K.; Nel, A.; Rosenberg, Z.; Soto-Torres, L. E.;

Hillier, S. L.; Brown, E. R. Safety, Uptake, and Use of a Dapivirine Vaginal Ring for HIV-1 Prevention in African Women (HOPE): An Open-Label, Extension Study. *Lancet HIV* **2021**, *8*, e87–e95.
145. Begley, R.; Lutz, J.; Rhee, M.; et al. In *GS-6207 Sustained Delivery Formulation Supports 6-Month Dosing Interval; Presented at: 23rd International AIDS Conference; July 6–10, 2020; Virtual*; 2020. https://www.natap.org/2020/IAC/IAC_51.htm.
146. Sun, L.; Peng, Y.; Yu, W.; Zhang, Y.; Liang, L.; Song, C.; Hou, J.; Qiao, Y.; Wang, Q.; Chen, J.; Wu, M.; Zhang, D.; Li, E.; Han, Z.; Zhao, Q.; Jin, X.; Zhang, B.; Huang, Z.; Chai, J.; Wang, J.-H.; Chang, J. Mechanistic Insight into Antiretroviral Potency of 2′-Deoxy-2′-β-fluoro-4′-azidocytidine (FNC) With a Long-Lasting Effect on HIV-1 Prevention. *J. Med. Chem.* **2020**, *63*, 8554–8566.
147. Pierson, T. C.; Diamond, M. S. The Continued Threat of Emerging Flaviviruses. *Nat. Microbiol.* **2020**, *5*, 796–812.
148. https://www.who.int/news-room/fact-sheets/detail/yellow-fever.
149. Monath, T. P.; Vasconcelos, P. F. C. Yellow fever. *J. Clin. Virol.* **2015**, *64*, 160–173.
150. Chen, L. H.; Wilson, M. E. Yellow Fever Control: Current Epidemiology and Vaccination Strategies. *Trop. Dis. Travel Med. Vac.* **2020**, *6*, 1.
151. Shearer, F. M.; Longbottom, J.; Browne, A. J.; Pigott, D. M.; Brady, O. J.; Kraemer, M. U. G.; Marinho, F.; Yactayo, S.; de Araújo, V. E. M.; da Nóbrega, A. A.; Fullman, N.; Ray, S. E.; Mosser, J. F.; Stanaway, J. D.; Lim, S. S.; Reiner, R. C., Jr.; Moyes, C. L.; Hay, S. I.; Golding, N. Existing and Potential Infection Risk Zones of Yellow Fever Worldwide: A Modelling Analysis. *Lancet Glob. Health* **2018**, *6*, e270–e278.
152. Gaythorpe, K. A.; Hamlet, A.; Jean, K.; Garkauskas Ramos, D.; Cibrelus, L.; Garske, T.; Ferguson, N. The Global Burden of Yellow Fever. *Elife* **2021**, *10*.
153. Paules, C. I.; Fauci, A. S. Yellow Fever—Once Again on the Radar Screen in the Americas. *N. Engl. J. Med.* **2017**, *376*, 1397–1399.
154. Lataillade, L. D. G. D.; Vazeille, M.; Obadia, T.; Madec, Y.; Mousson, L.; Kamgang, B.; Chen, C.-H.; Failloux, A.-B.; Yen, P.-S. Risk of Yellow Fever Virus Transmission in the Asia-Pacific Region. *Nat. Commun.* **2020**, *11*, 5801.
155. Iwamura, T.; Guzman-Holst, A.; Murray, K. A. Accelerating Invasion Potential of Disease Vector Aedes aegypti Under Climate Change. *Nat. Commun.* **2020**, *11*, 2130.
156. Khan, S. U.; Ogden, N. H.; Fazil, A. A.; Gachon, P. H.; Dueymes, G. U.; Greer, A. L.; Ng, V. Current and Projected Distributions of *Aedes aegypti* and *Ae. albopictus* in Canada and the U.S. *Environ. Health Perspect.* **2020**, *128*, 057007.
157. Lindsey, N. P.; Schroeder, B. A.; Miller, E. R.; Braun, M. M.; Hinckley, A. F.; Marano, N.; Slade, B. A.; Barnett, E. D.; Brunette, G. W.; Horan, K.; Staples, J. E.; Kozarsky, P. E.; Hayes, E. B. Adverse Event Reports Following Yellow Fever Vaccination. *Vaccine* **2008**, *26*, 6077–6082.
158. Zandi, K.; Amblard, F.; Amichai, S.; Bassit, L.; Tao, S. J.; Jiang, Y.; Zhou, L. H.; Russell, O. O.; Mengshetti, S.; Schinazi, R. F. Nucleoside Analogs With Antiviral Activity Against Yellow Fever Virus. *Antimicrob. Agents Chemother.* **2019**, *63*.
159. de Freitas, C. S.; Higa, L. M.; Sacramento, C. Q.; Ferreira, A. C.; Reis, P. A.; Delvecchio, R.; Monteiro, F. L.; Barbosa-Lima, G.; James Westgarth, H.; Vieira, Y. R.; Mattos, M.; Rocha, N.; Hoelz, L. V. B.; Leme, R. P. P.; Bastos, M. M.; Rodrigues, G. O. L.; Lopes, C. E. M.; Queiroz-Junior, C. M.; Lima, C. X.; Costa, V. V.; Teixeira, M. M.; Bozza, F. A.; Bozza, P. T.; Boechat, N.; Tanuri, A.; Souza, T. M. L. Yellow Fever Virus is Susceptible to Sofosbuvir Both In Vitro and In Vivo. *PLoS Negl. Trop. Dis.* **2019**, (13), e0007072.
160. Mendes, É. A.; Pilger, D. R. B. D.; Santos Nastri, A. C. D. S.; Malta, F. D. M.; Pascoalino, B. D. S.; Carneiro D'Albuquerque, L. A.; Balan, A.; Freitas, L. H.

G. D.; Durigon, E. L.; Carrilho, F. J.; Rebello Pinho, J. R. Sofosbuvir Inhibits Yellow Fever Virus In Vitro and in Patients With Acute Liver Failure. *Ann. Hepatol.* **2019**, *18*, 816–824.
161. Julander, J. G.; Bantia, S.; Taubenheim, B. R.; Minning, D. M.; Kotian, P.; Morrey, J. D.; Smee, D. F.; Sheridan, W. P.; Babu, Y. S. BCX4430, a Novel Nucleoside Analog, Effectively Treats Yellow Fever in a Hamster Model. *Antimicrob. Agents Chemother.* **2014**, *58*, 6607–6614.
162. Konkolova, E.; Dejmek, M.; Hřebabecký, H.; Šála, M.; Böserle, J.; Nencka, R.; Boura, E. Remdesivir Triphosphate can Efficiently Inhibit the RNA-Dependent RNA Polymerase From Various Flaviviruses. *Antiviral Res.* **2020**, *182*, 104899.
163. Fioravanti, R.; Desideri, N.; Carta, A.; Atzori, E. M.; Delogu, I.; Collu, G.; Loddo, R. Inhibitors of Yellow Fever Virus Replication Based on 1,3,5-Triphenyl-4,5-Dihydropyrazole Scaffold: Design, Synthesis and Antiviral Evaluation. *Eur. J. Med. Chem.* **2017**, *141*, 15–25.
164. Cannalire, R.; Tarantino, D.; Piorkowski, G.; Carletti, T.; Massari, S.; Felicetti, T.; Barreca, M. L.; Sabatini, S.; Tabarrini, O.; Marcello, A.; Milani, M.; Cecchetti, V.; Mastrangelo, E.; Manfroni, G.; Querat, G. Broad Spectrum Anti-flavivirus Pyridobenzothiazolones Leading to Less Infective Virions. *Antiviral Res.* **2019**, *167*, 6–12.
165. Gwon, Y.-D.; Strand, M.; Lindqvist, R.; Nilsson, E.; Saleeb, M.; Elofsson, M.; Överby, A. K.; Evander, M. Antiviral Activity of Benzavir-2 Against Emerging Flaviviruses. *Viruses* **2020**, *12*, 351.
166. https://www.who.int/news-room/fact-sheets/detail/poliomyelitis.
167. https://www.who.int/news-room/feature-stories/detail/two-out-of-three-wild-poliovirus-strains-eradicated.
168. Chumakov, K.; Ehrenfeld, E.; Agol, V. I.; Wimmer, E. Polio Eradication at the Crossroads. *Lancet Glob. Health* **2021**, *9*, e1172–e1175.
169. Kew, O.; Morris-Glasgow, V.; Landaverde, M.; Burns, C.; Shaw, J.; Garib, Z.a.; André, J.; Blackman, E.; Freeman, C. J.; Jorba, J.; Sutter, R.; Tambini, G.; Venczel, L.; Pedreira, C.; Laender, F.; Shimizu, H.; Yoneyama, T.; Miyamura, T.; van der Avoort, H.; Oberste, M. S.; Kilpatrick, D.; Cochi, S.; Pallansch, M.; de Quadros, C. Outbreak of Poliomyelitis in Hispaniola Associated with Circulating Type 1 Vaccine-Derived Poliovirus. *Science* **2002**, *296*, 356–359.
170. https://www.who.int/emergencies/disease-outbreak-news/item/circulating-vaccine-derived-poliovirus-type-2-global-update.
171. Macklin, G.; Diop, O. M.; Humayun, A.; Shahmahmoodi, S.; El-Sayed, Z. A.; Triki, H.; Rey, G.; Avagyan, T.; Grabovac, V.; Jorba, J.; Farag, N.; Mach, O. Update on Immunodeficiency-Associated Vaccine-Derived Polioviruses—Worldwide, July 2018-December 2019. *MMWR Morb. Mortal. Wkly Rep.* **2020**, *69*, 913–917.
172. Sutter, R. W.; Modlin, J. F.; Zaffran, M. Completing Polio Eradication: The Case for Antiviral Drugs. *J. Infect. Dis.* **2016**, *215*, 333–334.
173. Oberste, M. S.; Moore, D.; Anderson, B.; Pallansch, M. A.; Pevear, D. C.; Collett, M. S. In Vitro Antiviral Activity of V-073 Against Polioviruses. *Antimicrob. Agents Chemother.* **2009**, *53*, 4501–4503.
174. Collett, M. S.; Hincks, J. R.; Benschop, K.; Duizer, E.; van der Avoort, H.; Rhoden, E.; Liu, H.; Oberste, M. S.; McKinlay, M. A.; Hartford, M. Antiviral Activity of Pocapavir in a Randomized, Blinded, Placebo-Controlled Human Oral Poliovirus Vaccine Challenge Model. *J. Infect. Dis.* **2016**, *215*, 335–343.
175. Rhoden, E.; Liu, H. M.; Wang-Chern, S. W.; Oberste, M. S. Anti-Poliovirus Activity of Protease Inhibitor AG-7404, and Assessment of In Vitro Activity in Combination With Antiviral Capsid Inhibitor Compounds. *Antiviral Res.* **2013**, *98*, 186–191.

176. Xing, Y.; Zuo, J.; Krogstad, P.; Jung, M. E. Synthesis and Structure-Activity Relationship (SAR) Studies of Novel Pyrazolopyridine Derivatives as Inhibitors of Enterovirus Replication. *J. Med. Chem.* **2018**, *61*, 1688–1703.
177. Zhang, Z.; Yang, E.; Hu, C.; Cheng, H.; Chen, C. Y.; Huang, D.; Wang, R.; Zhao, Y.; Rong, L.; Vignuzzi, M.; Shen, H.; Shen, L.; Chen, Z. W. Cell-Based High-Throughput Screening Assay Identifies 2′,2′-Difluoro-2′-deoxycytidine Gemcitabine as a Potential Antipoliovirus Agent. *ACS Infect. Dis.* **2017**, *3*, 45–53.
178. Madia, V. N.; Messore, A.; Pescatori, L.; Saccoliti, F.; Tudino, V.; De Leo, A.; Scipione, L.; Fiore, L.; Rhoden, E.; Manetti, F.; Oberste, M. S.; Di Santo, R.; Costi, R. In Vitro Antiviral Activity of New Oxazoline Derivatives as Potent Poliovirus Inhibitors. *J. Med. Chem.* **2019**, *62*, 798–810.
179. Manns, M. P.; Buti, M.; Gane, E.; Pawlotsky, J.-M.; Razavi, H.; Terrault, N.; Younossi, Z. Hepatitis C Virus Infection. *Nat. Rev. Dis. Primers* **2017**, *3*, 17006.
180. Stanciu, C.; Muzica, C. M.; Girleanu, I.; Cojocariu, C.; Sfarti, C.; Singeap, A. M.; Huiban, L.; Chiriac, S.; Cuciureanu, T.; Trifan, A. An Update on Direct Antiviral Agents for the Treatment of Hepatitis C. *Expert Opin. Pharmacother.* **2021**, 1–13.
181. Smolders, E. J.; Jansen, A. M. E.; Ter Horst, P. G. J.; Rockstroh, J.; Back, D. J.; Burger, D. M. Viral Hepatitis C Therapy: Pharmacokinetic and Pharmacodynamic Considerations: A 2019 Update. *Clin. Pharmacokinet.* **2019**, *58*, 1237–1263.
182. Heo, Y. A.; Deeks, E. D. Sofosbuvir/Velpatasvir/Voxilaprevir: A Review in Chronic Hepatitis C. *Drugs* **2018**, *78*, 577–587.
183. Martinello, M.; Bhagani, S.; Gane, E.; Orkin, C.; Cooke, G.; Dore, G. J.; Petoumenos, K.; Applegate, T. L.; Tu, E.; Marks, P.; Pagani, N.; Grebely, J.; Nelson, M.; Matthews, G. V. Shortened Therapy of Eight Weeks With Paritaprevir/Ritonavir/Ombitasvir and Dasabuvir is Highly Effective in People With Recent HCV Genotype 1 Infection. *J. Viral Hepat.* **2018**, *25*, 1180–1188.
184. Do, A.; Reau, N. S. Chronic Viral Hepatitis: Current Management and Future Directions. *Hepatol. Commun.* **2020**, *4*, 329–341.
185. Brown, R. S., Jr.; Buti, M.; Rodrigues, L.; Chulanov, V.; Chuang, W. L.; Aguilar, H.; Horváth, G.; Zuckerman, E.; Carrion, B. R.; Rodriguez-Perez, F.; Urbánek, P.; Abergel, A.; Cohen, E.; Lovell, S. S.; Schnell, G.; Lin, C. W.; Zha, J.; Wang, S.; Trinh, R.; Mensa, F. J.; Burroughs, M.; Felizarta, F. Glecaprevir/Pibrentasvir for 8 Weeks in Treatment-Naïve Patients With Chronic HCV Genotypes 1-6 and Compensated Cirrhosis: The Expedition-8 Trial. *J. Hepatol.* **2020**, *72*, 441–449.
186. Martinello, M.; Gane, E.; Hellard, M.; Sasadeusz, J.; Shaw, D.; Petoumenos, K.; Applegate, T.; Grebely, J.; Maire, L.; Marks, P.; Dore, G. J.; Matthews, G. V. Sofosbuvir and Ribavirin for 6 Weeks is Not Effective Among People With Recent hepatitis C Virus Infection: The DARE-C II Study. *Hepatology* **2016**, *64*, 1911–1921.
187. Deterding, K.; Spinner, C. D.; Schott, E.; Welzel, T. M.; Gerken, G.; Klinker, H.; Spengler, U.; Wiegand, J.; Schulze Zur Wiesch, J.; Pathil, A.; Cornberg, M.; Umgelter, A.; Zöllner, C.; Zeuzem, S.; Papkalla, A.; Weber, K.; Hardtke, S.; von der Leyen, H.; Koch, A.; von Witzendorff, D.; Manns, M. P.; Wedemeyer, H. Ledipasvir Plus Sofosbuvir Fixed-Dose Combination for 6 Weeks in Patients With Acute Hepatitis C Virus Genotype 1 Monoinfection (HepNet Acute HCV IV): An Open-Label, Single-Arm, Phase 2 Study. *Lancet Infect. Dis.* **2017**, *17*, 215–222.
188. Rockstroh, J. K.; Bhagani, S.; Hyland, R. H.; Yun, C.; Dvory-Sobol, H.; Zheng, W.; Brainard, D. M.; Ingiliz, P.; Lutz, T.; Boesecke, C.; Nelson, M. Ledipasvir-Sofosbuvir for 6 Weeks to Treat Acute Hepatitis C Virus Genotype 1 or 4 Infection in Patients With HIV Coinfection: An Open-Label, Single-Arm Trial. *Lancet Gastroenterol. Hepatol.* **2017**, *2*, 347–353.

189. Casey, J. L.; Feld, J. J.; MacParland, S. A. Restoration of HCV-Specific Immune Responses With Antiviral Therapy: A Case for DAA Treatment in Acute HCV Infection. *Cell* **2019**, *8*.
190. Piecha, F.; Gänßler, J. M.; Ozga, A. K.; Wehmeyer, M. H.; Dietz, J.; Kluwe, J.; Laschtowitz, A.; von Felden, J.; Sterneck, M.; Jordan, S.; Pischke, S.; Lohse, A. W.; Schulze Zur Wiesch, J. Treatment and Re-treatment Results of HCV Patients in the DAA Era. *PLoS One* **2020**, *15*, e0232773.
191. Bourlière, M.; Gordon, S. C.; Flamm, S. L.; Cooper, C. L.; Ramji, A.; Tong, M.; Ravendhran, N.; Vierling, J. M.; Tran, T. T.; Pianko, S.; Bansal, M. B.; de Lédinghen, V.; Hyland, R. H.; Stamm, L. M.; Dvory-Sobol, H.; Svarovskaia, E.; Zhang, J.; Huang, K. C.; Subramanian, G. M.; Brainard, D. M.; McHutchison, J. G.; Verna, E. C.; Buggisch, P.; Landis, C. S.; Younes, Z. H.; Curry, M. P.; Strasser, S. I.; Schiff, E. R.; Reddy, K. R.; Manns, M. P.; Kowdley, K. V.; Zeuzem, S. Sofosbuvir, Velpatasvir, and Voxilaprevir for Previously Treated HCV Infection. *New Engl. J. Med.* **2017**, *376*, 2134–2146.
192. Sezaki, H.; Suzuki, F.; Hosaka, T.; Fujiyama, S.; Kawamura, Y.; Akuta, N.; Kobayashi, M.; Suzuki, Y.; Saitoh, S.; Arase, Y.; Ikeda, K.; Kobayashi, M.; Kumada, H. Initial- and Re-treatment Effectiveness of Glecaprevir and Pibrentasvir for Japanese Patients With Chronic Hepatitis C Virus-Genotype 1/2/3 Infections. *J. Gastroenterol.* **2019**, *54*, 916–927.
193. Meanwell, N. A.; Georg, G. I.; Wang, S. T. Nobel Prize in Physiology or Medicine. *J. Med. Chem.* **2020**, *2020* (63), 13197–13204.
194. Dore, G. J.; Bajis, S. Hepatitis C Virus Elimination: Laying the Foundation for Achieving 2030 Targets. *Nat. Rev. Gastroenterol. Hepatol.* **2021**, *18*, 91–92.
195. https://www.who.int/news-room/fact-sheets/detail/hepatitis-c.
196. Ramdas, V.; Talwar, R.; Banerjee, M.; Joshi, A. A.; Das, A. K.; Walke, D. S.; Borhade, P.; Dhayagude, U.; Loriya, R.; Gote, G.; Bommakanti, A.; Sivaram, A.; Agarwal, G.; Goswami, A.; Nigade, P.; Mehta, M.; Patil, V.; Modi, D.; Kumar, H.; Mallurwar, S.; Dash, A.; Modi, F.; Kuldharan, S.; Srivastava, P.; Singh, M.; Narasimham, L.; Gundu, J.; Sharma, S.; Kamboj, R. K.; Palle, V. P. Discovery and Characterization of Potent Pan-Genotypic HCV NS5A Inhibitors Containing Novel Tricyclic Central Core Leading to Clinical Candidate. *Journal of Medicinal Chemistry* **2019**, *62*, 10563–10582.
197. Liu, B.; Gai, K.; Qin, H.; Wang, J.; Liu, X.; Cao, Y.; Lu, Q.; Lu, D.; Chen, D.; Shen, H.; Song, W.; Mei, J.; Wang, X.; Xu, H.; Zhang, Y. Discovery of a Silicon-Containing Pan-Genotype Hepatitis C Virus NS5A Inhibitor. *J. Med. Chem.* **2020**, *63*, 5312–5323.
198. Kazmierski, W. M.; Baskaran, S.; Walker, J. T.; Miriyala, N.; Meesala, R.; Beesu, M.; Adjabeng, G.; Grimes, R. M.; Hamatake, R.; Leivers, M. R.; Crosby, R.; Xia, B.; Remlinger, K. GSK2818713, a Novel Biphenylene Scaffold-Based Hepatitis C NS5A Replication Complex Inhibitor with Broad Genotype Coverage. *J. Med. Chem.* **2020**, *63*, 4155–4170.
199. Randolph, J. T.; Voight, E. A.; Greszler, S. N.; Uno, B. E.; Newton, J. N.; Gleason, K. M.; Stolarik, D.; Van Handel, C.; Bow, D. A. J.; DeGoey, D. A. Prodrug Strategies to Improve the Solubility of the HCV NS5A Inhibitor Pibrentasvir (ABT-530). *J. Med. Chem.* **2020**, *63*, 11034–11044.
200. Sun, L. Q.; Mull, E.; D'Andrea, S.; Zheng, B.; Hiebert, S.; Gillis, E.; Bowsher, M.; Kandhasamy, S.; Baratam, V. R.; Puttaswamy, S.; Pulicharla, N.; Vishwakrishnan, S.; Reddy, S.; Trivedi, R.; Sinha, S.; Sivaprasad, S.; Rao, A.; Desai, S.; Ghosh, K.; Anumula, R.; Kumar, A.; Rajamani, R.; Wang, Y. K.; Fang, H.; Mathur, A.; Rampulla, R.; Zvyaga, T. A.; Mosure, K.; Jenkins, S.; Falk, P.; Tagore, D. M.; Chen, C.; Rendunchintala, K.; Loy, J.; Meanwell, N. A.;

McPhee, F.; Scola, P. M. Discovery of BMS-986144, a Third-Generation, Pan-Genotype NS3/4A Protease Inhibitor for the Treatment of Hepatitis C Virus Infection. *J. Med. Chem.* **2020**, *63*, 14740–14760.

201. Nageswara Rao, D.; Zephyr, J.; Henes, M.; Chan, E. T.; Matthew, A. N.; Hedger, A. K.; Conway, H. L.; Saeed, M.; Newton, A.; Petropoulos, C. J.; Huang, W.; Kurt Yilmaz, N.; Schiffer, C. A.; Ali, A. Discovery of Quinoxaline-Based P1–P3 Macrocyclic NS3/4A Protease Inhibitors With Potent Activity Against Drug-Resistant Hepatitis C Virus Variants. *J. Med. Chem.* **2021**, *64*, 11972–11989.
202. Wang, G.; Dyatkina, N.; Prhavc, M.; Williams, C.; Serebryany, V.; Hu, Y.; Huang, Y.; Wan, J.; Wu, X.; Deval, J.; Fung, A.; Jin, Z.; Tan, H.; Shaw, K.; Kang, H.; Zhang, Q.; Tam, Y.; Stoycheva, A.; Jekle, A.; Smith, D. B.; Beigelman, L. Synthesis and Anti-HCV Activities of 4′-Fluoro-2′-Substituted Uridine Triphosphates and Nucleotide Prodrugs: Discovery of 4′-Fluoro-2′-C-methyluridine 5′-Phosphoramidate Prodrug (AL-335) for the Treatment of Hepatitis C Infection. *J. Med. Chem.* **2019**, *62*, 4555–4570.
203. Mengshetti, S.; Zhou, L. H.; Sari, O.; De Schutter, C.; Zhang, H. W.; Cho, J. H.; Tao, S. J.; Bassit, L. C.; Verma, K.; Domaoal, R. A.; Ehteshami, M.; Jiang, Y.; Ovadia, R.; Kasthuri, M.; Russell, O. I.; McBrayer, T.; Whitaker, T.; Pattassery, J.; Pascual, M. L.; Uher, L.; Lin, B. Y.; Lee, S.; Amblard, F.; Coats, S. J.; Schinazi, R. F. Discovery of a Series of 2'-Alpha-Fluoro,2'-Beta-bromo-ribonucleosides and Their Phosphoramidate Prodrugs as Potent Pan-Genotypic Inhibitors of Hepatitis C Virus. *J. Med. Chem.* **2019**, *62*, 1859–1874.
204. Wang, G.; Dyatkina, N.; Prhavc, M.; Williams, C.; Serebryany, V.; Hu, Y.; Huang, Y.; Wu, X.; Chen, T.; Huang, W.; Rajwanshi, V. K.; Deval, J.; Fung, A.; Jin, Z.; Stoycheva, A.; Shaw, K.; Gupta, K.; Tam, Y.; Jekle, A.; Smith, D. B.; Beigelman, L. Synthesis and Anti-HCV Activity of Sugar-Modified Guanosine Analogues: Discovery of AL-611 as an HCV NS5B Polymerase Inhibitor for the Treatment of Chronic Hepatitis C. *J. Med. Chem.* **2020**, *63*, 10380–10395.
205. Good, S. S.; Moussa, A.; Zhou, X. J.; Pietropaolo, K.; Sommadossi, J. P. Preclinical Evaluation of AT-527, a Novel Guanosine Nucleotide Prodrug With Potent, Pan-Genotypic Activity Against Hepatitis C Virus. *PLoS One* **2020**, *15*, e0227104.
206. Chong, P. Y.; Shotwell, J. B.; Miller, J.; Price, D. J.; Maynard, A.; Voitenleitner, C.; Mathis, A.; Williams, S.; Pouliot, J. J.; Creech, K.; Wang, F.; Fang, J.; Zhang, H.; Tai, V. W. F.; Turner, E.; Kahler, K. M.; Crosby, R.; Peat, A. J. Design of N-Benzoxaborole Benzofuran GSK8175—Optimization of Human Pharmacokinetics Inspired by Metabolites of a Failed Clinical HCV Inhibitor. *J. Med. Chem.* **2019**, *62*, 3254–3267.
207. Jiang, X.; Tan, J.; Wang, Y.; Chen, J.; Li, J.; Jiang, Z.; Quan, Y.; Jin, J.; Li, Y.; Cen, S.; Li, Y.; Peng, Z.; Li, Z. 2-((4-Arylpiperazin-1-yl)methyl)benzonitrile Derivatives as Orally Available Inhibitors of Hepatitis C Virus with a Novel Mechanism of Action. *J. Med. Chem.* **2020**, *63*, 5972–5989.
208. Rolt, A.; Talley, D. C.; Park, S. B.; Hu, Z.; Dulcey, A.; Ma, C.; Irvin, P.; Leek, M.; Wang, A. Q.; Stachulski, A. V.; Xu, X.; Southall, N.; Ferrer, M.; Liang, T. J.; Marugan, J. J. Discovery and Optimization of a 4-Aminopiperidine Scaffold for Inhibition of Hepatitis C Virus Assembly. *J. Med. Chem.* **2021**, *64*, 9431–9443.
209. Yang, Y.; Cao, L.; Gao, H.; Wu, Y.; Wang, Y.; Fang, F.; Lan, T.; Lou, Z.; Rao, Y. Discovery, Optimization, and Target Identification of Novel Potent Broad-Spectrum Antiviral Inhibitors. *J. Med. Chem.* **2019**, *62*, 4056–4073.
210. Singh, R.; Singh, K. P.; Cherian, S.; Saminathan, M.; Kapoor, S.; Manjunatha Reddy, G. B.; Panda, S.; Dhama, K. Rabies—Epidemiology, Pathogenesis, Public Health Concerns and Advances in Diagnosis and Control: A Comprehensive Review. *Vet. Q.* **2017**, *37*, 212–251.

211. https://www.who.int/news-room/fact-sheets/detail/rabies.
212. Jackson, A. C. Diabolical Effects of Rabies Encephalitis. *J. Neurovirol.* **2016**, *22*, 8–13.
213. Kessels, J. A.; Recuenco, S.; Navarro-Vela, A. M.; Deray, R.; Vigilato, M.; Ertl, H.; Durrheim, D.; Rees, H.; Nel, L. H.; Abela-Ridder, B.; Briggs, D. Pre-Exposure Rabies Prophylaxis: A Systematic Review. *Bull. World Health Organ.* **2017**, *95*, 210–219c.
214. Kessels, J.; Tarantola, A.; Salahuddin, N.; Blumberg, L.; Knopf, L. Rabies Post-Exposure Prophylaxis: A Systematic Review on Abridged Vaccination Schedules and the Effect of Changing Administration Routes During a Single Course. *Vaccine* **2019**, *37*, A107–A117.
215. Smith, S. P.; Wu, G.; Fooks, A. R.; Ma, J.; Banyard, A. C. Trying to Treat the Untreatable: Experimental Approaches to Clear Rabies Virus Infection From the CNS. *J. Gen. Virol.* **2019**, *100*, 1171–1186.
216. Du Pont, V.; Plemper, R. K.; Schnell, M. J. Status of Antiviral Therapeutics Against Rabies Virus and Related Emerging Lyssaviruses. *Curr. Opin. Virol.* **2019**, *35*, 1–13.
217. Appolinario, C. M.; Jackson, A. C. Antiviral Therapy for Human Rabies. *Antivir. Ther.* **2015**, *20*, 1–10.
218. Anindita, P. D.; Sasaki, M.; Okada, K.; Ito, N.; Sugiyama, M.; Saito-Tarashima, N.; Minakawa, N.; Shuto, S.; Otsuguro, S.; Ichikawa, S.; Matsuda, A.; Maenaka, K.; Orba, Y.; Sawa, H. Ribavirin-Related Compounds Exert In Vitro Inhibitory Effects Toward Rabies Virus. *Antiviral Res.* **2018**, *154*, 1–9.
219. Yamada, K.; Noguchi, K.; Komeno, T.; Furuta, Y.; Nishizono, A. Efficacy of Favipiravir (T-705) in Rabies Postexposure Prophylaxis. *J. Infect. Dis.* **2016**, *213*, 1253–1261.
220. Banyard, A. C.; Mansfield, K. L.; Wu, G.; Selden, D.; Thorne, L.; Birch, C.; Koraka, P.; Osterhaus, A.; Fooks, A. R. Re-Evaluating the Effect of Favipiravir Treatment on Rabies Virus Infection. *Vaccine* **2019**, *37*, 4686–4693.
221. Yamada, K.; Noguchi, K.; Kimitsuki, K.; Kaimori, R.; Saito, N.; Komeno, T.; Nakajima, N.; Furuta, Y.; Nishizono, A. Reevaluation of the Efficacy of Favipiravir Against Rabies Virus Using In Vivo Imaging Analysis. *Antiviral Res.* **2019**, *172*, 104641.
222. Kali, S.; Jallet, C.; Azebi, S.; Cokelaer, T.; Da Fonseca, J. P.; Wu, Y.; Barbier, J.; Cintrat, J. C.; Gillet, D.; Tordo, N. Broad Spectrum Compounds Targeting Early Stages of Rabies Virus (RABV) Infection. *Antiviral Res.* **2021**, *188*, 105016.
223. Pont, V. D.; Wirblich, C.; Yoon, J.-J.; Cox, R. M.; Schnell, M. J.; Plemper, R. K.; Dutch, R. E. Identification and Characterization of a Small-Molecule Rabies Virus Entry Inhibitor. *J. Virol.* **2020**, *94*, e00320–e00321.
224. Tang, H.-B.; Lu, Z.-L.; Wei, X.-K.; Zhong, T.-Z.; Zhong, Y.-Z.; Ouyang, L.-X.; Luo, Y.; Xing, X.-W.; Liao, F.; Peng, K.-K.; Deng, C.-Q.; Minamoto, N.; Luo, T. R. Viperin Inhibits Rabies Virus Replication Via Reduced Cholesterol and Sphingomyelin and is Regulated Upstream by TLR4. *Sci. Rep.* **2016**, *6*, 30529.
225. Gizzi, A. S.; Grove, T. L.; Arnold, J. J.; Jose, J.; Jangra, R. K.; Garforth, S. J.; Du, Q.; Cahill, S. M.; Dulyaninova, N. G.; Love, J. D.; Chandran, K.; Bresnick, A. R.; Cameron, C. E.; Almo, S. C. A Naturally Occurring Antiviral Ribonucleotide Encoded by the Human Genome. *Nature* **2018**, *558*, 610–614.
226. Patel, A. M.; Marsh, E. N. G. The Antiviral Enzyme, Viperin, Activates Protein Ubiquitination by the E3 Ubiquitin Ligase, TRAF6. *J. Am. Chem. Soc.* **2021**, *143*, 4910–4914.
227. Durrheim, D. N. Measles Eradication-Retreating is Not an Option. *Lancet Infect. Dis.* **2020**, *20*, e138–e141.
228. Durrheim, D. N.; Andrus, J. K.; Tabassum, S.; Bashour, H.; Githanga, D.; Pfaff, G. A Dangerous Measles Future Looms Beyond the COVID-19 Pandemic. *Nat. Med.* **2021**, *27*, 360–361.

229. Rana, M. S.; Alam, M. M.; Ikram, A.; Salman, M.; Mere, M. O.; Usman, M.; Umair, M.; Zaidi, S. S. Z.; Arshad, Y. Emergence of Measles During the COVID-19 Pandemic Threatens Pakistan's Children and the Wider Region. *Nat. Med.* **2021**, *27*, 1127–1128.
230. Laventhal, N.; Tarini, B. A.; Lantos, J. Ethical Issues in Neonatal and Pediatric Clinical Trials. *Pediatr. Clin. North Am.* **2012**, *59*, 1205–1220.
231. Ferren, M.; Horvat, B.; Mathieu, C. Measles Encephalitis: Towards New Therapeutics. *Viruses* **2019**, *11*, 1017.
232. Plemper, R. K.; Doyle, J.; Sun, A.; Prussia, A.; Cheng, L.-T.; Rota, P. A.; Liotta, D. C.; Snyder, J. P.; Compans, R. W. Design of a Small-Molecule Entry Inhibitor With Activity Against Primary Measles Virus Strains. *Antimicrob. Agents Chemother.* **2005**, *49*, 3755–3761.
233. Sun, A.; Prussia, A.; Zhan, W.; Murray, E. E.; Doyle, J.; Cheng, L.-T.; Yoon, J.-J.; Radchenko, E. V.; Palyulin, V. A.; Compans, R. W.; Liotta, D. C.; Plemper, R. K.; Snyder, J. P. Nonpeptide Inhibitors of Measles Virus Entry. *J. Med. Chem.* **2006**, *49*, 5080–5092.
234. Lo, M. K.; Jordan, R.; Arvey, A.; Sudhamsu, J.; Shrivastava-Ranjan, P.; Hotard, A. L.; Flint, M.; McMullan, L. K.; Siegel, D.; Clarke, M. O.; Mackman, R. L.; Hui, H. C.; Perron, M.; Ray, A. S.; Cihlar, T.; Nichol, S. T.; Spiropoulou, C. F. GS-5734 and Its Parent Nucleoside Analog Inhibit Filo-, Pneumo-, and Paramyxoviruses. *Sci. Rep.* **2017**, 7, 43395.
235. Lo, M. K.; Jordan, P. C.; Stevens, S.; Tam, Y.; Deval, J.; Nichol, S. T.; Spiropoulou, C. F. Susceptibility of Paramyxoviruses and Filoviruses to Inhibition by 2'-Monofluoro- and 2'-Difluoro-4'-Azidocytidine Analogs. *Antiviral Res.* **2018**, *153*, 101–113.
236. Ndungu, J. M.; Krumm, S. A.; Yan, D.; Arrendale, R. F.; Reddy, G. P.; Evers, T.; Howard, R.; Natchus, M. G.; Saindane, M. T.; Liotta, D. C.; Plemper, R. K.; Snyder, J. P.; Sun, A. Non-nucleoside Inhibitors of the Measles Virus RNA-Dependent RNA Polymerase: Synthesis, Structure–Activity Relationships, and Pharmacokinetics. *J. Med. Chem.* **2012**, *55*, 4220–4230.
237. Krumm, S. A.; Yan, D.; Hovingh, E. S.; Evers, T. J.; Enkirch, T.; Reddy, G. P.; Sun, A.; Saindane, M. T.; Arrendale, R. F.; Painter, G.; Liotta, D. C.; Natchus, M. G.; von Messling, V.; Plemper, R. K. An Orally Available, Small-Molecule Polymerase inhibitor Shows Efficacy Against a Lethal Morbillivirus Infection in a Large Animal Model. *Sci. Transl. Med.* **2014**, *6*, 232ra252.
238. Wittwer, K.; Anderson, D. E.; Pfeffermann, K.; Cox, R. M.; Wolf, J. D.; Santibanez, S.; Mankertz, A.; Plesker, R.; Sticher, Z. M.; Kolkykhalov, A. A.; Natchus, M. G.; Pfaller, C. K.; Plemper, R. K.; von Messling, V. Small-Molecule Polymerase Inhibitor Protects Non-Human Primates From Measles and Reduces Shedding. *Nat. Commun.* **2021**, *12*, 5233.
239. Cox, R. M.; Sourimant, J.; Govindarajan, M.; Natchus, M. G.; Plemper, R. K. Therapeutic Targeting of Measles Virus Polymerase With ERDRP-0519 Suppresses All RNA Synthesis Activity. *PLoS Pathog.* **2021**, *17*.
240. Cox, R. M.; Sourimant, J.; Toots, M.; Yoon, J.-J.; Ikegame, S.; Govindarajan, M.; Watkinson, R. E.; Thibault, P.; Makhsous, N.; Lin, M. J.; Marengo, J. R.; Sticher, Z.; Kolykhalov, A. A.; Natchus, M. G.; Greninger, A. L.; Lee, B.; Plemper, R. K. Orally Efficacious Broad-Spectrum Allosteric Inhibitor of Paramyxovirus Polymerase. *Nat. Microbiol.* **2020**, *5*, 1232–1246.
241. Krumm, S. A.; Ndungu, J. M.; Yoon, J. J.; Dochow, M.; Sun, A.; Natchus, M.; Snyder, J. P.; Plemper, R. K. Potent Host-Directed Small-Molecule Inhibitors of Myxovirus RNA-Dependent RNA-Polymerases. *PLoS One* **2011**, *6*, e20069.
242. Grafen, A.; Schumacher, F.; Chithelen, J.; Kleuser, B.; Beyersdorf, N.; Schneider-Schaulies, J. Use of Acid Ceramidase and Sphingosine Kinase Inhibitors as Antiviral Compounds Against Measles Virus Infection of Lymphocytes In Vitro. *Front. Cell Dev. Biol.* **2019**, 7.

243. Gorbalenya, A. E.; Baker, S. C.; Baric, R. S.; de Groot, R. J.; Drosten, C.; Gulyaeva, A. A.; Haagmans, B. L.; Lauber, C.; Leontovich, A. M.; Neuman, B. W.; Penzar, D.; Perlman, S.; Poon, L. L. M.; Samborskiy, D. V.; Sidorov, I. A.; Sola, I.; Ziebuhr, J.; Coronaviridae Study Group of the International Committee on Taxonomy of. V. The Species Severe Acute Respiratory Syndrome-Related Coronavirus: Classifying 2019-nCoV and Naming it SARS-CoV-2. *Nat. Microbiol.* **2020**, *5*, 536–544.
244. Su, S.; Wong, G.; Shi, W.; Liu, J.; Lai, A. C. K.; Zhou, J.; Liu, W.; Bi, Y.; Gao, G. F. Epidemiology, Genetic Recombination, and Pathogenesis of Coronaviruses. *Trends Microbiol.* **2016**, *24*, 490–502.
245. Hu, B.; Guo, H.; Zhou, P.; Shi, Z.-L. Characteristics of SARS-CoV-2 and COVID-19. *Nat. Rev. Microbiol.* **2021**, *19*, 141–154.
246. Cui, J.; Li, F.; Shi, Z. L. Origin and Evolution of Pathogenic Coronaviruses. *Nat. Rev. Microbiol.* **2019**, *17*, 181–192.
247. https://www.who.int/emergencies/disease-outbreak-news/item/2021-DON333.
248. https://covid19.who.int/.
249. Gerges Harb, J.; Noureldine, H. A.; Chedid, G.; Eldine, M. N.; Abdallah, D. A.; Chedid, N. F.; Nour-Eldine, W. SARS, MERS and COVID-19: Clinical Manifestations and Organ-System Complications: A Mini Review. *Pathog. Dis.* **2020**, *78*.
250. Woo, P. C.; Lau, S. K.; Lam, C. S.; Lau, C. C.; Tsang, A. K.; Lau, J. H.; Bai, R.; Teng, J. L.; Tsang, C. C.; Wang, M.; Zheng, B. J.; Chan, K. H.; Yuen, K. Y. Discovery of Seven Novel Mammalian and Avian Coronaviruses in the Genus Deltacoronavirus Supports Bat Coronaviruses as the Gene Source of Alphacoronavirus and Betacoronavirus and Avian Coronaviruses as the Gene Source of Gammacoronavirus and Deltacoronavirus. *J. Virol.* **2012**, *86*, 3995–4008.
251. Boni, M. F.; Lemey, P.; Jiang, X.; Lam, T. T.-Y.; Perry, B. W.; Castoe, T. A.; Rambaut, A.; Robertson, D. L. Evolutionary Origins of the SARS-CoV-2 Sarbecovirus Lineage Responsible for the COVID-19 Pandemic. *Nat. Microbiol.* **2020**, *5*, 1408–1417.
252. Dolgin, E. The Race for Antiviral Drugs to Beat COVID—and the Next Pandemic. *Nature* **2021**, *592*, 340–343.
253. Nascimento Junior, J. A. C.; Santos, A. M.; Quintans-Júnior, L. J.; Walker, C. I. B.; Borges, L. P.; Serafini, M. R. SARS, MERS and SARS-CoV-2 (COVID-19) Treatment: A Patent Review. *Expert Opin. Ther. Pat.* **2020**, *30*, 567–579.
254. Artese, A.; Svicher, V.; Costa, G.; Salpini, R.; Di Maio, V. C.; Alkhatib, M.; Ambrosio, F. A.; Santoro, M. M.; Assaraf, Y. G.; Alcaro, S.; Ceccherini-Silberstein, F. Current Status of Antivirals and Druggable Targets of SARS CoV-2 and Other Human Pathogenic Coronaviruses. *Drug Resist. Updat.* **2020**, *53*, 100721.
255. Zhang, L.; Lin, D.; Sun, X.; Curth, U.; Drosten, C.; Sauerhering, L.; Becker, S.; Rox, K.; Hilgenfeld, R. Crystal Structure of SARS-CoV-2 Main Protease Provides a Basis for Design of Improved α-Ketoamide Inhibitors. *Science* **2020**, *368*, 409–412.
256. Zhang, L.; Lin, D.; Kusov, Y.; Nian, Y.; Ma, Q.; Wang, J.; von Brunn, A.; Leyssen, P.; Lanko, K.; Neyts, J.; de Wilde, A.; Snijder, E. J.; Liu, H.; Hilgenfeld, R. α-Ketoamides as Broad-Spectrum Inhibitors of Coronavirus and Enterovirus Replication: Structure-Based Design, Synthesis, and Activity Assessment. *J. Med. Chem.* **2020**, *63*, 4562–4578.
257. Hu, Y.; Ma, C.; Szeto, T.; Hurst, B.; Tarbet, B.; Wang, J. Boceprevir, Calpain Inhibitors II and XII, and GC-376 Have Broad-Spectrum Antiviral Activity Against Coronaviruses. *ACS Infect. Dis.* **2021**, 7, 586–597.

258. Dampalla, C. S.; Zheng, J.; Perera, K. D.; Wong, L.-Y. R.; Meyerholz, D. K.; Nguyen, H. N.; Kashipathy, M. M.; Battaile, K. P.; Lovell, S.; Kim, Y.; Perlman, S.; Groutas, W. C.; Chang, K.-O. Postinfection Treatment With a Protease Inhibitor Increases Survival of Mice With a Fatal SARS-CoV-2 Infection. *Proc. Natl. Acad. Sci. U. S. A.* **2021**, *118*.
259. Hoffman, R. L.; Kania, R. S.; Brothers, M. A.; Davies, J. F.; Ferre, R. A.; Gajiwala, K. S.; He, M.; Hogan, R. J.; Kozminski, K.; Li, L. Y.; Lockner, J. W.; Lou, J.; Marra, M. T.; Mitchell, L. J.; Murray, B. W.; Nieman, J. A.; Noell, S.; Planken, S. P.; Rowe, T.; Ryan, K.; Smith, G. J.; Solowiej, J. E.; Steppan, C. M.; Taggart, B. Discovery of Ketone-Based Covalent Inhibitors of Coronavirus 3CL Proteases for the Potential Therapeutic Treatment of COVID-19. *J. Med. Chem.* **2020**, *63*, 12725–12747.
260. Gao, Y.; Yan, L.; Huang, Y.; Liu, F.; Zhao, Y.; Cao, L.; Wang, T.; Sun, Q.; Ming, Z.; Zhang, L.; Ge, J.; Zheng, L.; Zhang, Y.; Wang, H.; Zhu, Y.; Zhu, C.; Hu, T.; Hua, T.; Zhang, B.; Yang, X.; Li, J.; Yang, H.; Liu, Z.; Xu, W.; Guddat, L. W.; Wang, Q.; Lou, Z.; Rao, Z. Structure of the RNA-Dependent RNA Polymerase From COVID-19 Virus. *Science* **2020**, *368*, 779–782.
261. Malin, J. J.; Suárez, I.; Priesner, V.; Fätkenheuer, G.; Rybniker, J. Remdesivir Against COVID-19 and Other Viral Diseases. *Clin. Microbiol. Rev.* **2020**, *34*.
262. Xia, S.; Liu, M.; Wang, C.; Xu, W.; Lan, Q.; Feng, S.; Qi, F.; Bao, L.; Du, L.; Liu, S.; Qin, C.; Sun, F.; Shi, Z.; Zhu, Y.; Jiang, S.; Lu, L. Inhibition of SARS-CoV-2 (Previously 2019-nCoV) Infection by a Highly Potent Pan-Coronavirus Fusion Inhibitor Targeting its Spike Protein That Harbors a High Capacity to Mediate Membrane Fusion. *Cell Res.* **2020**, *30*, 343–355.
263. Lee, J. Y.; Shin, Y. S.; Jeon, S.; Lee, S. I.; Noh, S.; Cho, J.-E.; Jang, M. S.; Kim, S.; Song, J. H.; Kim, H. R.; Park, C. M. Design, Synthesis and Biological Evaluation of 2-Aminoquinazolin-4(3H)-one Derivatives as Potential SARS-CoV-2 and MERS-CoV Treatments. *Bioorg. Med. Chem. Lett.* **2021**, *39*, 127885.
264. Xu, J.; Shi, P.-Y.; Li, H.; Zhou, J. Broad Spectrum Antiviral Agent Niclosamide and Its Therapeutic Potential. *ACS Infect. Dis.* **2020**, *6*, 909–915.
265. Pillaiyar, T.; Laufer, S. Kinases as Potential Therapeutic Targets for Anti-coronaviral Therapy. *J. Med. Chem.* **2021**.
266. Mellott, D. M.; Tseng, C.-T.; Drelich, A.; Fajtová, P.; Chenna, B. C.; Kostomiris, D. H.; Hsu, J.; Zhu, J.; Taylor, Z. W.; Kocurek, K. I.; Tat, V.; Katzfuss, A.; Li, L.; Giardini, M. A.; Skinner, D.; Hirata, K.; Yoon, M. C.; Beck, S.; Carlin, A. F.; Clark, A. E.; Beretta, L.; Maneval, D.; Hook, V.; Frueh, F.; Hurst, B. L.; Wang, H.; Raushel, F. M.; O'Donoghue, A. J.; de Siqueira-Neto, J. L.; Meek, T. D.; McKerrow, J. H. A Clinical-Stage Cysteine Protease Inhibitor Blocks SARS-CoV-2 Infection of Human and Monkey Cells. *ACS Chem. Biol.* **2021**, *16*, 642–650.
267. Firpo, M. R.; Mastrodomenico, V.; Hawkins, G. M.; Prot, M.; Levillayer, L.; Gallagher, T.; Simon-Loriere, E.; Mounce, B. C. Targeting Polyamines Inhibits Coronavirus Infection by Reducing Cellular Attachment and Entry. *ACS Infect. Dis.* **2021**, 7, 1423–1432.

CHAPTER THREE

Broad spectrum antiviral nucleosides—Our best hope for the future

Katherine L. Seley-Radtke*, Joy E. Thames, and Charles D. Waters III
Department of Chemistry & Biochemistry, University of Maryland, Baltimore County, Baltimore, MD, United States
*Corresponding author: e-mail address: kseley@umbc.edu

Contents

1. Introduction

Nucleosides have long served as the primary scaffold for treating many diseases, including those caused by cancers, parasites, bacteria, but most notably, against many different families of viruses.[1–4] Because the naturally occurring nucleosides are the building blocks found in our DNA and RNA, and also play critical roles in numerous biological processes, even small modifications to the nucleoside scaffold can have profound therapeutic effects. This observation provides impetus for designing modified nucleoside analogues. Modifications to their structure can be designed in and/or refined, based on the key interactions identified in the binding site of target enzymes. As the field has progressed and new information has become available about nucleoside structure, enzyme recognition, and biological activity, unique and more complex modifications have been pursued, including multiple modifications on the same scaffold.[2–4] As a result of these modifications,

Annual Reports in Medicinal Chemistry, Volume 57
ISSN 0065-7743
https://doi.org/10.1016/bs.armc.2021.09.001
Copyright © 2021 Elsevier Inc.
All rights reserved.

there are currently more than 30 nucleoside/tide analogues on the market approved for use in treating viruses, cancers, parasites, as well as bacterial and fungal infections, with many more in preclinical and clinical trials.[1–4]

In that regard, we now find ourselves in the middle of a global pandemic caused by the deadly coronavirus, SARS-CoV-2, the zoonotic virus responsible for COVID-19. Never has it been more urgent for researchers to revisit old nucleosides, as well as to explore new and more novel possibilities.[5–7] In addition, given the strong likelihood that this will not be the last pandemic we face, given the number of zoonotic and other diseases circulating globally, it is critical that we find new ways to target future outbreaks.[8] As a result, the current focus for many researchers has turned to the development of therapeutics that have the potential for serving as broad-spectrum inhibitors that can target numerous viruses, both within a particular family, as well as to span across multiple viral families so that we can build an arsenal in preparation for the next outbreak.[9–12] This is not as out of reach as it may seem. As detailed below, many nucleosides have been shown to inhibit multiple viruses due to the conserved nature of many viral enzyme binding sites. Thus, it is somewhat surprising that up until very recently, many researchers focused more on "one bug one drug", rather than trying to target multiple viruses given those similarities.[8] This approach is now changing due to the realization that we need to be proactive rather than reactive to combat emerging and reemerging infectious diseases.[8]

In general, most nucleoside drugs are direct acting antivirals that target the viral polymerases.[1–4,13] Since they resemble the naturally occurring nucleosides, they are taken up by cells, then sequentially phosphorylated by cellular and/or viral kinases to ultimately form the triphosphate, which is typically the active form of the drug.[1–3,13] The triphosphate is recognized by the polymerase and subsequently incorporated in the growing nucleic acid chain, but through one of several different possible mechanisms, the drug acts as a chain terminator, thereby stopping the virus from replicating.[1,13,14] This occurs typically because the nucleoside lacks the necessary 3'-hydroxyl group needed for attachment of the next incoming nucleotide (obligate chain terminator), or if it has a 3'-hydroxyl but also has other substituents that can cause steric issues, the chain either stops immediately (pseudo-obligate chain terminator) or is able to add some additional nucleotides, albeit at an increasingly slower rate until ultimately no more can be added (delayed chain terminator).[1,13,14] There are several other mechanisms that have seen some promise, and those will be discussed further on in this chapter.

2. Early antiviral nucleosides—Obligate chain terminators

The earliest antiviral nucleoside was idoxuridine (Fig. 1), which was approved by the FDA in 1962.[15] It was initially developed as an anticancer drug, but was later approved for use against the ocular form of herpes simplex virus (HSV), HSV keratitis.[15] It works by mimicking thymidine, and once incorporated into viral DNA, leads to inhibition of thymidylate phosphorylase and the polymerase.[15–17] It is not used much anymore, and has been replaced by another nucleoside analogue, trifluridine (Fig. 1).[18] Somewhat related in terms of structure, Brivudine (Fig. 1), another pyrimidine substitute analogue, is widely used in many European countries, but not until recently in the US, for treatment of varicella zoster virus (VZV), which causes herpes zoster, better known as shingles.[14,19,20] It has also been used to treat HSV-1, but not HSV-2.[19] Another older nucleoside that is now being investigated against SARS-CoV-2 is 6-azauridine (Fig. 1). This triazole analogue has shown broad-spectrum activity against a number of different coronaviruses, flaviviruses, including West Nile virus, thus it is not surprising it is now being looked at again.[21–23]

Probably the most well-known of the early FDA-approved antiviral nucleosides is Zidovudine (AZT) shown in Fig. 2, which was the first nucleoside to be widely used to treat HIV and AIDS.[2,16,24–26] Shortly thereafter, Zalcitibine (ddC), Didanosine (ddI) and stavudine (d4T) were developed (Fig. 2), however ddI and ddC were plagued with various toxicity issues and not used extensively after initial excitement.[26–28] All of these are considered obligate chain terminators and work by inhibiting the HIV reverse transcriptase (RT).[1,13]

Other early nucleoside drugs were the arabinose nucleosides vidarabine (Ara-A) and cytarabine (Ara-C) shown in Fig. 3, both of which are

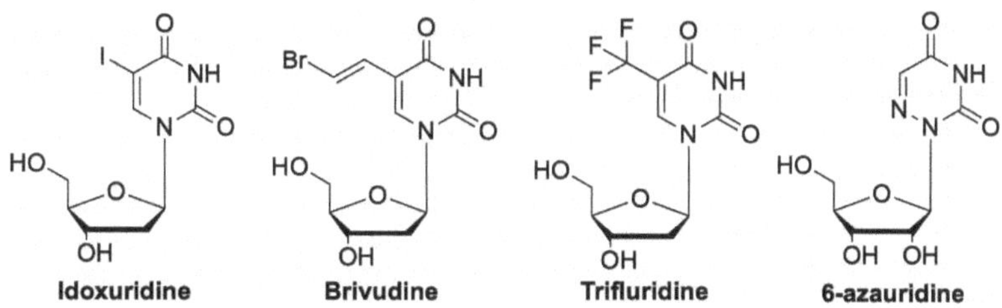

Fig. 1 Early modified nucleoside analogues idoxuridine, brivudine, trifluridine, and 6-azauridine.

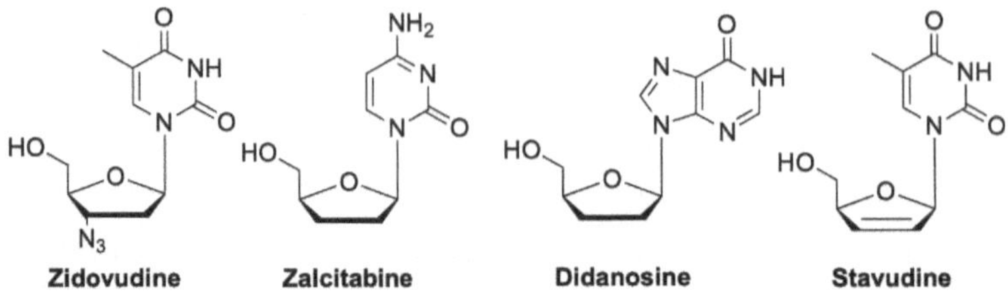

Fig. 2 Other early modified nucleoside analogues zidovudine, zalcitabine, didanosine, and stavudine.

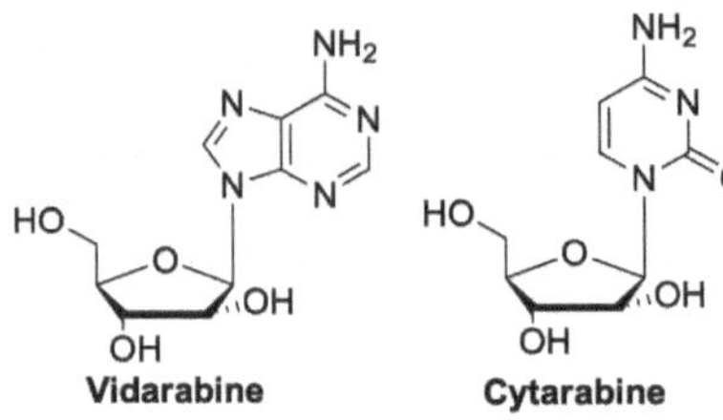

Fig. 3 Vidarabine and cytarabine, competitive inhibitors of DNA polymerases.

competitive inhibitors of DNA polymerases.[1,13] They are incorporated but then due to the steric configuration of the 2'-hydroxyl on the sugar moiety cause chain termination. Ara-A was used clinically for herpes, but both Ara-A and Ara-C were also shown to inhibit vaccinia (VV) and VZV viruses.[15,29] Ara-C went on to be used primarily as an anticancer drug, due to its toxicity.[17,30]

3. Acyclic obligate chain terminator nucleosides

As an extreme approach to obligate chain termination, Professor Antonín Holý developed the acyclic nucleosides and nucleoside phosphonates which lack not only the hydroxyls on the 2'- and 3'-carbons, but also the carbons themselves.[16,31,32] The acyclic nucleosides that are FDA-approved include acyclovir, penciclovir, and ganciclovir, while the phosphonates include adefovir, cidofovir, and tenofovir.[2,16,31] In addition, due to issues with poor oral bioavailability, various prodrugs have been subsequently approved including valacyclovir, valganciclovir, famciclovir, tenofovir disoproxil fumarate (TDF), adefovir dipivoxil, and most recently, tenofovir alafenamide (TAF).[1,2,16] Each will be discussed in more detail below but each work by the same mechanisms – competitive inhibition of the viral polymerases.[16]

Acyclovir and penciclovir—the carbocyclic version of acyclovir—and their respective prodrugs valacyclovir and famciclovir (Fig. 4), are used primarily to treat various forms of HSV, while ganciclovir and its prodrug valganciclovir are approved for use against human cytomegalovirus (HCMV).[2,16,33–38] Interestingly, HP-083, the fleximer version of acyclovir, and the related fleximer HP-100 (Fig. 5) as well as their prodrugs that were developed in our laboratories some years ago, are not active against HSV or HCMV, but rather are potent inhibitors of Ebola (EBOV), Marburg and Sudan viruses of the filovirus class, as well as SARS-CoV-1, MERS and human coronaviruses, in addition to a number of flaviviruses, including dengue (DENV), yellow fever (YFV) and tickborne encephalitis virus (TBEV).[39–41] Notably, acyclovir is not active against any of those at concentrations up to 1000 μM.[40] Not surprisingly, these compounds have attracted widespread attention for development as potential broad-spectrum inhibitors and are also currently being investigated for activity against SARS-CoV-2.

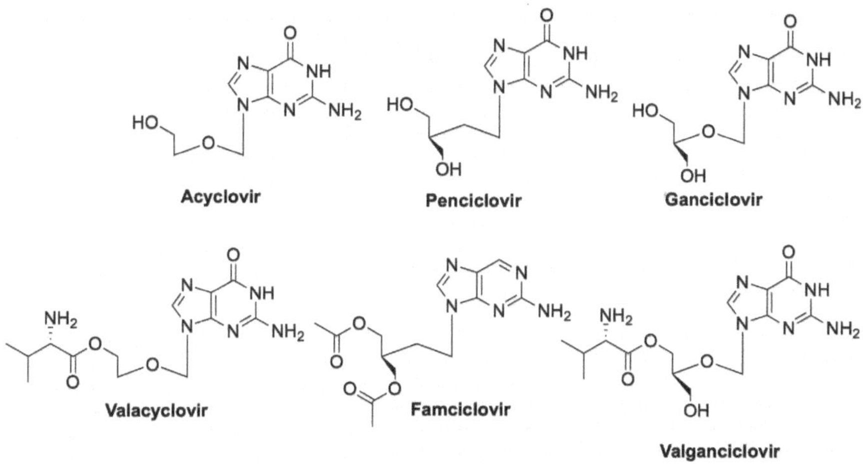

Fig. 4 Approved acyclic nucleosides acyclovir, penciclovir, ganciclovir and their prodrugs valacyclovir, famciclovir, valganciclovir.

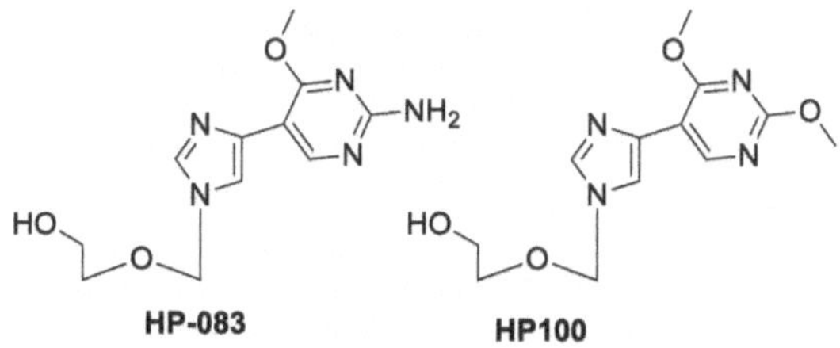

Fig. 5 Fleximer analogues of acyclovir, HP-083 and HP-100.

In terms of the acyclic phosphonates, cidofovir (Fig. 6) has the broadest spectrum of antiviral activity, with inhibition being noted against both DNA and RNA viruses and numerous different families of viruses.[12,32,42–47] Activity exhibited against DNA viruses by cidofovir includes HBV, papillomavirus, polyomavirus, adenovirus, numerous herpes viruses including HSV, HCMV, Epstein Barr virus (EBV) and VZV.[12,32,42–47] In addition, cidofovir has been explored extensively against various poxviruses, such as VV, molluscum contagiosum virus (MCV), variola virus, cowpox, monkeypox, and camelpox.[42,44,45,47] Related to this, brincidofovir (Fig. 6), the lipid prodrug form of cidofovir, was recently stockpiled by the U.S. government for potential outbreaks or terrorist attacks of smallpox.[47–50] As mentioned earlier, not just limited to DNA viruses, cidofovir also works against retroviruses such as HIV, and the corresponding feline and simian forms of HIV, FIV and SIV respectively.[12] As a result, cidofovir and brincidofovir should certainly be revisited for potential use as broad-spectrum antiviral agents.

While cidofovir certainly has the widest range of activities, one of the most important nucleosides for the antiviral field is tenofovir (Fig. 7).[1,2] Like cidofovir, tenofovir is an acyclic phosphonate and lacks the 3'-OH, thus works as an obligate chain terminator.[1,2] Tenofovir was initially

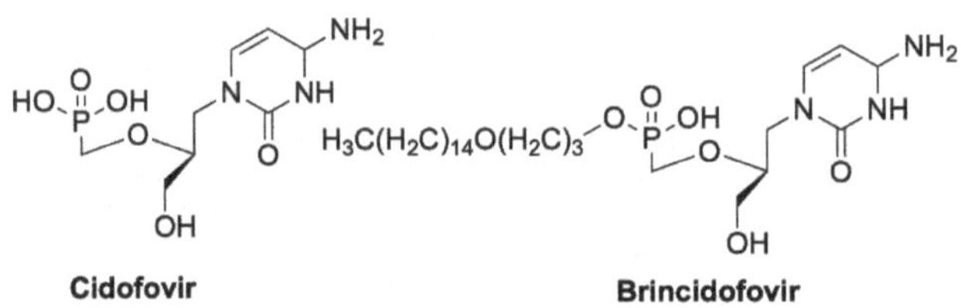

Fig. 6 Phosphonate nucleoside cidofovir and its lipid prodrug brincidofovir.

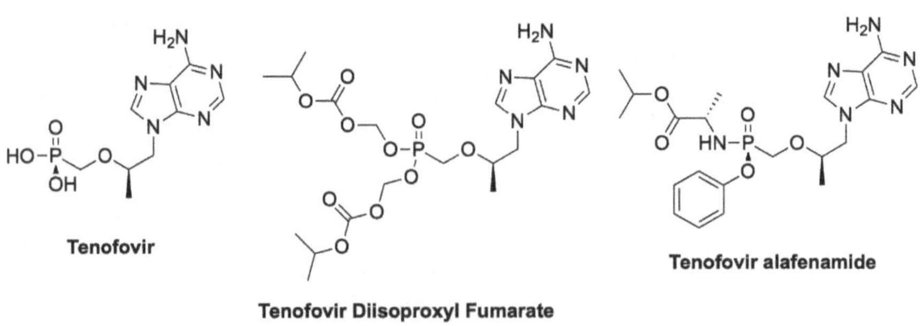

Fig. 7 Phosphonate nucleoside tenofovir and its two prodrug forms, tenofovir disoproxil fumarate and tenofovir alafenamide.

approved as a treatment for HIV, but was later also approved for treatment of HBV, and represents the nucleoside backbone in nearly every combination drug cocktail that has been developed in the past two decades.[27,51–53] Due to issues with bioavailability, tenofovir was developed as a prodrug, with the first form being tenofovir disoproxil fumarate (TDF, Fig. 7).[1,27,53] Recently however, a newer form was introduced, which possesses a McGuigan ProTide moiety, tenofovir alafenamide (TAF, Fig. 7).[1,2,7,54] This change in prodrug drastically lowered the required dose from 300 mgs/kg to 25 mgs//kg with far less side effects due to more rapid absorption into cells, which lessens the amount of phosphonate in the plasma.[55]

Perhaps it is important at this point, to deviate and explain a bit about the impact that the discovery and development of the McGuigan ProTides has had on medicinal chemistry, particularly on nucleoside drugs.[56–59] There are two inherent problems with nucleoside drugs—one, delivery across the cell membrane and two, phosphorylation once inside the cell.[56–59] In terms of delivery, nucleosides as a class are typically polar, so it leads to issues with poor permeability.[56–59] By masking the polar hydroxyl group at the 5'-position with a non-polar moiety, it allows the drug to cross more readily.[56–59] As mentioned previously, typically the active form of the nucleoside is the triphosphate, however you cannot directly administer the triphosphate due to the extreme polarity.[56–59] Moreover, the first phosphorylation step is typically rate limiting due to the acute selectivity of most kinases.[56,58,59] Once the first phosphorylation has occurred, the second and third tend to be more facile, although not always. As a solution to both problems, a number of years ago the late Professor Chris McGuigan and his team developed what they dubbed the McGuigan ProTide (Fig. 8).[56–59]

As shown in Fig. 8, there are 3 basic components to the ProTide—the ester, the aromatic moiety, and the amino acid. The ProTide renders the nucleoside less polar, thus it is able to cross the cell membrane.[56–59]

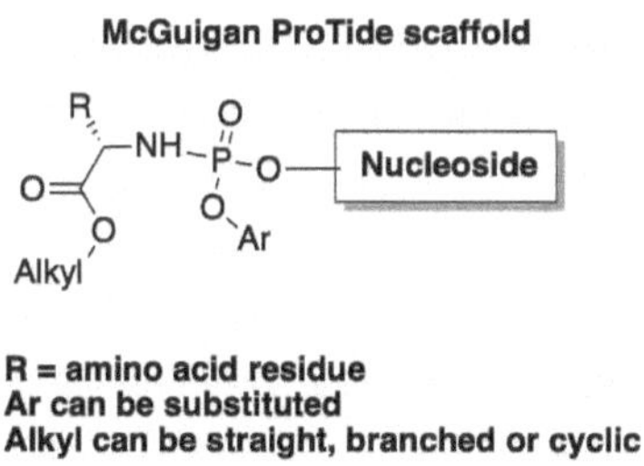

Fig. 8 General structure of the McGuigan ProTide.

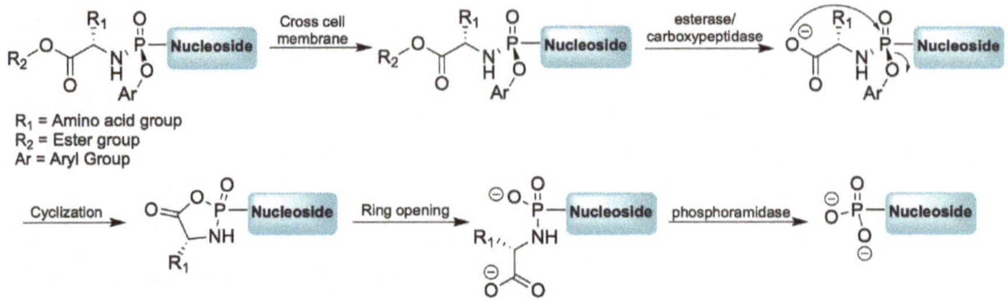

Fig. 9 Mechanism of McGuigan ProTide metabolism.

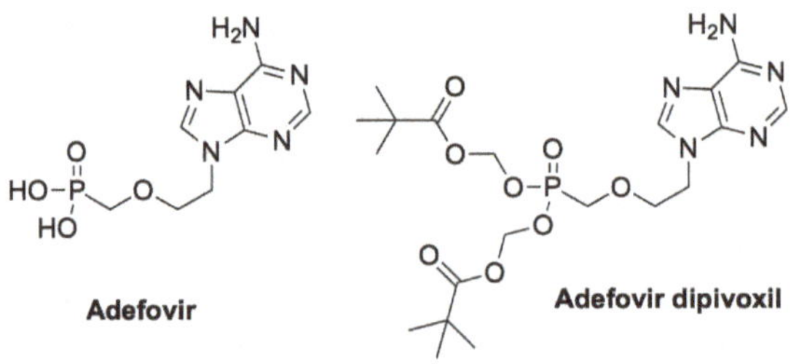

Fig. 10 Other approved phosphonate nucleosides, adefovir and its prodrug adefovir dipivoxil.

Once inside, each piece of the ProTide is sequentially removed by two cellular enzymes as well as spontaneous mechanistic steps (Fig. 9). This in turn, reveals the monophosphate form of the nucleoside, thus bypassing that first phosphorylation step and overcoming the rate limiting barrier to activation for the drug.[56–59] After extensive structure activity relationship (SAR) studies on the various components of this clever system, many nucleoside analogues are now being pursued and even marketed, for example, the aforementioned TAF as the McGuigan ProTide form.[55,56]

Prior to the development of TDF and TAF, the other FDA approved nucleoside phosphonate was adefovir and its prodrug adefovir dipivoxil (Fig. 10), which are used to treat chronic HBV.[60–62] Like tenofovir, adefovir works by competitive inhibition of the viral reverse transcriptase and subsequent chain termination.[1,16] In general, these inhibitors are known as nucleoside reverse transcriptase inhibitors or NRTIs.[1,16,63]

4. Other obligate chain terminator antivirals

Returning to antivirals possessing a cyclic sugar moiety, other important antiviral NRTIs of note that target HBV are entecavir, and three L-nucleoside analogues—telbivudine, lamivudine (3TC), and its closely

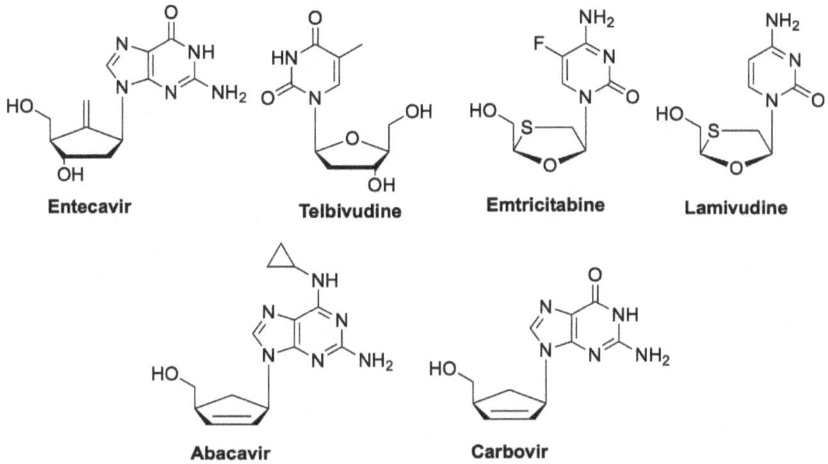

Fig. 11 Nucleoside reverse transcriptase inhibitors entecavir, telbuvidine, emtricitabine, and lamuvidine (L-nucleosides), as well as abacavir and carbovir.

related analogue emtricitabine (FTC) (Fig. 11).[1,2,43,64,65] Interestingly, entecavir introduced the idea of rigidifying the sugar of the nucleoside with the exocyclic double bond, something that still remains quite novel in terms of scaffold modifications.[2] Also of note, the discovery that L-nucleosides could be active was an important finding, as many believed that enzymes would not recognize the unnatural form of the drug.[2,62,66–73] Two other FDA-approved NRTIs, abacavir and carbovir (Fig. 11), are carbocyclic nucleosides used to treat HIV.[24,74–76] Notably, abacavir is one of the few drugs available for use in children with HIV.[76]

5. Antivirals that act by lethal mutagenesis

All of the nucleosides discussed to this point have been NRTIs or polymerase inhibitors that act by chain termination, however another mechanism of action has been explored that doesn't involve chain termination. Lethal mutagenesis is a type of antiviral therapy that essentially increases the viral mutation rate by incorporating nucleosides that act as mutagens, to the point of destroying the viability of the virus.[10,77]

One approved nucleoside that has been shown to work by this mechanism is ribavirin, although ribavirin has also been shown to work by other mechanisms and in general, its mechanism of action is not well understood.[6] Ribavirin (Fig. 12) was originally approved for the treatment of severe respiratory syncytial virus (RSV) infection, it has also been used for treatment of

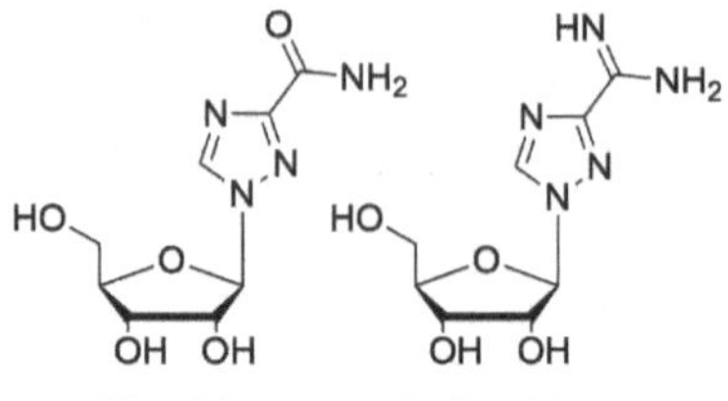

Fig. 12 Two nucleosides that work by lethal mutagenesis ribavirin and taribavirin.

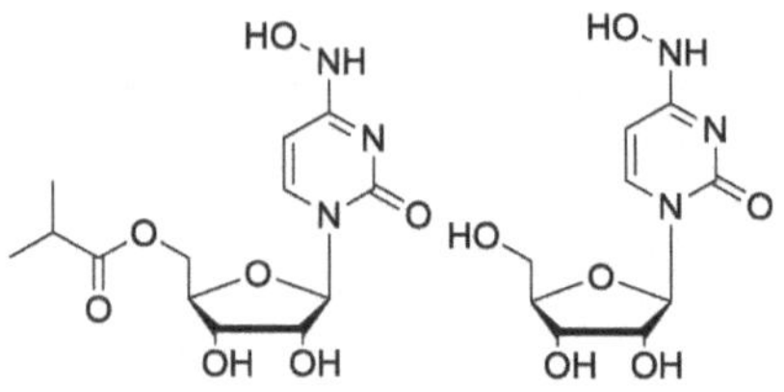

Fig. 13 NHC and its prodrug molnupiravir, another nucleoside that works by lethal mutagenesis.

Lassa fever virus infection, as well as influenza A and B.[78–80] Most recently it has been approved for use against HBV and HCV but has been tested against many other viruses including coronaviruses, however without much success in the latter case.[6,81,82] More recently a prodrug of ribavirin, taribavirin (also known as viramidine) has entered Phase III human trials (Fig. 12). Notably, taribavirin has been reported to target the liver better than ribavirin, which is important because as previously mentioned, both ribavirin and taribavirin have been shown to be highly effective against HCV and HBV.[83–86]

Another nucleoside analogue that also works by lethal mutagenesis that has recently gained significant attention against SARS-CoV-2 is molnupiravir (Fig. 13).[10] Molnupiravir is a prodrug of N-hydroxycytidine (NHC, Fig. 13) and is reported to increase mutations of G to A and C to U in replicating coronaviruses.[77] So far it is showing great promise and is currently in Phase II/III clinical trials. Interestingly, the U.S. Department of Health and Human Services has committed to purchasing $1.2B worth of molnupiravir should it ultimately be approved by the FDA.[87]

Similarly, the heterocyclic base favipiravir (Fig. 14) also works by lethal mutagenesis. Although favipiravir is administered as the base, it is subsequently converted to the nucleoside following ribosylation, and then undergoes normal phosphorylation by the various kinases, thus providing the triphosphate active form.[6,88,89]

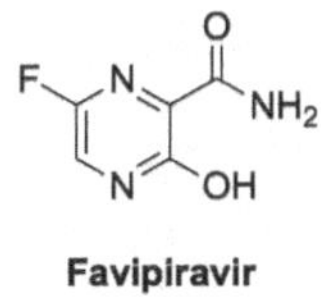

Fig. 14 Favipiravir.

6. Newer antiviral therapeutics—And new mechanisms of action

As more and more information has become known about emerging and reemerging viruses due to increased access to crystal structures and various enzymes involved in viral replication, multiple modifications to the nucleoside scaffold have become quite standard.[2,3] Examples of these complex nucleoside scaffolds include such drugs such as sofosbuvir and the closely related uprifosbuvir, remdesivir and the parent nucleoside GS-441524, EFdA (also known as Islatravir), NITD008, AT-527, and galidesivir, an immunocillin analogue.[1–3]

Sofosbuvir is a nucleoside analogue that features an alanine McGuigan ProTide moiety as well as a di-substitution at the C-2 position (Fig. 15).[90–92] Sofosbuvir is used to treat HCV, and what makes sofosbuvir stand out among nucleoside analogues is that it is not just a treatment, but rather, a cure.[91,92] For example, HIV patients must take their drug cocktails for their lifetime in order to maintain undetectable levels of viral load, but Sofosbuvir clears the virus completely in 8-12 weeks.[91] This was a remarkable breakthrough in antiviral therapy and has set a new standard for researchers to achieve that same effect in other viruses with their nucleoside analogues. Sofosbuvir, and the closely related uprifosbuvir (Fig. 15), are both delayed chain terminators.[1,93] Uprifosbuvir differs from Sofosbuvir at the C-2 position, featuring a chlorine atom instead of the fluorine, as well as

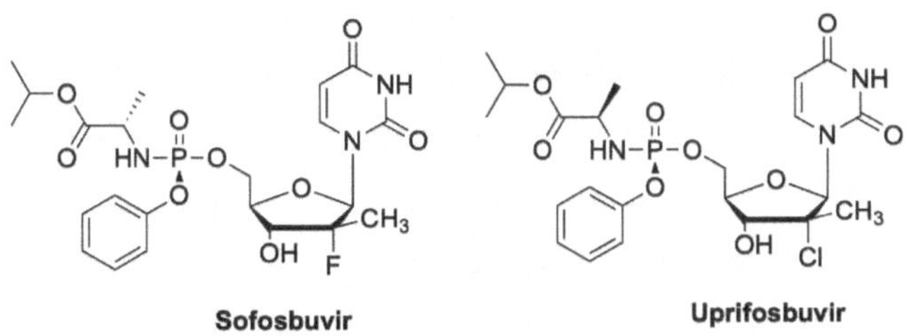

Fig. 15 Delayed chain terminators, sofosbuvir and uprifosbuvir.

the configuration of the methyl group on the ProTide moiety.[93,94] Both have shown broad-spectrum activity against a number of viruses including ZIKV, YFV and chikungunya viruses, as well as more recently, SARS-CoV-2. Uprifosbuvir is currently in Phase III clinical trials.[90,93–96]

Similarly, remdesivir (RDV, Fig. 16), currently the only drug approved for emergency use against SARS-CoV-2 is also a broad-spectrum inhibitor and also contains a McGuigan ProTide moiety like sofosbuvir, uprifosbuvir and TAF, however RDV's ester moiety differs than TAF and sofosbuvir's.[3,97] RDV was first under development for Ebola, including five compassionate cases, but was later pushed aside due to the development of two monoclonal antibodies that showed significant activity compared to RDV.[6,98,99] RDV is not ideal as it has to be administered in a hospital setting, which limits its widespread usage and effectiveness. RDV was later shown to have activity against SARS-1 and MERS, thus it wasn't surprising that it was immediately investigated for SARS-2.[6,98]

Interestingly, the parent nucleoside of remdesivir, GS-441524 (Fig. 16), has been used for some time for coronaviruses in cats.[98] Current efforts involve development a nasal spray as well as an oral form of RDV, both of which would have a major impact on making this drug more accessible globally.[6,98,100] Regardless, RDV has shown significant broad-spectrum activity against a number of other viruses, therefore certainly has promise for future use in fighting other emerging RNA viruses.[9] Remdesivir is also a delayed chain terminator that appears to involve incorporation, followed by pausing and then restarting.[6,9,98] More recently, RDV has been shown to also work by a second mechanism, as a translocation RT inhibitor, which is currently being studied more in depth.[101]

A new nucleoside analogue that has recently emerged on the scene due to its excellent activity against SARS-CoV-2 is AT-527 (Fig. 17).[102] AT-527 is a guanosine analogue being developed by Atea Pharmaceuticals and was recently licensed by Merck to take through clinical trials.[103] AT-527 has

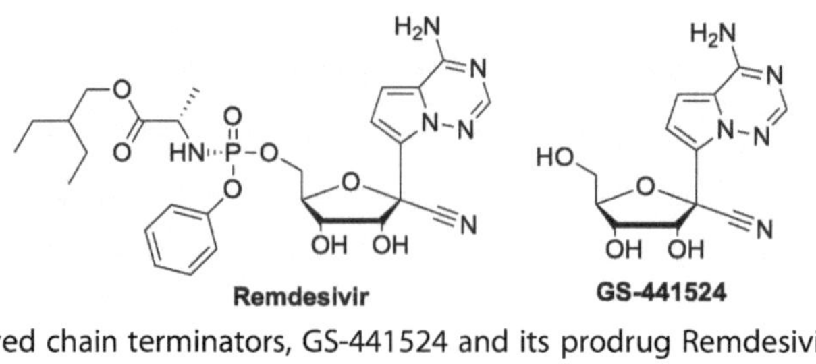

Fig. 16 Delayed chain terminators, GS-441524 and its prodrug Remdesivir.

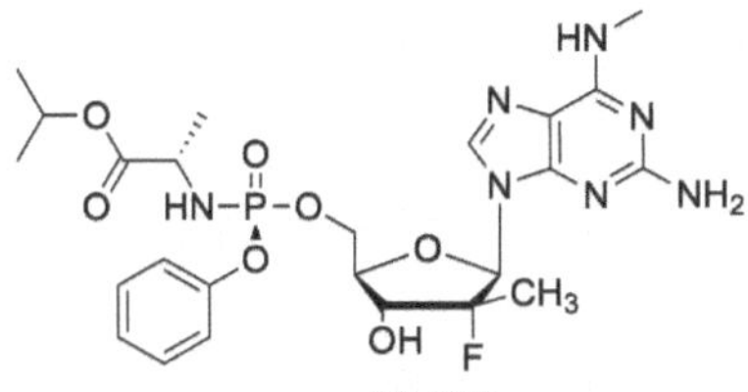

Fig. 17 Double prodrug AT-527.

the C-2' modification found in sofosbuvir as well as the same McGuigan alanine moiety, but is a double prodrug due to the N-methyl group on the heterocyclic base.[102,104,105]

AT-527 has shown significant promise and has numerous advantages over RDV, including being orally bioavailable and safe, thus representing a direct acting antiviral suitable for easy and early administration, as well as for pre- or post-exposure prophylaxis.[102–105] Additionally, AT-527 has exhibited no inhibition of human polymerases, no mitochondrial toxicity in vivo, no toxicity, cardiotoxicity, or neurological effects in rats or monkeys at doses up to 1000 mg/kg, a 10 fold safety profile over RDV for the 550 mg BID dose in humans, a 7 fold higher concentrations in human nasal and lung cells at 72 h, and a much longer half-life.[102–105]

Like other nucleoside analogues, AT-527 contains a McGuigan ProTide moiety, which delivers the monophosphate.[102,104,105] In contrast to other ProTides such as RDV, sofosbuvir and tenofovir, AT-527 exhibits much greater concentrations of the triphosphate, the active form of the drug, in the lungs.[102,105] Most notably however, AT-527 inhibits SARS-CoV-2 in a very unique manner—one molecule of AT-527 is incorporated into the growing strand, while another gets stuck in the active site, both of which lead to chain termination, while a third molecule inhibits the Nidovirus RdRp-Associated Nucleotidyltransferase (NiRAN) binding site, which is adjacent to the RdRp binding site.[102] This unique dual mechanism of action is highly synergistic and is likely what is responsible for its superior potency.[102]

Another nucleoside that is not as well-known as sofosbuvir or RDV, is Islatravir or EFdA (Fig. 18). EFdA is an adenosine analogue that has also been shown to work by multiple mechanisms.[106,107] EFdA has shown remarkable activity against HIV, including resistant strains.[106–108] What is even more remarkable, is that unlike other nucleoside analogues that typically work by a single mechanism, EFdA has been shown to work by a number of mechanisms. Similar to RDV, EFdA can function as a delayed chain

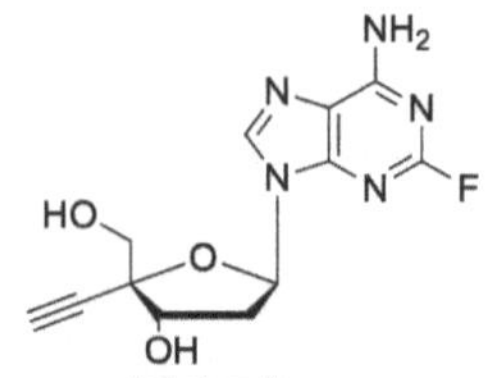

Fig. 18 Adenosine analogue islatravir.

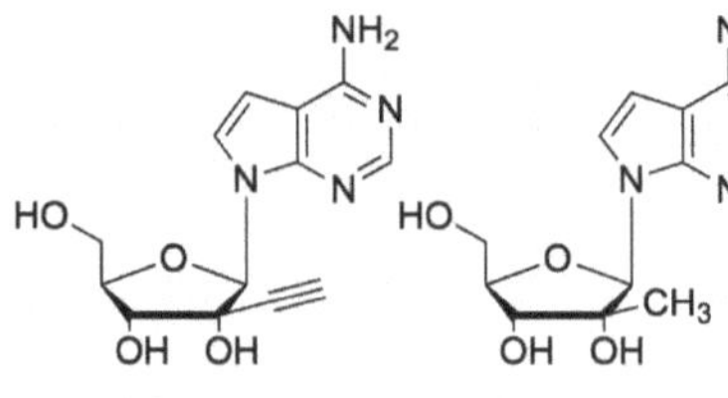

Fig. 19 Modified C2′ nucleosides NITD008 and MK-608.

terminator due to the presence of the 4'-triple bond, which causes steric issues.[106,107] Also like RDV, EFdA can act as a translocation RT inhibitor that essentially slows DNA synthesis, thereby acting as a de facto immediate, or pseudo obligate chain terminator.[106,107] Finally, EFdA-MP can also be misincorporated by RT, leading to mismatched primers that are extremely hard to extend, and are protected from excision by proofreading enzymes.[106,107] This multifaceted mechanism of action is likely to be important for combating the development of viral resistance, something that has plagued medicinal chemists for decades.[106,107]

Similar to EFdA, another nucleoside analogue that features a triple bond, albeit at the C2′-position, that initially showed promise is NITD008 (Fig. 19).[109–112] NITD008 and the closely related MK-608 (Fig. 19), have both exhibit broad-spectrum inhibition against a number of viruses.[109–115] NITD008 has shown potent activity against flaviviruses such as HCV, DENV, TBEV, ZIKV, West Nile virus, and YFV, as well as more exotic viruses such as Powassan virus, Kyasanur forest disease, which is a tickborne virus, and Omsk hemorrhagic fever virus.[110–115] In contrast, MK-608 appears to have only shown activity against DENV, TBEV, ZIKV and poliovirus. Both are delayed chain terminators, both are 7-deaza analogues and both are di-substituted at the C-2'-position, however NITD008 possesses the aforementioned triple bond in the "up" configuration, while MK-608 has a methyl group.[3] Unfortunately neither have been pursued

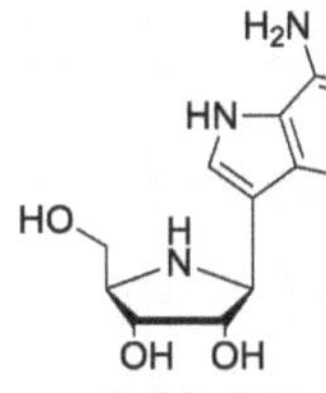

Fig. 20 Galidesivir, one of the immucillin nucleosides.

extensively after NITD008 proved too toxic in preclinical trials and MK-608 failed in early human clinical trials.[116,117]

Finally, this brings us to galidesivir, also known as Immucillin A or BX-4430. Galidesivir is interesting structurally due to the presence of a nitrogen in the sugar ring, rather than the typical furanose oxygen found in most nucleoside analogues, or the methylene group found in carbocyclic nucleosides.[2] This class of nucleosides are known as immucillins and they have shown potent activity against a number of viruses. Galidesivir is an adenosine analogue that has been investigated for use against numerous viruses from various different families.[118–122] It has shown activity against filoviruses including Ebola and Marburg, against flaviviruses such as YFV and ZIKV, as well as broad spectrum activity against various negative- and positive-sense RNA viruses, including coronaviruses and arenaviruses.[11] Because of this broad-spectrum activity, it is now being investigated against SARS-CoV-2, and like many of the nucleosides revisited herein, is an excellent example of how nucleosides can be effective against many viruses[123] (Fig. 20).

In summary, while this review is by no means comprehensive, rest assured, there are many other nucleoside analogues out there that are being investigated not only as treatments against one virus, but now against multiple viruses. The stark realization brought about by the SARS-CoV-2 pandemic and the over 4.55 M deaths as of September 2021 shows us just how unprepared we were to fight emerging viruses.[124] It has never been clear that we must be able to develop and stockpile small molecule broad-spectrum inhibitors to fight emerging and reemerging infectious diseases, particularly zoonotic ones given the vast number of viruses circulating in animals. In that regard, nucleoside inhibitors will indoubtedly remain our best hope just as they have for decades.

Acknowledgments

The authors thank Ms. Christianna Kutz for her excellent editing of the manuscript.

References

1. De Clercq, E.; Li, G. Approved Antiviral Drugs Over the Past 50 Years. *Clin. Microbiol. Rev.* **2016**, *29*, 695–747.
2. Seley-Radtke, K. L.; Yates, M. K. The Evolution of Nucleoside Analogue Antivirals: A Review for Chemists and Non-Chemists. Part 1: Early Structural Modifications to the Nucleoside Scaffold. *Antivir. Res.* **2018**, *154*, 66–86.
3. Yates, M. K.; Seley-Radtke, K. L. The Evolution of Antiviral Nucleoside Analogues: A Review for Chemists and Non-chemists. Part II: Complex Modifications to the Nucleoside Scaffold. *Antivir. Res.* **2019**, *162*, 5–21.
4. Seley-Radtke, K. Flexibility-Not Just for Yoga Anymore! *Antivir. Chem. Chemother.* **2018**, *26*. 2040206618756788.
5. Park, S.-J.; Yu, K.-M.; Kim, Y.-I.; Kim, S.-M.; Kim, E.-H.; Kim, S.-G.; Kim, E. J.; Casel, M. A. B.; Rollon, R.; Jang, S.-G.; Lee, M.-H.; Chang, J.-H.; Song, M.-S.; Jeong, H. W.; Choi, Y.; Chen, W.; Shin, W.-J.; Jung, J. U.; Choi, Y. K. Antiviral Efficacies of FDA-Approved Drugs Against SARS-CoV-2 Infection in Ferrets. *mBio* **2020**, *11*, e01114-20.
6. Simonis, A.; Theobald, S. J.; Fätkenheuer, G.; Rybniker, J.; Malin, J. J. A Comparative Analysis of Remdesivir and Other Repurposed Antivirals Against SARS-CoV-2. *EMBO Mol. Med.* **2021**, *13*, e13105.
7. Zanella, I.; Zizioli, D.; Castelli, F.; Quiros-Roldan, E. Tenofovir, Another Inexpensive, Well-Known and Widely Available Old Drug Repurposed for SARS-COV-2 Infection. *Pharmaceuticals* **2021**, *14*, 454.
8. Dolgin, E. The Race for Antiviral Drugs to Beat COVID—and the Next Pandemic. *Nature* **2021**, *592*, 340–343.
9. Sheahan, T. P.; Sims, A. C.; Graham, R. L.; Menachery, V. D.; Gralinski, L. E.; Case, J. B.; Leist, S. R.; Pyrc, K.; Feng, J. Y.; Trantcheva, I.; Bannister, R.; Park, Y.; Babusis, D.; Clarke, M. O.; Mackman, R. L.; Spahn, J. E.; Palmiotti, C. A.; Siegel, D.; Ray, A. S.; Cihlar, T.; Jordan, R.; Denison, M. R.; Baric, R. S. Broad-Spectrum Antiviral GS-5734 Inhibits Both Epidemic and Zoonotic Coronaviruses. *Sci. Transl. Med.* **2017**, *9*.
10. Sheahan, T. P.; Sims, A. C.; Zhou, S.; Graham, R. L.; Pruijssers, A. J.; Agostini, M. L.; Leist, S. R.; Schäfer, A.; Dinnon, K. H.; Stevens, L. J.; Chappell, J. D.; Lu, X.; Hughes, T. M.; George, A. S.; Hill, C. S.; Montgomery, S. A.; Brown, A. J.; Bluemling, G. R.; Natchus, M. G.; Saindane, M.; Kolykhalov, A. A.; Painter, G.; Harcourt, J.; Tamin, A.; Thornburg, N. J.; Swanstrom, R.; Denison, M. R.; Baric, R. S. An Orally Bioavailable Broad-Spectrum Antiviral Inhibits SARS-CoV-2 in Human Airway Epithelial Cell Cultures and Multiple Coronaviruses in Mice. *Sci. Transl. Med.* **2020**, *12*, eabb5883.
11. Taylor, R.; Kotian, P.; Warren, T.; Panchal, R.; Bavari, S.; Julander, J.; Dobo, S.; Rose, A.; El-Kattan, Y.; Taubenheim, B.; Babu, Y.; Sheridan, W. P. BCX4430—A Broad-Spectrum Antiviral Adenosine Nucleoside Analog Under Development for the Treatment of Ebola Virus Disease. *J. Infect. Publ. Health* **2016**, *9*, 220–226.
12. Painter, W.; Robertson, A.; Trost, L. C.; Godkin, S.; Lampert, B.; Painter, G. First Pharmacokinetic and Safety Study in Humans of the Novel Lipid Antiviral Conjugate CMX001, a Broad-Spectrum Oral Drug Active against Double-Stranded DNA Viruses. *Antimicrob. Agents Chemother.* **2012**, *56*, 2726–2734.
13. De Clercq, E.; Neyts, J. Antiviral Agents Acting as DNA or RNA Chain Terminators. *Handb. Exp. Pharmacol.* **2009**, 53–84.
14. Deval, J. Antimicrobial Strategies: Inhibition of Viral Polymerases by 3'-Hydroxyl Nucleosides. *Drugs* **2009**, *69*, 151–166.
15. Wilhelmus, K. R. Antiviral Treatment and Other Therapeutic Interventions for Herpes Simplex Virus Epithelial Keratitis. *Cochrane Database Systemat. Rev.* **2015**.

16. De Clercq, E.; Field, H. J. Antiviral Prodrugs—the Development of Successful Prodrug Strategies for Antiviral Chemotherapy. *Brit. J. Pharmacol.* **2006**, *147*, 1–11.
17. Lauter, C. B.; Bailey, E. J.; Lerner, A. M. Assessment of Cytosine Arabinoside as an Antiviral Agent in Humans. *Antimicrob. Agents Chemother.* **1974**, *6*, 598–602.
18. O'Brien, W. J.; Taylor, J. L. Therapeutic Response of Herpes Simplex Virus-Induced Corneal Edema to Trifluridine in Combination With Immunosuppressive Agents. *Invest. Ophthalmol. Vis. Sci.* **1991**, *32*, 2455–2461.
19. De Clercq, E. Discovery and Development of BVDU (Brivudin) as a Therapeutic for the Treatment of Herpes Zoster. *Biochem. Pharmacol.* **2004**, *68*, 2301–2315.
20. Desgranges, C.; Razaka, G.; Rabaud, M.; Bricaud, H.; Balzarini, J.; de Clercq, E. Phosphorolysis of (E)-5-(2-bromovinyl)-2'-deoxyuridine (BVDU) and Other 5-Substituted-2'-deoxyuridines by Purified Human Thymidine Phosphorylase and Intact Blood Platelets. *Biochem. Pharmacol.* **1983**, *32*, 3583–3590.
21. Morrey, J. D.; Smee, D. F.; Sidwell, R. W.; Tseng, C. Identification of Active Antiviral Compounds Against a New York Isolate of West Nile Virus. *Antiviral Res.* **2002**, *55*, 107–116.
22. Pyrc, K.; Bosch, B. J.; Berkhout, B.; Jebbink, M. F.; Dijkman, R.; Rottier, P.; Van Der Hoek, L. Inhibition of Human Coronavirus NL63 Infection at Early Stages of the Replication Cycle. *Antimicrob. Agents Chemother.* **2006**, *50*, 2000–2008.
23. Crance, J. M.; Scaramozzino, N.; Jouan, A.; Garin, D. Interferon, Ribavirin, 6-Azauridine and Glycyrrhizin: Antiviral Compounds Active Against Pathogenic Flaviviruses. *Antivir. Res.* **2003**, *58*, 73–79.
24. Parker, W. B.; White, E. L.; Shaddix, S. C.; Ross, L. J.; Buckheit, R. W., Jr.; Germany, J. M.; Secrist, J. A., 3rd; Vince, R.; Shannon, W. M. Mechanism of Inhibition of Human Immunodeficiency Virus Type 1 Reverse Transcriptase and Human DNA Polymerases Alpha, Beta, and Gamma by the 5'-Triphosphates of Carbovir, 3'-azido-3'-deoxythymidine, 2',3'-Dideoxyguanosine and 3'-Deoxythymidine. A Novel RNA Template for the Evaluation of Antiretroviral Drugs. *J. Biol. Chem.* **1991**, *266*, 1754–1762.
25. Furman, P. A.; Fyfe, J. A.; St Clair, M. H.; Weinhold, K.; Rideout, J. L.; Freeman, G. A.; Lehrman, S. N.; Bolognesi, D. P.; Broder, S.; Mitsuya, H. Phosphorylation of 3'-azido-3'-Deoxythymidine and Selective Interaction of the 5'-Triphosphate With Human Immunodeficiency Virus Reverse Transcriptase. *Proc. Natl. Acad. Sci. U. S. A.* **1986**, *83*, 8333.
26. Yarchoan, R.; Thomas, R.; Allain, J.-P.; McAtee, N.; Dubinsky, R.; Mitsuya, H.; Lawley, T.; Safai, B.; Myers, C.; Perno, C.; Klecker, R.; Wills, R.; Fischl, M.; McNeely, M. C.; Pluda, J.; Leuther, M.; Collins, J.; Broder, S. Phase I Studies of 2',3'-Dideoxycytidine in Severe Human Immunodeficiency Virus Infection as a Single Agent and Alternating With Zidovudine (AZT). *Lancet* **1988**, *331*, 76–81.
27. Pruvost, A.; Negredo, E.; Benech, H.; Theodoro, F.; Puig, J.; Grau, E.; García, E.; Moltó, J.; Grassi, J.; Clotet, B. Measurement of Intracellular Didanosine and Tenofovir Phosphorylated Metabolites and Possible Interaction of the Two Drugs in Human Immunodeficiency Virus-Infected Patients. *Antimicrob. Agents Chemother.* **2005**, *49*, 1907–1914.
28. Mitsuya, H.; Broder, S. Inhibition of the In Vitro Infectivity and Cytopathic Effect of Human T-Lymphotrophic Virus Type III/Lymphadenopathy-Associated Virus (HTLV-III/LAV) by 2',3'-Dideoxynucleosides. *Proc. Natl. Acad. Sci. U. S. A.* **1986**, *83*, 1911–1915.
29. Whitley, R. J.; Tucker, B. C.; Kinkel, A. W.; Barton, N. H.; Pass, R. F.; Whelchel, J. D.; Cobbs, C. G.; Diethelm, A. G.; Buchanan, R. A. Pharmacology, Tolerance, and Antiviral Activity of Vidarabine Monophosphate in Humans. *Antimicrob. Agents Chemother.* **1980**, *18*, 709–715.

30. Chhikara, B. S.; Parang, K. Development of Cytarabine Prodrugs and Delivery Systems for Leukemia Treatment. *Expert Opin. Drug Deliv.* **2010**, 7, 1399–1414.
31. De Clercq, E. The Holý Trinity: The Acyclic Nucleoside Phosphonates. *Adv. Pharmacol.* **2013**, *67*, 293–316.
32. Tichý, T.; Andrei, G.; Dračínský, M.; Holý, A.; Balzarini, J.; Snoeck, R.; Krečmerová, M. New Prodrugs of Adefovir and Cidofovir. *Bioorg. Med. Chem.* **2011**, *19*, 3527–3539.
33. Shaw, M.; Gurr, W.; Watts, P.; Littler, E.; Field, H. Ganciclovir and Penciclovir, but Not Acyclovir, Induce Apoptosis in Herpes Simplex Virus Thymidine Kinasetransformed Baby Hamster Kidney Cells. *Antivir. Chem. Chemother.* **2001**, *12*, 175–186.
34. Chou, T.; Hong, B. Ganciclovir Ophthalmic Gel 0.15% for the Treatment of Acute Herpetic Keratitis: Background, Effectiveness, Tolerability, Safety, and Future Applications. *Ther. Clin. Risk Manage.* **2014**, *665*.
35. Worrall, G. Acyclovir in Recurrent Herpes labialis. *BMJ* **1996**, *312*, 6.
36. Spruance, S. L.; Nett, R.; Marbury, T.; Wolff, R.; Johnson, J.; Spaulding, T. Acyclovir Cream for Treatment of Herpes Simplex Labialis: Results of Two Randomized, Double-Blind, Vehicle-Controlled, Multicenter Clinical Trials. *Antimicrob. Agents Chemother.* **2002**, *46*, 2238–2243.
37. Rafailidis, P. I.; Mavros, M. N.; Kapaskelis, A.; Falagas, M. E. Antiviral Treatment for Severe EBV Infections in Apparently Immunocompetent Patients. *J. Clin. Virol.* **2010**, *49*, 151–157.
38. Gershburg, E.; Pagano, J. S. Epstein-Barr Virus Infections: Prospects for Treatment. *J. Antimicrob. Chemother.* **2005**, *56*, 277–281.
39. Yates, M. K.; Raje, M. R.; Chatterjee, P.; Spiropoulou, C. F.; Bavari, S.; Flint, M.; Soloveva, V.; Seley-Radtke, K. L. Flex-Nucleoside Analogues—Novel Therapeutics Against Filoviruses. *Bioorg. Med. Chem. Lett.* **2017**, *27*, 2800–2802.
40. Peters, H. L.; Jochmans, D.; de Wilde, A. H.; Posthuma, C. C.; Snijder, E. J.; Neyts, J.; Seley-Radtke, K. L. Design, Synthesis and Evaluation of a Series of Acyclic Fleximer Nucleoside Analogues With Anti-Coronavirus Activity. *Bioorg. Med. Chem. Lett.* **2015**, *25*, 2923–2926.
41. Thames, J. E.; Waters, C. D.; Valle, C.; Bassetto, M.; Aouadi, W.; Martin, B.; Selisko, B.; Falat, A.; Coutard, B.; Brancale, A.; Canard, B.; Decroly, E.; Seley-Radtke, K. L. Synthesis and Biological Evaluation of Novel Flexible Nucleoside Analogues That Inhibit Flavivirus Replication In Vitro. *Bioorg. Med. Chem.* **2020**, *28*, 115713.
42. Magee, W. C.; Hostetler, K. Y.; Evans, D. H. Mechanism of Inhibition of Vaccinia Virus DNA Polymerase by Cidofovir Diphosphate. *Antimicrob. Agents Chemother.* **2005**, *49*, 3153–3162.
43. Jockusch, S.; Tao, C.; Li, X.; Anderson, T. K.; Chien, M.; Kumar, S.; Russo, J. J.; Kirchdoerfer, R. N.; Ju, J. A Library of Nucleotide Analogues Terminate RNA Synthesis Catalyzed by Polymerases of Coronaviruses That Cause SARS and COVID-19. *Antivir. Res.* **2020**, *180*, 104857.
44. De Clercq, E. Cidofovir in the Treatment of Poxvirus Infections. *Antivir. Res.* **2002**, *55*, 1–13.
45. Scagliarini, A.; McInnes, C. J.; Gallina, L.; Dal Pozzo, F.; Scagliarini, L.; Snoeck, R.; Prosperi, S.; Sales, J.; Gilray, J. A.; Nettleton, P. F. Antiviral Activity of HPMPC (Cidofovir) Against orf Virus Infected Lambs. *Antivir. Res.* **2007**, *73*, 169–174.
46. De Clercq, E. Clinical Potential of the Acyclic Nucleoside Phosphonates Cidofovir, Adefovir, and Tenofovir in Treatment of DNA Virus and Retrovirus Infections. *Clin. Microbiol. Rev.* **2003**, *16*, 569–596.
47. Bradbury, J. Orally Available Cidofovir Derivative Active Against Smallpox. *Lancet* **2002**, *359*, 1041.

48. Marty, F. M.; Winston, D. J.; Chemaly, R. F.; Mullane, K. M.; Shore, T. B.; Papanicolaou, G. A.; Chittick, G.; Brundage, T. M.; Wilson, C.; Morrison, M. E.; Foster, S. A.; Nichols, W. G.; Boeckh, M. J. A Randomized, Double-Blind, Placebo-Controlled Phase 3 Trial of Oral Brincidofovir for Cytomegalovirus Prophylaxis in Allogeneic Hematopoietic Cell Transplantation. *Biol. Blood Marrow Transplant* **2019**, *25*, 369–381.
49. Lanier, R.; Trost, L.; Tippin, T.; Lampert, B.; Robertson, A.; Foster, S.; Rose, M.; Painter, W.; O'Mahony, R.; Almond, M.; Painter, G. Development of CMX001 for the Treatment of Poxvirus Infections. *Viruses* **2010**, *2*, 2740–2762.
50. Florescu, D. F.; Keck, M. A. Development of CMX001 (Brincidofovir) for the Treatment of Serious Diseases or Conditions Caused by dsDNA Viruses. *Expert Rev. Anti Infect. Ther.* **2014**, *12*, 1171–1178.
51. Pan, C. Q.; Duan, Z.; Dai, E.; Zhang, S.; Han, G.; Wang, Y.; Zhang, H.; Zou, H.; Zhu, B.; Zhao, W.; Jiang, H. Tenofovir to Prevent Hepatitis B Transmission in Mothers With High Viral Load. *New Engl. J. Med.* **2016**, *374*, 2324–2334.
52. Charlton, M. R.; Alam, A.; Shukla, A.; Dashtseren, B.; Lesmana, C. R. A.; Duger, D.; Payawal, D. A.; Duy Cuong, D.; Jargalsaikhan, G.; Cua, I. H. Y.; Sollano, J. D.; Singh, K. R.; Madan, K.; Win, K. M.; Kyi, K. P.; Tun, K. S.; Salih, M.; Rastogi, M.; Saraf, N.; Thuy, P. T. T.; Hien, P. T. D.; Gani, R. A.; Mohamed, R.; Tanwandee, T.; Piratvisuth, T.; Sukeepaisarnjaroen, W.; Naing, W.; Hashmi, Z. Y. An Expert Review on the Use of Tenofovir Alafenamide for the Treatment of Chronic Hepatitis B Virus Infection in Asia. *J. Gastroenterol.* **2020**, *55*, 811–823.
53. Tao, X.; Lu, Y.; Zhou, Y.; Zhang, L.; Chen, Y. Efficacy and Safety of the Regimens Containing Tenofovir Alafenamide Versus Tenofovir Disoproxil Fumarate in Fixed-Dose Single-Tablet Regimens for Initial Treatment of HIV-1 Infection: A Meta-Analysis of Randomized Controlled Trials. *Int. J. Infect. Dis.* **2020**, *93*, 108–117.
54. Kutlu, Ö. Can Tenofovir Diphosphate be a Candidate Drug for Sars-Cov2? First Clinical Perspective. *Int. J. Clin. Pract.* **2021**, *75*.
55. Ray, A. S.; Fordyce, M. W.; Hitchcock, M. J. M. Tenofovir Alafenamide: A Novel Prodrug of Tenofovir for the Treatment of Human Immunodeficiency Virus. *Antivir. Res.* **2016**, *125*, 63–70.
56. Mehellou, Y.; Rattan, H. S.; Balzarini, J. The ProTide Prodrug Technology: From the Concept to the Clinic. *J. Med. Chem.* **2018**, *61*, 2211–2226.
57. McGuigan, C.; Pathirana, R. N.; Balzarini, J.; De Clercq, E. Intracellular Delivery of Bioactive AZT Nucleotides by Aryl Phosphate Derivatives of AZT. *J. Med. Chem.* **1993**, *36*, 1048–1052.
58. McGuigan, C.; Harris, S. A.; Daluge, S. M.; Gudmundsson, K. S.; McLean, E. W.; Burnette, T. C.; Marr, H.; Hazen, R.; Condreay, L. D.; Johnson, L.; De Clercq, E.; Balzarini, J. Application of Phosphoramidate Pronucleotide Technology to Abacavir Leads to a Significant Enhancement of Antiviral Potency. *J. Med. Chem.* **2005**, *48*, 3504–3515.
59. Cahard, D.; McGuigan, C.; Balzarini, J. Aryloxy Phosphoramidate Triesters as Pro-Tides. *Mini Rev. Med. Chem.* **2004**, *4*, 371–381.
60. Marcellin, P.; Chang, T. T.; Lim, S. G.; Tong, M. J.; Sievert, W.; Shiffman, M. L.; Jeffers, L.; Goodman, Z.; Wulfsohn, M. S.; Xiong, S.; Fry, J.; Brosgart, C. L. Adefovir Dipivoxil for the Treatment of Hepatitis B e Antigen-Positive Chronic Hepatitis B. *N. Engl. J. Med.* **2003**, *348*, 808–816.
61. Manolakopoulos, S.; Bethanis, S.; Koutsounas, S.; Goulis, J.; Vlachogiannakos, J.; Christias, E.; Saveriadis, A.; Pavlidis, C.; Triantos, C.; Christidou, A.; Papatheodoridis, G.; Karamanolis, D.; Tzourmakliotis, D. Long-Term Therapy With Adefovir Dipivoxil in Hepatitis B e Antigen-Negative Patients Developing Resistance to Lamivudine. *Aliment. Pharmacol. Ther.* **2008**, *27*, 266–273.

62. Chan, H. L.; Heathcote, E. J.; Marcellin, P.; Lai, C. L.; Cho, M.; Moon, Y. M.; Chao, Y. C.; Myers, R. P.; Minuk, G. Y.; Jeffers, L.; Sievert, W.; Bzowej, N.; Harb, G.; Kaiser, R.; Qiao, X. J.; Brown, N. A. Treatment of Hepatitis B e Antigen Positive Chronic Hepatitis With Telbivudine or Adefovir: A Randomized Trial. *Ann. Intern. Med.* **2007**, *147*, 745–754.
63. Fisher, E. J.; Chaloner, K.; Cohn, D. L.; Grant, L. B.; Alston, B.; Brosgart, C. L.; Schmetter, B.; El-Sadr, W. M.; Sampson, J. The Safety and Efficacy of Adefovir Dipivoxil in Patients With Advanced HIV Disease: A Randomized, Placebo-Controlled Trial. *Aids* **2001**, *15*, 1695–1700.
64. Lai, C. L.; Shouval, D.; Lok, A. S.; Chang, T. T.; Cheinquer, H.; Goodman, Z.; DeHertogh, D.; Wilber, R.; Zink, R. C.; Cross, A.; Colonno, R.; Fernandes, L. Entecavir Versus Lamivudine for Patients With HBeAg-Negative Chronic Hepatitis B. *N. Engl. J. Med.* **2006**, *354*, 1011–1020.
65. Sims, K. A.; Woodland, A. M. Entecavir: A New Nucleoside Analog for the Treatment of Chronic Hepatitis B Infection. *Pharmacotherapy* **2006**, *26*, 1745–1757.
66. Khalighinejad, P.; Alavian, S. M.; Fesharaki, M. G.; Jalilianhasanpour, R. Lamivudine's Efficacy and Safety in Preventing Mother-to-Child Transmission of Hepatitis B: A Meta-Analysis. *Turk. J. Gastroenterol.* **2018**.
67. Lai, C. L.; Leung, N.; Teo, E. K.; Tong, M.; Wong, F.; Hann, H. W.; Han, S.; Poynard, T.; Myers, M.; Chao, G.; Lloyd, D.; Brown, N. A. A 1-Year Trial of Telbivudine, Lamivudine, and the Combination in Patients With Hepatitis B e Antigen-Positive Chronic Hepatitis B. *Gastroenterology* **2005**, *129*, 528–536.
68. Lai, C. L.; Gane, E.; Liaw, Y. F.; Hsu, C. W.; Thongsawat, S.; Wang, Y.; Chen, Y.; Heathcote, E. J.; Rasenack, J.; Bzowej, N.; Naoumov, N. V.; Di Bisceglie, A. M.; Zeuzem, S.; Moon, Y. M.; Goodman, Z.; Chao, G.; Constance, B. F.; Brown, N. A. Telbivudine Versus Lamivudine in Patients With Chronic Hepatitis B. *N. Engl. J. Med.* **2007**, *357*, 2576–2588.
69. Peiris, J. S.; Chu, C. M.; Cheng, V. C.; Chan, K. S.; Hung, I. F.; Poon, L. L.; Law, K. I.; Tang, B. S.; Hon, T. Y.; Chan, C. S.; Chan, K. H.; Ng, J. S.; Zheng, B. J.; Ng, W. L.; Lai, R. W.; Guan, Y.; Yuen, K. Y. Clinical Progression and Viral Load in a Community Outbreak of Coronavirus-Associated SARS Pneumonia: A Prospective Study. *Lancet* **2003**, *361*, 1767–1772.
70. Osborn, M. K. Safety and Efficacy of Telbivudine for the Treatment of Chronic Hepatitis B. *Ther. Clin. Risk Manag.* **2009**, *5*, 789–798.
71. Wohl, D.; Clarke, A.; Maggiolo, F.; Garner, W.; Laouri, M.; Martin, H.; Quirk, E. Patient-Reported Symptoms Over 48 Weeks Among Participants in Randomized, Double-Blind, Phase III Non-inferiority Trials of Adults With HIV on Co-formulated Bictegravir, Emtricitabine, and Tenofovir Alafenamide Versus Co-Formulated Abacavir, Dolutegravir. *Pat. Pat. Center. Outcomes Res.* **2018**, *11*, 561–573.
72. Deeks, E. D. Bictegravir/Emtricitabine/Tenofovir Alafenamide: A Review in HIV-1 Infection. *Drugs* **2018**, *78*, 1817–1828.
73. Sculier, D.; Wandeler, G.; Yerly, S.; Marinosci, A.; Stoeckle, M.; Bernasconi, E.; Braun, D. L.; Vernazza, P.; Cavassini, M.; Buzzi, M.; Metzner, K. J.; Decosterd, L. A.; Günthard, H. F.; Schmid, P.; Limacher, A.; Egger, M.; Calmy, A. Efficacy and Safety of Dolutegravir Plus Emtricitabine Versus Standard ART for the Maintenance of HIV-1 Suppression: 48-Week Results of the Factorial, Randomized, Non-Inferiority SIMPL'HIV Trial. *PLoS Med.* **2020**, *17*, e1003421.
74. Jackson, A.; Moyle, G.; Dickinson, L.; Back, D.; Khoo, S.; Taylor, J.; Gedela, K.; Abongomera, G.; Gazzard, B.; Boffito, M. Pharmacokinetics of Abacavir and Its Anabolite Carbovir Triphosphate Without and With Darunavir/Ritonavir or Raltegravir in HIV-Infected Subjects. *Antivir. Ther.* **2011**, *17*, 19–24.

75. Yeom, Y. H.; Remmel, R. P.; Huang, S. H.; Hua, M.; Vince, R.; Zimmerman, C. L. Pharmacokinetics and Bioavailability of Carbovir, a Carbocyclic Nucleoside Active Against Human Immunodeficiency Virus, in Rats. *Antimicrob. Agents Chemother.* **1989**, *33*, 171–175.
76. Mega, T. A.; Usamo, F. B.; Negera, G. Z. Abacavir Versus Zidovudine-Based Regimens for Treatment of HIV-Infected Children in Resource Limited Settings: A Retrospective Cohort Study. *BMC Pediat.* **2020**, *20*.
77. Kabinger, F.; Stiller, C.; Schmitzová, J.; Dienemann, C.; Kokic, G.; Hillen, H. S.; Höbartner, C.; Cramer, P. Mechanism of Molnupiravir-Induced SARS-CoV-2 Mutagenesis. *Nat. Struct. Mol. Biol.* **2021**.
78. Eberhardt, K. A.; Mischlinger, J.; Jordan, S.; Groger, M.; Günther, S.; Ramharter, M. Ribavirin for the Treatment of Lassa Fever: A Systematic Review and Meta-Analysis. *Int. J. Infect. Dis.* **2019**, *87*, 15–20.
79. Pelaez, A.; Lyon, G. M.; Force, S. D.; Ramirez, A. M.; Neujahr, D. C.; Foster, M.; Naik, P. M.; Gal, A. A.; Mitchell, P. O.; Lawrence, E. C. Efficacy of Oral Ribavirin in Lung Transplant Patients With Respiratory Syncytial Virus Lower Respiratory Tract Infection. *J. Heart Lung Transplant.* **2009**, *28*, 67–71.
80. Smee, D. F.; Wong, M.-H.; Bailey, K. W.; Sidwell, R. W. Activities of Oseltamivir and Ribavirin Used Alone and in Combination Against Infections in Mice With Recent Isolates of Influenza a (H1N1) and B Viruses. *Antivir. Chem. Chemother.* **2006**, *17*, 185–192.
81. Liu, C. J.; Chen, P. J.; Lai, M. Y.; Kao, J. H.; Jeng, Y. M.; Chen, D. S. Ribavirin and interferon is effective for hepatitis C virus clearance in hepatitis B and C dually infected patients. *Hepatology* **2003**, *37*, 568–576.
82. Eslami, G.; Mousaviasl, S.; Radmanesh, E.; Jelvay, S.; Bitaraf, S.; Simmons, B.; Wentzel, H.; Hill, A.; Sadeghi, A.; Freeman, J.; Salmanzadeh, S.; Esmaeilian, H.; Mobarak, M.; Tabibi, R.; Jafari Kashi, A. H.; Lotfi, Z.; Talebzadeh, S. M.; Wickramatillake, A.; Momtazan, M.; Hajizadeh Farsani, M.; Marjani, S.; Mobarak, S. The Impact of Sofosbuvir/Daclatasvir or Ribavirin in Patients With Severe COVID-19. *J. Antimicrob. Chemother.* **2020**, *75*, 3366–3372.
83. Gish, R. G. Treating HCV With Ribavirin Analogues and Ribavirin-Like Molecules. *J. Antimicrob. Chemother.* **2006**, *57*, 8–13.
84. Poordad, F.; Lawitz, E.; Shiffman, M. L.; Hassanein, T.; Muir, A. J.; Bacon, B. R.; Heise, J.; Halliman, D.; Chun, E.; Hammond, J. Virologic Response Rates of Weight-Based Taribavirin Versus Ribavirin in Treatment-Naive Patients With Genotype 1 Chronic Hepatitis C. *Hepatology* **2010**, *52*, 1208–1215.
85. Lin, C. C.; Luu, K.; Lourenco, D.; Yeh, L. T. Pharmacokinetics and Metabolism of [14C]viramidine in Rats and Cynomolgus Monkeys. *Antimicrob. Agents Chemother.* **2003**, *47*, 2458–2463.
86. Lin, C. C.; Yeh, L. T.; Vitarella, D.; Hong, Z. Viramidine, a Prodrug of Ribavirin, Shows Better Liver-Targeting Properties and Safety Profiles Than Ribavirin in Animals. *Antivir. Chem. Chemother.* **2003**, *14*, 145–152.
87. Zimmer, C. *A Pill to Treat Covid-19? The U.S. Is Betting on It*; The New York Times, 2021.
88. Ghasemnejad-Berenji, M.; Pashapour, S. Favipiravir and COVID-19: A Simplified Summary. *Drug Res.* **2021**, *71*, 166–170.
89. Furuta, Y.; Gowen, B. B.; Takahashi, K.; Shiraki, K.; Smee, D. F.; Barnard, D. L. Favipiravir (T-705), a Novel Viral RNA Polymerase Inhibitor. *Antivir. Res.* **2013**, *100*, 446–454.
90. Chien, M.; Anderson, T. K.; Jockusch, S.; Tao, C.; Li, X.; Kumar, S.; Russo, J. J.; Kirchdoerfer, R. N.; Ju, J. Nucleotide Analogues as Inhibitors of SARS-CoV-2 Polymerase, a Key Drug Target for COVID-19. *J. Proteome Res.* **2020**, *19*, 4690–4697.

91. Heo, Y.-A.; Deeks, E. D. Sofosbuvir/Velpatasvir/Voxilaprevir: A Review in Chronic Hepatitis C. *Drugs* **2018**, *78*, 577–587.
92. Mathur, P.; Kottilil, S.; Wilson, E. Sofosbuvir/Velpatasvir/Voxilaprevir: A Highly Effective Option for Retreatment of Hepatitis C in Difficult-to-Treat Patients. *Antivir. Ther.* **2019**, *24*, 1–10.
93. Lawitz, E.; Gane, E.; Feld, J. J.; Buti, M.; Foster, G. R.; Rabinovitz, M.; Burnevich, E.; Katchman, H.; Tomasiewicz, K.; Lahser, F.; Jackson, B.; Shaughnessy, M.; Klopfer, S.; Yeh, W. W.; Robertson, M. N.; Hanna, G. J.; Barr, E.; Platt, H. L. Efficacy and Safety of a Two-Drug Direct-Acting Antiviral Agent Regimen Ruzasvir 180 mg and Uprifosbuvir 450 mg for 12 weeks in Adults With Chronic Hepatitis C Virus Genotype 1, 2, 3, 4, 5 or 6. *J. Viral Hepat.* **2019**, *26*, 1127–1138.
94. Wyles, D.; Wedemeyer, H.; Ben-Ari, Z.; Gane, E. J.; Hansen, J. B.; Jacobson, I. M.; Laursen, A. L.; Luetkemeyer, A.; Nahass, R.; Pianko, S.; Zeuzem, S.; Jumes, P.; Huang, H. C.; Butterton, J.; Robertson, M.; Wahl, J.; Barr, E.; Joeng, H. K.; Martin, E.; Serfaty, L. Grazoprevir, Ruzasvir, and Uprifosbuvir for Hepatitis C Virus After NS5A Treatment Failure. *Hepatology* **2017**, *66*, 1794–1804.
95. Siqueira-Batista, R.; De Souza Bayão, T.; Do Carmo Cupertino, M.; Alfred Joseph Mayers, N.; Patrícia Gomes, A. Sofosbuvir Use for Yellow Fever: A New Perspective Treatment. *Pathog. Glob. Health* **2019**, *113*, 207–208.
96. Mumtaz, N.; Jimmerson, L. C.; Bushman, L. R.; Kiser, J. J.; Aron, G.; Reusken, C.; Koopmans, M. P. G.; van Kampen, J. J. A. Cell-Line Dependent Antiviral Activity of Sofosbuvir Against Zika Virus. *Antivir. Res.* **2017**, *146*, 161–163.
97. Warren, T. K.; Jordan, R.; Lo, M. K.; Ray, A. S.; Mackman, R. L.; Soloveva, V.; Siegel, D.; Perron, M.; Bannister, R.; Hui, H. C.; Larson, N.; Strickley, R.; Wells, J.; Stuthman, K. S.; Van Tongeren, S. A.; Garza, N. L.; Donnelly, G.; Shurtleff, A. C.; Retterer, C. J.; Gharaibeh, D.; Zamani, R.; Kenny, T.; Eaton, B. P.; Grimes, E.; Welch, L. S.; Gomba, L.; Wilhelmsen, C. L.; Nichols, D. K.; Nuss, J. E.; Nagle, E. R.; Kugelman, J. R.; Palacios, G.; Doerffler, E.; Neville, S.; Carra, E.; Clarke, M. O.; Zhang, L.; Lew, W.; Ross, B.; Wang, Q.; Chun, K.; Wolfe, L.; Babusis, D.; Park, Y.; Stray, K. M.; Trancheva, I.; Feng, J. Y.; Barauskas, O.; Xu, Y.; Wong, P.; Braun, M. R.; Flint, M.; McMullan, L. K.; Chen, S. S.; Fearns, R.; Swaminathan, S.; Mayers, D. L.; Spiropoulou, C. F.; Lee, W. A.; Nichol, S. T.; Cihlar, T.; Bavari, S. Therapeutic Efficacy of the Small Molecule GS-5734 Against Ebola Virus in Rhesus Monkeys. *Nature* **2016**, *531*, 381–385.
98. Amirian, E. S.; Levy, J. K. Current Knowledge About the Antivirals Remdesivir (GS-5734) and GS-441524 as Therapeutic Options for Coronaviruses. *One Health* **2020**, *9*, 100128.
99. Agostini, M. L.; Andres, E. L.; Sims, A. C.; Graham, R. L.; Sheahan, T. P.; Lu, X.; Smith, E. C.; Case, J. B.; Feng, J. Y.; Jordan, R.; Ray, A. S.; Cihlar, T.; Siegel, D.; Mackman, R. L.; Clarke, M. O.; Baric, R. S.; Denison, M. R. Coronavirus Susceptibility to the Antiviral Remdesivir (GS-5734) Is Mediated by the Viral Polymerase and the Proofreading Exoribonuclease. *mBio* **2018**, *9*.
100. Wu, K. *Gilead to Test a Version of Remdesivir That Can Be Inhaled*; The New York Times, 2020.
101. Bravo, J. P. K.; Dangerfield, T. L.; Taylor, D. W.; Johnson, K. A. Remdesivir is a Delayed Translocation Inhibitor of SARS-CoV-2 Replication. *Mol. Cell* **2021**, *81*, 1548–1552.e4.
102. Good, S. S.; Westover, J.; Jung, K. H.; Zhou, X. J.; Moussa, A.; La Colla, P.; Collu, G.; Canard, B.; Sommadossi, J. P. AT-527, a Double Prodrug of a Guanosine Nucleotide Analog, Is a Potent Inhibitor of SARS-CoV-2 In Vitro and a Promising Oral Antiviral for Treatment of COVID-19. *Antimicrob. Agents Chemother.* **2021**, *65*.
103. Beasley, D. *Inside the Race to Find a COVID-19 Treatment Pill*; Reuters, 2021.

104. Good, S. S.; Moussa, A.; Zhou, X.-J.; Pietropaolo, K.; Sommadossi, J.-P. Preclinical Evaluation of AT-527, a Novel Guanosine Nucleotide Prodrug With Potent, Pan-Genotypic Activity Against Hepatitis C Virus. *PLoS One* **2020**, *15*, e0227104.
105. Berliba, E.; Bogus, M.; Vanhoutte, F.; Berghmans, P.-J.; Good, S. S.; Moussa, A.; Pietropaolo, K.; Murphy, R. L.; Zhou, X.-J.; Sommadossi, J.-P. Safety, Pharmacokinetics, and Antiviral Activity of AT-527, a Novel Purine Nucleotide Prodrug, in Hepatitis C Virus-Infected Subjects With or Without Cirrhosis. *Antimicrob. Agents Chemother.* **2019**, *63*.
106. Michailidis, E.; Ryan, E. M.; Hachiya, A.; Kirby, K. A.; Marchand, B.; Leslie, M. D.; Huber, A. D.; Ong, Y. T.; Jackson, J. C.; Singh, K.; Kodama, E. N.; Mitsuya, H.; Parniak, M. A.; Sarafianos, S. G. Hypersusceptibility Mechanism of Tenofovir-Resistant HIV to EFdA. *Retrovirology* **2013**, *10*, 65.
107. Michailidis, E.; Huber, A. D.; Ryan, E. M.; Ong, Y. T.; Leslie, M. D.; Matzek, K. B.; Singh, K.; Marchand, B.; Hagedorn, A. N.; Kirby, K. A.; Rohan, L. C.; Kodama, E. N.; Mitsuya, H.; Parniak, M. A.; Sarafianos, S. G. 4′-Ethynyl-2-fluoro-2′-deoxyadenosine (EFdA) Inhibits HIV-1 Reverse Transcriptase With Multiple Mechanisms. *J. Biol. Chem.* **2014**, *289*, 24533–24548.
108. Stoddart, C. A.; Galkina, S. A.; Joshi, P.; Kosikova, G.; Moreno, M. E.; Rivera, J. M.; Sloan, B.; Reeve, A. B.; Sarafianos, S. G.; Murphey-Corb, M.; Parniak, M. A. Oral Administration of the Nucleoside EFdA (4′-Ethynyl-2-Fluoro-2′-Deoxyadenosine) Provides Rapid Suppression of HIV Viremia in Humanized Mice and Favorable Pharmacokinetic Properties in Mice and the Rhesus Macaque. *Antimicrob. Agents Chemother.* **2015**, *59*, 4190–4198.
109. Deng, C. L.; Yeo, H.; Ye, H. Q.; Liu, S. Q.; Shang, B. D.; Gong, P.; Alonso, S.; Shi, P. Y.; Zhang, B. Inhibition of Enterovirus 71 by Adenosine Analog NITD008. *J. Virol.* **2014**, *88*, 11915–11923.
110. Deng, Y. Q.; Zhang, N. N.; Li, C. F.; Tian, M.; Hao, J. N.; Xie, X. P.; Shi, P. Y.; Qin, C. F. Adenosine Analog NITD008 Is a Potent Inhibitor of Zika Virus. *Open Forum Infect. Dis.* **2016**, *3*, ofw175.
111. Enosi Tuipulotu, D.; Fumian, T. M.; Netzler, N. E.; Mackenzie, J. M.; White, P. A. The Adenosine Analogue NITD008 has Potent Antiviral Activity Against Human and Animal Caliciviruses. *Viruses* **2019**, *11*.
112. Lo, M. K.; Shi, P. Y.; Chen, Y. L.; Flint, M.; Spiropoulou, C. F. In Vitro Antiviral Activity of Adenosine Analog NITD008 Against Tick-Borne Flaviviruses. *Antivir. Res.* **2016**, *130*, 46–49.
113. Schul, W.; Liu, W.; Xu, H. Y.; Flamand, M.; Vasudevan, S. G. A Dengue Fever Viremia Model in Mice Shows Reduction in Viral Replication and Suppression of the Inflammatory Response After Treatment With Antiviral Drugs. *J. Infect. Dis.* **2007**, *195*, 665–674.
114. Wu, R.; Smidansky, E. D.; Oh, H. S.; Takhampunya, R.; Padmanabhan, R.; Cameron, C. E.; Peterson, B. R. Synthesis of a 6-Methyl-7-Deaza Analogue of Adenosine That Potently Inhibits Replication of Polio and Dengue Viruses. *J. Med. Chem.* **2010**, *53*, 7958–7966.
115. Eyer, L.; Valdés, J. J.; Gil, V. A.; Nencka, R.; Hřebabecký, H.; Šála, M.; Salát, J.; Černý, J.; Palus, M.; De Clercq, E.; Růžek, D. Nucleoside Inhibitors of Tick-Borne Encephalitis Virus. *Antimicrob. Agents Chemother.* **2015**, *59*, 5483–5493.
116. Nelson, J.; Roe, K.; Orillo, B.; Shi, P. Y.; Verma, S. Combined Treatment of Adenosine Nucleoside Inhibitor NITD008 and Histone Deacetylase Inhibitor Vorinostat Represents an Immunotherapy Strategy to Ameliorate West Nile Virus Infection. *Antivir. Res.* **2015**, *122*, 39–45.
117. Vittori, S.; Dal Ben, D.; Lambertucci, C.; Marucci, G.; Volpini, R.; Cristalli, G. Antiviral Properties of Deazaadenine Nucleoside Derivatives. *Curr. Med. Chem.* **2006**, *13*, 3529–3552.

118. Westover, J. B.; Mathis, A.; Taylor, R.; Wandersee, L.; Bailey, K. W.; Sefing, E. J.; Hickerson, B. T.; Jung, K. H.; Sheridan, W. P.; Gowen, B. B. Galidesivir Limits Rift Valley Fever Virus Infection and Disease in Syrian Golden Hamsters. *Antivir. Res.* **2018**, *156*, 38–45.
119. Julander, J. G.; Bantia, S.; Taubenheim, B. R.; Minning, D. M.; Kotian, P.; Morrey, J. D.; Smee, D. F.; Sheridan, W. P.; Babu, Y. S. BCX4430, a Novel Nucleoside Analog, Effectively Treats Yellow Fever in a Hamster Model. *Antimicrob. Agents Chemother.* **2014**, *58*, 6607–6614.
120. Lim, S.-Y.; Osuna, C. E.; Best, K.; Taylor, R.; Chen, E.; Yoon, G.; Kublin, J. L.; Schalk, D.; Schultz-Darken, N.; Capuano, S.; Safronetz, D.; Luo, M.; Maclennan, S.; Mathis, A.; Babu, Y. S.; Sheridan, W. P.; Perelson, A. S.; Whitney, J. B. A Direct-Acting Antiviral Drug Abrogates Viremia in Zika Virus–Infected Rhesus macaques. *Sci. Transl. Med.* **2020**, (12), eaau9135.
121. De Clercq, E. New Nucleoside Analogues for the Treatment of Hemorrhagic Fever Virus Infections. *Chem. Asian J.* **2019**, *14*, 3962–3968.
122. Evans, G. B.; Tyler, P. C.; Schramm, V. L. Immucillins in Infectious Diseases. *ACS Infect. Dis.* **2018**, *4*, 107–117.
123. Elfiky, A.; Ribavirin, A. Remdesivir, Sofosbuvir, Galidesivir, and Tenofovir Against SARS-CoV-2 RNA Dependent RNA Polymerase (RdRp): A Molecular Docking Study. *Life Sci.* **2020**, *253*, 117592.
124. https://coronavirus.jhu.edu/map.html.

CHAPTER FOUR

The search for antivirals to treat alphavirus infections

Marcella Bassetto[a] and Andrea Brancale[b,*]
[a]Department of Chemistry, Swansea University, Swansea, United Kingdom
[b]School of Pharmacy and Pharmaceutical Sciences, Cardiff University, Cardiff, United Kingdom
*Corresponding author: e-mail address: brancalea@cardiff.ac.uk

Contents

1. Introduction

Alphaviruses are positive strand RNA viruses, transmitted mainly by mosquitoes, that are responsible for different human diseases.[1] Although these illnesses are not generally considered to be life-threatening, several can evolve into very invalidating conditions that can often become chronic. For this reason, alphavirus infections can represent a significant economic and social burden, for which neither a specific antiviral therapy nor vaccines are available. Alphaviruses are widely distributed worldwide, although the specific viruses' location is associated to the presence of the vector responsible for transmission. A common classification of the different viruses in this family divides them in two groups: Old World (OW) and New World (NW) viruses. This classification gives us a rough indication of the geographical distribution of these viruses, although with some notable exceptions such as Chikungunya virus (CHIKV), which is now established also in Europe and the US. In addition, viruses belonging to do these different groups also have a different set of symptoms: OW viral infections are associated with fever and arthralgia, while NW viral infections are associated

Annual Reports in Medicinal Chemistry, Volume 57
ISSN 0065-7743
https://doi.org/10.1016/bs.armc.2021.09.002
Copyright © 2021 Elsevier Inc.
All rights reserved.

Table 1 Alphaviruses that affect humans.

	Virus name	Abbreviation	Symptoms
Old World	Chikungunya	CHIKV	Fever, rash, arthralgia
	Sindbis	SINV	
	O′nyong-nyong	ONNV	
	Ross River	RRV	
	Semliki forest	SFV	Generally non-pathogenic
New World	Venezuelan equine encephalitis	VEEV	Fever, encephalitis
	Eastern equine encephalitis	EEEV	
	Western equine encephalitis	WEEV	
	Mayaro	MAYV	
	Una	UNAV	Fever, arthralgia

with encephalitis and other neurological disorders.[1] Some of the most representative alphaviruses are reported in Table 1.

Alphaviruses are a member of the *Togaviridae* family and they share a simple genome, on a positive strand of RNA. All viruses in the group encode five structural proteins, and four non-structural proteins. Structural proteins include a capsid protein (CP), three envelope proteins (E1, E2, E3), and the 6K protein, which is involved in the viral maturation process. The four, multifunctional, non-structural proteins (nsP1-4) are responsible for the viral replication.[2] Fig. 1 gives a graphic summary of the genomic organization, and protein functions.

The alphaviruses life cycle starts with a virus particle invading the host cell through a clathrin-mediated endocytosis that is facilitated by the envelope protein E2. Once inside the cell, the endosome pH becomes more acidic, resulting in the release of the viral nucleocapsid in the cytoplasm. The viral (+)-RNA is then partially translated to form the P1234 polyprotein, which is then processed in the individual nsP1-4 proteins, through a series of proteolytic events, mediated by the viral protease (nsP2). The resulting nsPs form the replication complex that can then generate the complementary negative strand of RNA, which, in turn, serves as a template for both the generation of the genomic (+)-RNA, and a subgenomic (+)-RNA (26S RNA), that will encode for the structural polyprotein. During the

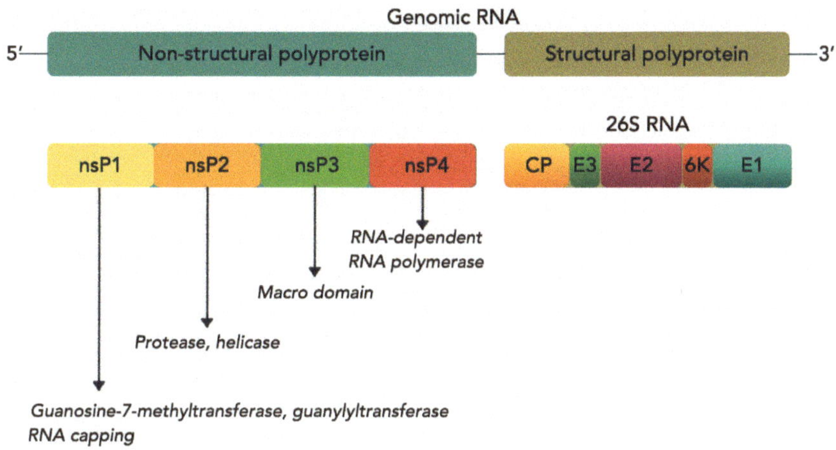

Fig. 1 Alphavirus genomic organization.

maturation stages, the CP is initially released, thanks to its autoprotease function, while the other structural proteins are generated by host-mediated events in the endoplasmic reticulum and in the Golgi apparatus. Finally, the envelope proteins are translocated to the plasma membrane, so that mature virus particles can form and be released.[2]

From a drug design point of view, all the described viral replication stages can potentially be disrupted to achieve an antiviral effect.[3] Targeting the viral proteins is often considered to be the preferential strategy, as it could potentially lead to potent, specific, and safe antivirals. However, the resulting compounds could generate viral resistance quickly, becoming ineffective. The alternative strategy is to target the host factors required in the viral replication. Compounds in this group should have a higher barrier to resistance, and they could also present a broad-spectrum profile, as viruses in the same family often rely on the same cellular pathways. The drawback of this strategy is the possibility of having compounds that interfere with the normal cellular processes, hence presenting some level of toxicity. In the context of alphavirus research, the recent and current literature shows how both approaches are actively pursued to generate novel antivirals. Indeed, many of the viral infections caused by this group of viruses are a clear unmet medical need, as discussed above, and although the anti-alphavirus field has now been very active for at least a decade, no specific antivirals have been, or are close to be, approved for the treatment of any alphavirus infection.[1]

In the last few years, several excellent reviews were published on this topic, giving often a very detailed overview of antiviral research on alphaviruses. For example, Ching et al. provided a thorough landscape of the anti-alphavirus drug design and discovery efforts up until 2017,[4] Abdelnabi et al. focused their attention to arthritogenic alphaviruses,[5] Kovacikova et al.,[6] and Hucke et al.,[7] independently, discussed small-molecules inhibitors of CHIKV, and Mello et al. looked even more specifically at antivirals for Mayaro virus (MAYV).[8]

In this chapter, we will briefly discuss the strategies that researchers are using in their quest of finding new antivirals to treat alphavirus infections, focusing primarily on the most recent publications in the field, to understand what drug discoveries approaches are currently used, and how.

2. Current approaches in the discovery of novel anti-alphavirus compounds

2.1 Drug repurposing strategies

Drug repurposing is often considered an attractive strategy, as it potentially allows to quickly identify a promising compound that could then be evaluated clinically very rapidly. The underlying idea is to find a different indication for an approved drug, or more often, a compound that has at least proven safe in a Phase I clinical trial. On the other hand, a repurposed compound could present some important limitations, as it represents a drug that was originally developed for a different target/disease, hence it could show selectivity problems. An approach that could limit this issue, in the case of antivirals, is to explore the activity of other approved antivirals, developed for similar viruses, or that have shown a broad-spectrum activity. This is the case, for example, of the identification of favipiravir (**1**), a compound initially developed, and approved in Japan, to treat influenza, as active against several RNA viruses, including different alphaviruses.[9] More recently, Julander et al. consolidated this observation by showing the activity of favipiravir against different clades of CHIKV, *in vivo*, with an efficacy inversely proportional to the disease severity caused by the different CHIKV strains.[10] Another example is described by Ferreira et al., who have shown the efficacy of sofosbuvir, a nucleotide prodrug approved for the treatment of hepatitis C, against CHIKV, *in vivo*.[11] Both favipiravir and sofosbuvir target the viral polymerase (nsP4), a target generally considered very attractive for the development of antivirals. Considerably more unusual is the work reported by Dey et al., where they studied the CHIKV 6K protein, and its ion channel function, and they found that the old drug amantadine showed some antiviral activity (IC_{50} of 29.51 μM) against CHIKV in Vero cells, also reporting a change in morphology of the viral particles treated with the drug.[12]

Viral proteins are not the only targets to be exploited in focused selections of potential repurposed drugs. Indeed, it is now becoming common to produce maps of the interactions between the viral proteins and the associated host factors, that could be used to target specific cellular pathways to produce an antiviral effect.[13,14] In other cases, specific observations can provide some new original insights. Kaur et al., starting from the knowledge that proteasome activity was important in CHIKV replication, evaluated the antiviral activity of the approved proteasome inhibitor bortezomib (**2**) in different cell lines.[15] Remarkably, the EC_{50} observed were all in the sub-micromolar range. However, the actual mechanism of action for this compound remains elusive. Similarly, Hwang et al., based on the observation that CHIKV envelope is high in cholesterol, and that cholesterol is also important in viral replication, decided to evaluate the anti-CHIKV activity of an agonist of the liver X receptor beta, an important regulator of cholesterol homeostasis. LXR-623 (**3**), originally developed to treat atherosclerosis, showed an interesting IC_{50} of 2.5 μM against CHIKV replication in primary human fibroblasts.[16] (Fig. 2).

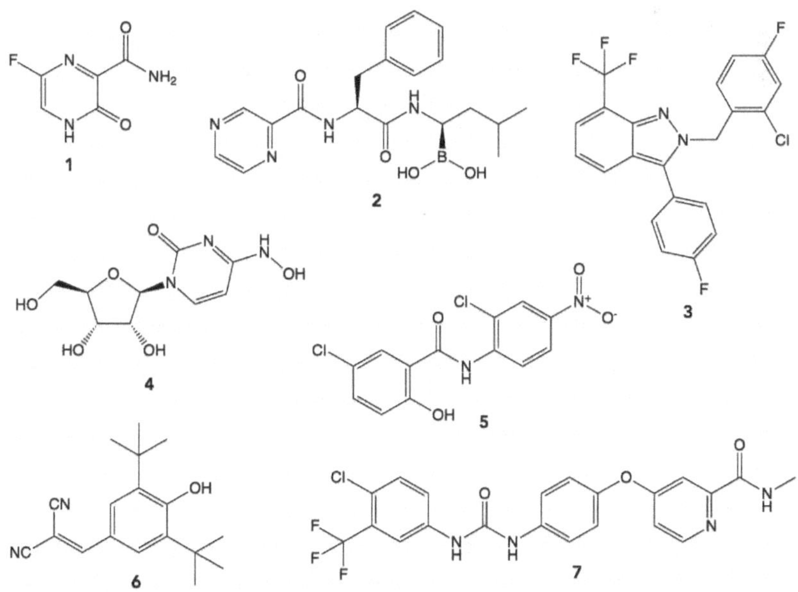

Fig. 2 Examples of repurposed compounds active on alphavirus replication.

Although cherry-picking single compounds and evaluate them in antiviral assays could sometimes prove to be very useful, it is more common to screen compound libraries to improve the chances of identifying novel, sometimes unexpected, antiviral molecules. Langendries et al., for example, initially developed a cell-based antiviral assay for MAYV, and they used it to screen a small library of compounds with known antiviral activities, identifying three active molecules: EIDD-1931(**4**) (the parent nucleoside of molnupiravir), favipiravir (**1**), and suramin. In particular, EID-1931, a nucleoside analog that would most likely act on the nsP4, showed a very interesting EC_{50} of 1.6 μM against viral replication, while showing no toxicity at >100 μM.[17] Suramin did not show a particularly potent activity (EC_{50} of 124 μM), however it is known to have an antiviral effect, *in vitro*, against other alphaviruses, and Albulescu et al. have also characterized its mechanism of action against CHIKV. Resistance studies have shown that the likely target for suramin is the E2 protein, disrupting the viral entry stage.[18]

Using libraries of compounds with antiviral activity might provide a shortcut, but wider repurposing campaigns that also include compounds with other therapeutic uses will certainly produce a more diverse set of hits, potentially with a broader antiviral spectrum. Mazzon et al. performed a sophisticated phenotypic screen that looked at the cellular pathways involved in viral entry, using the LOPAC (Sigma Aldrich) library of pharmacologically active compounds, in search for broad spectrum antivirals. Their protocol initially included an assay against an alphavirus (SFV), followed by an assay against a flavivirus. The 12 compounds identified proved to be active against a diverse panel of viruses, from different families. The two most interesting compounds identified were niclosamide (**5**) and tyrphostin A9 (**6**), that inhibit infection by SFV (IC_{50} of 3 and 2.9 μM respectively), SINV (IC_{50} of 2.8 and 0.8 μM), and RRV (IC_{50} of 0.82 and 0.6 μM).[19] In another report, a library of FDA-approved compounds was used by Lundberg et al. in a high-throughput luciferase reporter assay for VEEV, to identify the kinase inhibitor sorafenib (**7**) as active not only against VEEV (EC_{50} of 6.2 μM), but also against EEEV (EC_{50} of 6.7 μM), SINV (EC_{50} of 1.3 μM), and CHIKV (EC_{50} of 0.16 μM).[20]

Another possible approach to repurposed drugs is to evaluate the compound library against a specific viral target, instead of using a cell-based assay. In this way, a new direct antiviral mechanism can be discovered for compounds that were developed for a completely different use or biological pathway. Using this approach, Ferreira-Ramos et al. have screened a library of 1220 compounds, mainly approved drugs, in a high-throughput ELISA-based assay to monitor the capping activity of the VEEV nsP1, finding 18 hits that showed a significant inhibition at 50 μM. Among those, the most promising compound was the dihydrofolate reductase inhibitor pyrimethamine, with an IC_{50} of 2.7 μM.[21]

In some cases, researchers that did not have access to a valid high-throughput assay, have used a virtual screening approach to assess the compound library *in silico* as a method to select a much smaller number of compounds to be evaluated in a standard assay. More examples if this approach will be discussed in Section 2.2, but in the case of the results reported by Fatma et al., and Tripathi et al., the compound libraries used included clinically evaluated compounds, suitable for repurposing. Fatma et al. identified three compounds that can inhibit the autoprotease activity of the CHIKV CP, and inhibit CHIKV replication in Vero cells. One of the compounds, eptifibatide acetate, a molecule developed to prevent blood clots, shows an EC_{50} of 4.01 μM in the cell-based assay.[22] Tripathi et al. focused their studies on the CHIKV nsP2, and they initially screened *in silico* a library of FDA-approved compounds. Two compounds were selected (novobiocin and telmisartan) and evaluated in different biochemical, biophysical, and cell-based assays, showing that both molecules were able to bind to the nsP2, inhibit its protease activity, and have a modest antiviral activity in a plaque reduction assay.[23]

As mentioned above, one of the possible shortcomings of a repurposed drug that acts on a new antiviral target, is that it will also present the original biological effect for which it was developed, with, most likely, a more potent activity against the primary target, versus the newly identified one. This could represent an obstacle, sometimes insurmountable, in the clinical development of the compound for the new antiviral indication. However, this does not mean that the results obtained from a repurposing approach are not valid. The data gained could instead represent a useful starting point for the functionalization of the compound, to optimize its newly identified biological property, while reducing, or abolishing, the original biological effect. Clearly, the newly designed compounds would be new chemical entities that will need to go through the whole drug development process, losing the clinical advantage that a repurposed drug offers. However, there could be still some benefits, as the starting point of this new hit-to-lead optimization, the approved drug, is a high quality, well characterized compound, with plenty of valuable data available. An example of this approach is represented by the work reported by DeBono et al. The group had previously identified the antiviral properties of mifepristone (**8**) against VEEV, which acts by reducing the nuclear accumulation of the CP. However, the antiprogesterone

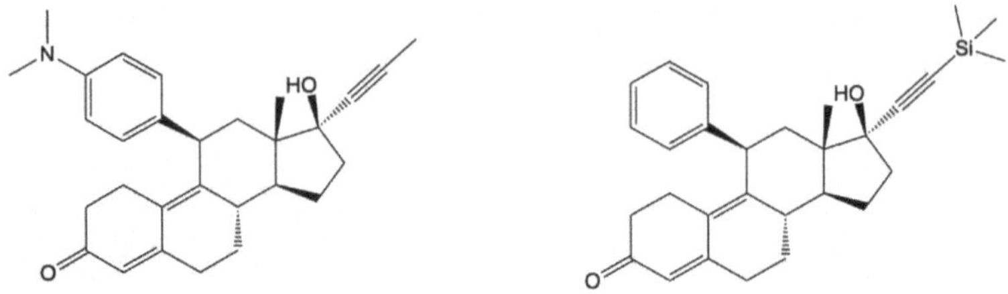

8 9

Fig. 3 Chemical modification of mifepristone (8) that lead to a more selective antiviral analog (9).

properties of the compound, which is approved for early pregnancy termination, make it unsuitable as an antiviral. For this reason, the researchers investigated the structure-activity relationships (SARs) of the compound to identify a new lead molecule (**9**) that showed an improved antiviral activity and a markedly reduced affinity for the progesterone receptor (Fig. 3).[24]

2.2 Medicinal chemistry strategies

Drug discovery is a highly multidisciplinary endeavor, and a key discipline is medicinal chemistry, which often dominates the early stages of the process: hit discovery, hit-to lead and lead optimization. A variety of tools are available to researchers to successfully navigate these phases, but the goal is always to design a compound with good efficacy, low toxicity, and a satisfactory set of properties (metabolic stability, water solubility, etc.) that will make the compound more likely to succeed in the clinic and become a useful therapeutic agent. This is clearly no different when we are looking at developing new antivirals for alphavirus infections. Indeed, the literature clearly shows how different groups are approaching this research. A good number of groups are using computer-aided drug design, especially in the context of hit identification, while classical SARs exploration strategies are preferred in the hit-to-lead stage, often supported by mechanism of action studies. In addition, following what can be seen also with the repurposing approach, hits are very often discovered by high-throughput screening, or direct evaluation of small, focused libraries of compounds. Yang et al. reported on the results of a phenotypic screening of a small library of substituted quinolones, where they identified a very interesting initial hit, that showed sub-micromolar activity against HCV. A series of analogs were then prepared through a systematic SARs exploration, leading to RYL-634 (**10**), a compound that not only showed activity in the original HCV assay, but proved to have a much broader spectrum, with antiviral activity against a diverse panel of viruses, including CHIKV (EC_{50} of 290 nM in Vero cells). Interestingly, they also investigated the possible biological target for the compound, providing good evidence that dihydroorotate dehydrogenase (hDHODH) inhibition is at least partially responsible for the observed antiviral activity.[25] Moesslacher et al. also reported on hit optimization efforts of a compound that was initially identified by high-throughput screening (**11**). The group built a SARs model by introducing logical modifications in the hit scaffold, resulting in several compounds that were able to inhibit CHIKV-induced cell death at low μM concentrations, in Vero cells.[26] In a different publication, they also describe the mechanism of action for this class of compounds, suggesting that they inhibit the viral mRNA capping, possibly interfering with the nsP1 protein.[27] A similar approach was also described by Ahmed et al., who described the optimization of a hit identified by high-throughput screening initially against VEEV, but which was then developed based on its antiviral activity against CHIKV (virus titer assay). Their lead compound (**12**) showed an EC_{90} of 270 nM in NHDF cells against CHIKV, and it also showed low/sub-micromolar activity against other alphaviruses (ONNV, MAYV, VEEV, RRV, UNAV), as well as other unrelated viruses (Dengue Virus and West Nile Virus).[28] Mode of action studies suggested that the biological target for the most active compound could be the hDHODH. Compound **12** was also evaluated *in vivo* against CHIKV, but modest results were obtained.[28] Gigante et al. also reported an extensive set of SAR studies around the original hit discovered through a biological screening against CHIKV in a cell-based assay.[29,30] The compounds seem to be active at the viral mRNA capping level, hence on the nsP1 protein, with resistance studies supporting this hypothesis.[31] One of the most interesting compounds of this family (**13**) showed a sub-micromolar antiviral activity against CHIKV (EC_{50} of 0.3 μM in Vero cells), but no activity against other alphaviruses was observed, despite showing some inhibition of the VEEV nsP1.[32] CHIKV nsP1 was also shown to be the target for the compounds Kovacikova et al. identified from a screening of adenosine analogs that were originally designed to be SAH hydrolase inhibitors. The two hits identified (6'-β-fluoro-homoaristeromycin and 6'-fluoro-homoneplanocin A) showed sub-micromolar activity in a cytopatic effect (CPE) reduction assay of CHIKV, in VeroE6 cells, and low micromolar activity against SFV.[33] Jonsson et al. reported a very interesting compound (**14**) with activity against VEEV both *in vitro* and *in vivo*.[34] The compound was derived by a hit identified by a high-throughput screening campaign,[35] and investigations on the mechanism of action on this family of molecules showed that they interfere with the viral RNA synthesis, possibly binding at the interface between nsP2 and nsP4.[36] The helicase domain of the VEEV nsP2 was instead indicated as the target of potent quinolinone analogs (**15**) reported by Haese et al. (IC_{90} of 0.77 μM in a CPE reduction assay, using NHDF cells), as a result of a large high-throughput screening campaign. Interestingly, the compound showed selectivity towards VEEV versus other alphaviruses.[37] (Fig. 4).

Virtual screening is now commonly used in the hit identification stages of drug discovery processes. Although with this approach the screening is performed on a computer, generating less accurate results compared to any other experimental method, it still represents a very attractive solution when a high-throughput assay is not readily available. Also, it is a considerably fast and economical technique, making it a very attractive tool to jump-start a new project, often on unexplored, more challenging, targets. Most virtual screening simulations include a molecular docking step, where the putative binding of the compounds from a virtual library onto the structure of the

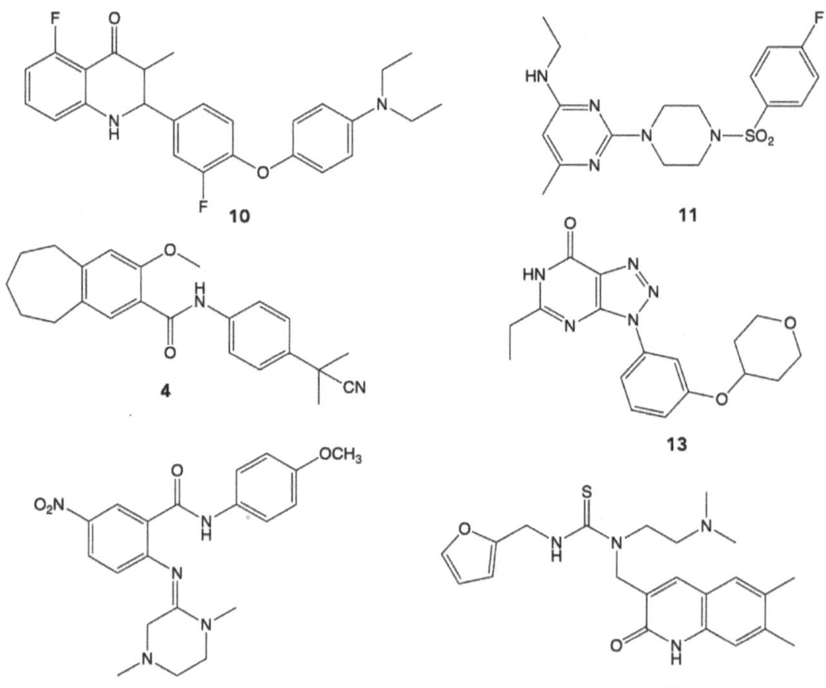

Fig. 4 Examples of anti-alphavirus compounds identified by experimental methods.

desired target is simulated and evaluated. One of the earliest examples of the application if these methods in the identification of novel alphavirus inhibitors is represented by the work reported by Bassetto et al. In the original work, the group performed a virtual screening simulation of a library of commercially available compounds, against the structure of the CHIKV nsP2. Some analogs of the original hit were prepared and evaluated, leading to a compound (**16**) with low micromolar activity in a CPE reduction assay of CHIKV, using Vero cells (EC_{50} of 3.2 μM).[38] The group has more recently expanded their work, extensively exploring the chemical space, unfortunately without significantly improving the antiviral activity for this class of compounds.[39,40] Nonetheless, this work was also followed up by Ivanova et al., who used different computer-aided approaches to design a new set of molecules, also active against CHIKV at a low micromolar level.[41] The same group also reported another compound, obtained by combining molecular docking and molecular dynamics, that inhibit CHIKV-NanoLuc replication (**17**, IC_{50} of 8.3 μM in BHK-21 cells).[42]

As mentioned before, the versatility of molecular modeling and structure-based drug design can be very useful when exploring less know targets, or new ways of engaging with a known protein. Battini et al., for example, have initially identified, *in silico*, a putative binding pocket for a small molecule at the interface between the CHIKV E1 and E2 proteins. They then used this information to run a virtual screening simulation on this pocket, using a library of commercially available compounds. The compounds were ranked according to a calculated consensus score, and the top 1% were visually examined to assess the interaction of the compounds with the key residues in the pocket. Finally, 7 compounds were evaluated against a reporter CHIKV variant (CHIKV-ZsGreen) in Vero cells. One compound proved to show some antiviral activity, and they then prepared a group of analogs to generate an initial SAR profile. The best compound (**18**) showed an EC_{50} of 1.6 μM, and it also showed to be active at the early stages of replication, probably interfering with viral entry. Furthermore, emergence of resistance through mutations in the envelope proteins further supported the initial design hypothesis.[43] Another very peculiar target was exploited by Shechter et al. The group focused their attention on the interaction between the VEEV CP and the host importin (IMP) α/β1 nuclear transport proteins. They initially identified the putative interaction pocket on IMPα2, and the short sequence of the VEEV CP that should bind into it, through a sophisticated computational approach. This allowed them to clearly define the site to be used in the docking simulation. A library of

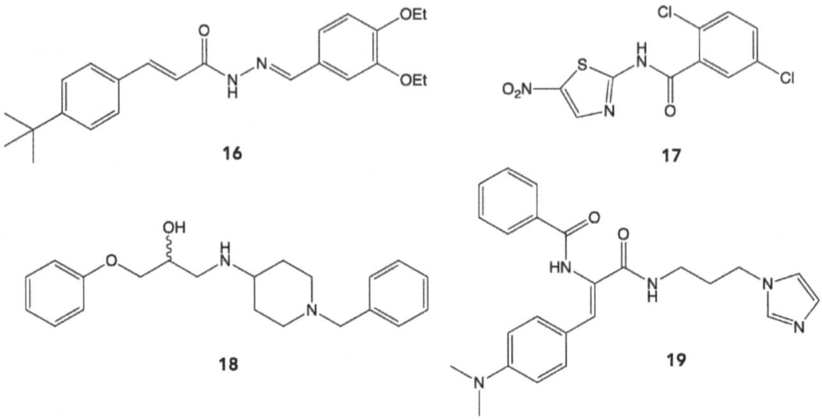

Fig. 5 Examples of anti-alphavirus compounds originally identified by *in silico* methods.

publicly available compounds was then virtually screened, ranked, and 84 of the top compounds were evaluated in a IMPα/β1:CP interaction assay. The four most active molecules were further evaluated, eventually leading to the lead compound (**19**) that showed a EC_{50} of 9.9 μM against VEEV replication in Vero cells. Furthermore, compound **19** proved to be inactive against a mutant VEEV that showed reduced affinity to IMPα/β1 (Fig. 5).[44]

Computer-aided drug design is also used at the hit-to-lead stage, although this is generally more effective if a structure of the hit has been co-crystallized with its target. Indeed, combining molecular modeling and structural biology is a very powerful way to identify high quality hits, which can then be optimized into very useful leads. An example of this approach can be seen in the work described by Zhang et al., that combined a virtual screening of a fragment library, using the CHIKV nsp3 macrodomain as target, with X-ray crystallography. In particular, the most promising fragments identified *in silico* were then evaluated for their binding to the macrodomain, and then soaked in the nsP3 crystal. This resulted in the identification of a fragment (2-pyrimidone-4-carboxylic acid) with a relevant activity, considering its size, against CHIKV replication (IC_{50} of 23 μM) in HFF cells. These results represent a promising starting point for the design of a very potent nsP3 macrodomain inhibitor.[45]

2.3 Natural compounds

Looking at nature to find a cure for a disease is how medicine has started thousands of years ago. Even nowadays, compounds from natural sources represent an important portion of the drugs we regularly use to treat a variety of conditions. Nature can readily offer very complex structures that could instead take a significant amount of time to be prepared following synthetic chemistry methods. This complexity could potentially lead to specific and potent biologically active compounds, albeit often with poor pharmacokinetic profiles. Nevertheless, the natural world still represents an important resource for drug discovery to this day. For this reason, it is not surprising that there are very active researchers in the field who are enthusiastically looking for anti-alphavirus compounds derived, very often, from plants.[46] The approach to search for a natural compound with antiviral activity is very often carried out by carefully selecting compounds with some interesting properties (e.g. anti-inflammatory, antioxidant), or that are traditionally considered to have curative effects, and test them in an antiviral assay, either as extracts or as purified molecules.[47] In the last few years, several groups have reported on the antiviral activity of a set of very diverse compounds, that have been cherry-picked to be evaluated against different alphaviruses. For example, silymarin and proanthocyanidin, both with antioxidant properties, were reported to be active against MAYV[48,49]; ginkgolic acid showed activity against MAYV, as well as UNAV and CHIKV[50]; quercetin, a compound that has already shown a variety of biological effects including antiviral properties, was reported to be active against ONNV.[51] For these compounds a mechanism of action is very often difficult to determine, as they possess a variety of biological properties. In some cases, however, natural compounds are selected to be evaluated as antivirals specifically because of their effect on some important cellular enzymes that could be somehow involved in viral replication. For example, Henss et al. found that silvestrol (**20**), an inhibitor of the host DEAD-box helicase eIF4A, was able to inhibit CHIKV replication at nanomolar concentrations (IC_{50} values of 1.89 and 5.06 nM in 293T and NIH3T3 cells, respectively). In addition, studies on the mode of action showed that silvestrol inhibited the translation of the viral mRNA that encodes the nsPs.[52] Hwang et al. also selected a natural compound with a clear biological activity: halofuginone (**21**), a synthetic derivative of the plant compound febrifugine, is a potent inhibitor of the host prolyl tRNA synthetase, and it can significantly inhibit CHIKV and ONNV replication in HFFs cells.[53]

Another approach commonly used to identify natural compounds with a desired biological activity is to screen focused libraries in specific biological assays. This strategy was successfully applied by Feiberlman et al. to screen a library of natural compounds against the CHIKV nsP1. Lobaric acid (**22**) proved to inhibit the viral mRNA capping and showed activity against

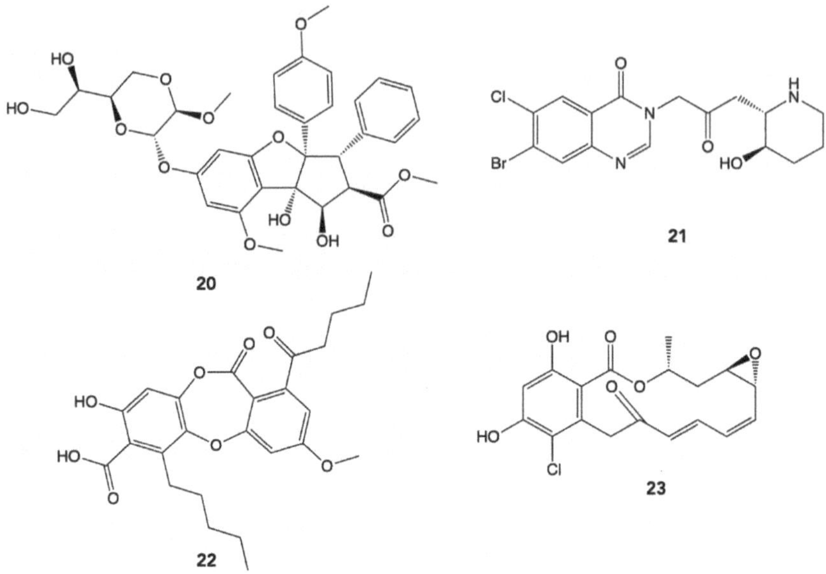

Fig. 6 Examples of natural compounds with anti-alphavirus activity.

CHIKV replication in BHK cells (EC_{50} of 9.9 μM).[54] Nam et al. performed a screening of 11,478 compounds in a CHIKV cell-based assay, and they found that radicicol (**23**), an inhibitor of the heat shock protein 90 (Hsp90), was able to completely abrogate plaque formation in infected Vero76 cells at a concentration between 0.2 μM and 2 μM (Fig. 6). From a mechanism point of view, resistance studies showed that mutations appeared on the nsP2 protein, on the domain that is believed to interact with Hsp90.[55]

3. Conclusions

Alphaviruses include a group of emerging ad re-emerging viruses that represent a clear unmet medical need. Scientists in the field, as discussed, clearly recognize this and an increasing number of research programs are now focused on the search for new antivirals that could be useful for the treatment of alphavirus infections. The recent examples reported in this chapter highlight the diversity of approaches that researchers are using to achieve this goal. Yet, these efforts have not delivered a clinically available drug. Repurposing approaches have indicated some promising compounds, and some of these may be clinically evaluated in the next few years.

However, these might not yet represent ideal drugs, as discussed above. The search for potent and specific anti-alphaviruses compounds should therefore start from drug design programs. Computer-aided drug design could provide a powerful set of tools, especially in combination with structure biology. However, a search on the Protein Data Bank (https://www.rcsb.org; accessed on the 7th of September 2021) shows that there are less than 300 crystal structures available from alphaviruses, with some important gaps in the proteins deposited. For example, the nsP4 protein, considered a highly important target for developing antiviral agents, has proven very difficult to crystallize. On this note, Cryo-Em methods may make an impact in this field in the near future, as has been the case in other contexts.[56] High-throughput screening approaches have also proven to be very useful in finding new hits, and medicinal chemists are rapidly transforming interesting hits into useful lead candidates. However, a very important obstacle is present in the development of new antiviral drugs for emerging and re-emerging diseases like the ones caused by alphaviruses. There is a clear, objective, difficulty to carry out a clinical trial to assess the clinical benefit of any potential anti-alphavirus drug, with challenges at every stage of the process, from protocol design to patient recruitment.[57] This would remain a significant obstacle even when, likely very soon, medicinal chemists and virologists will be able to identify new, much needed, antiviral compounds with a clear clinical potential to treat alphavirus infections.

References

1. Azar, S. R.; Campos, R. K.; Bergren, N. A.; Camargos, V. N.; Rossi, S. L. Epidemic Alphaviruses: Ecology, Emergence and Outbreaks. *Microorganisms* **2020**, *8* (8).
2. Shin, G.; Yost, S. A.; Miller, M. T.; Elrod, E. J.; Grakoui, A.; Marcotrigiano, J. Structural and Functional Insights Into Alphavirus Polyprotein Processing and Pathogenesis. *Proc. Natl. Acad. Sci. U. S. A.* **2012**, *109* (41), 16534–16539.
3. Abdelnabi, R.; Jacobs, S.; Delang, L.; Neyts, J. Antiviral Drug Discovery Against Arthritogenic Alphaviruses: Tools and Molecular Targets. *Biochem. Pharmacol.* **2020**, *174*, 113777.
4. Ching, K. C.; Ng, L. F. P.; Chai, C. L. L. A Compendium of Small Molecule Direct-Acting and Host-Targeting Inhibitors as Therapies Against Alphaviruses. *J. Antimicrob. Chemother.* **2017**, *72* (11), 2973–2989.
5. Abdelnabi, R.; Delang, L. Antiviral Strategies against Arthritogenic Alphaviruses. *Microorganisms* **2020**, *8* (9).
6. Kovacikova, K.; van Hemert, M. J. Small-Molecule Inhibitors of Chikungunya Virus: Mechanisms of Action and Antiviral Drug Resistance. *Antimicrob. Agents Chemother.* **2020**, *64* (12).
7. Hucke, F. I. L.; Bugert, J. J. Current and Promising Antivirals Against Chikungunya Virus. *Front. Public Health* **2020**, *8*, 618624.
8. Mello, M. V. P.; Domingos, T. F. S.; Ferreira, D. F.; Ribeiro, M. M. J.; Ribeiro, T. P.; Rodrigues, C. R.; Souza, A. M. T. Antiviral Drug Discovery and Development for

Mayaro Fever—What Do We Have So Far? *Mini Rev. Med. Chem.* **2020**, *20* (10), 921–928.
9. Furuta, Y.; Gowen, B. B.; Takahashi, K.; Shiraki, K.; Smee, D. F.; Barnard, D. L. Favipiravir (T-705), a Novel Viral RNA Polymerase Inhibitor. *Antiviral Res.* **2013**, *100* (2), 446–454.
10. Julander, J. G.; Dagley, A.; Gebre, M.; Komeno, T.; Nakajima, N.; Smee, D. F.; Furuta, Y. Strain-Dependent Disease and Response to Favipiravir Treatment in Mice Infected With Chikungunya Virus. *Antiviral Res.* **2020**, *182*, 104904.
11. Ferreira, A. C.; Reis, P. A.; de Freitas, C. S.; Sacramento, C. Q.; Villas Bôas Hoelz, L.; Bastos, M. M.; Mattos, M.; Rocha, N.; Gomes de Azevedo Quintanilha, I.; da Silva Gouveia Pedrosa, C.; Rocha Quintino Souza, L.; Correia Loiola, E.; Trindade, P.; Rangel Vieira, Y.; Barbosa-Lima, G.; de Castro Faria Neto, H. C.; Boechat, N.; Rehen, S. K.; Brüning, K.; Bozza, F. A.; Bozza, P. T.; Souza, T. M. L. Beyond Members of the Flaviviridae Family, Sofosbuvir Also Inhibits Chikungunya Virus Replication. *Antimicrob. Agents Chemother.* **2019**, *63* (2).
12. Dey, D.; Siddiqui, S. I.; Mamidi, P.; Ghosh, S.; Kumar, C. S.; Chattopadhyay, S.; Ghosh, S.; Banerjee, M. The Effect of Amantadine on an Ion Channel Protein From Chikungunya Virus. *PLoS Negl. Trop. Dis.* **2019**, *13* (7), e0007548.
13. Carey, B. D.; Bakovic, A.; Callahan, V.; Narayanan, A.; Kehn-Hall, K. New World Alphavirus Protein Interactomes From a Therapeutic Perspective. *Antiviral Res.* **2019**, *163*, 125–139.
14. Contu, L.; Balistreri, G.; Domanski, M.; Uldry, A. C.; Mühlemann, O. Characterisation of the Semliki Forest Virus-Host Cell Interactome Reveals the Viral Capsid Protein as an Inhibitor of Nonsense-Mediated mRNA Decay. *PLoS Pathog.* **2021**, *17* (5), e1009603.
15. Kaur, P.; Lello, L. S.; Utt, A.; Dutta, S. K.; Merits, A.; Chu, J. J. H. Bortezomib Inhibits Chikungunya Virus Replication by Interfering With Viral Protein Synthesis. *PLoS Negl. Trop. Dis.* **2020**, *14* (5), e0008336.
16. Hwang, J.; Wang, Y.; Fikrig, E. Inhibition of Chikungunya Virus Replication in Primary Human Fibroblasts by Liver X Receptor Agonist. *Antimicrob. Agents Chemother.* **2019**, *63* (9).
17. Langendries, L.; Abdelnabi, R.; Neyts, J.; Delang, L. Repurposing Drugs for Mayaro Virus: Identification of EIDD-1931, Favipiravir and Suramin as Mayaro Virus Inhibitors. *Microorganisms* **2021**, *9* (4).
18. Albulescu, I. C.; White-Scholten, L.; Tas, A.; Hoornweg, T. E.; Ferla, S.; Kovacikova, K.; Smit, J. M.; Brancale, A.; Snijder, E. J.; van Hemert, M. J. Suramin Inhibits Chikungunya Virus Replication by Interacting With Virions and Blocking the Early Steps of Infection. *Viruses* **2020**, *12* (3).
19. Mazzon, M.; Ortega-Prieto, A. M.; Imrie, D.; Luft, C.; Hess, L.; Czieso, S.; Grove, J.; Skelton, J. K.; Farleigh, L.; Bugert, J. J.; Wright, E.; Temperton, N.; Angell, R.; Oxenford, S.; Jacobs, M.; Ketteler, R.; Dorner, M.; Marsh, M. Identification of Broad-Spectrum Antiviral Compounds by Targeting Viral Entry. *Viruses* **2019**, *11* (2).
20. Lundberg, L.; Brahms, A.; Hooper, I.; Carey, B.; Lin, S. C.; Dahal, B.; Narayanan, A.; Kehn-Hall, K. Repurposed FDA-Approved Drug Sorafenib Reduces Replication of Venezuelan Equine Encephalitis Virus and Other Alphaviruses. *Antiviral Res.* **2018**, *157*, 57–67.
21. Ferreira-Ramos, A. S.; Li, C.; Eydoux, C.; Contreras, J. M.; Morice, C.; Quérat, G.; Gigante, A.; Pérez Pérez, M. J.; Jung, M. L.; Canard, B.; Guillemot, J. C.; Decroly, E.; Coutard, B. Approved Drugs Screening Against the nsP1 Capping Enzyme of Venezuelan Equine Encephalitis Virus Using An Immuno-Based Assay. *Antiviral Res.* **2019**, *163*, 59–69.
22. Fatma, B.; Kumar, R.; Singh, V. A.; Nehul, S.; Sharma, R.; Kesari, P.; Kuhn, R. J.; Tomar, S. Alphavirus Capsid Protease Inhibitors as Potential Antiviral Agents for Chikungunya Infection. *Antiviral Res.* **2020**, *179*, 104808.

23. Tripathi, P. K.; Soni, A.; Singh Yadav, S. P.; Kumar, A.; Gaurav, N.; Raghavendhar, S.; Sharma, P.; Sunil, S.; Ashish; Jayaram, B.; Patel, A. K. Evaluation of Novobiocin and Telmisartan for Anti-CHIKV Activity. *Virology* **2020**, *548*, 250–260.
24. DeBono, A.; Thomas, D. R.; Lundberg, L.; Pinkham, C.; Cao, Y.; Graham, J. D.; Clarke, C. L.; Wagstaff, K. M.; Shechter, S.; Kehn-Hall, K.; Jans, D. A. Novel RU486 (mifepristone) Analogues With Increased Activity Against Venezuelan Equine Encephalitis Virus But Reduced Progesterone Receptor Antagonistic Activity. *Sci. Rep* **2019**, *9* (1), 2634.
25. Yang, Y.; Cao, L.; Gao, H.; Wu, Y.; Wang, Y.; Fang, F.; Lan, T.; Lou, Z.; Rao, Y. Discovery, Optimization, and Target Identification of Novel Potent Broad-Spectrum Antiviral Inhibitors. *J. Med. Chem.* **2019**, *62* (8), 4056–4073.
26. Moesslacher, J.; Battisti, V.; Delang, L.; Neyts, J.; Abdelnabi, R.; Pürstinger, G.; Urban, E.; Langer, T. Identification of 2-(4-(Phenylsulfonyl)piperazine-1-yl)pyrimidine Analogues as Novel Inhibitors of Chikungunya Virus. *ACS Med. Chem. Lett.* **2020**, *11* (5), 906–912.
27. Abdelnabi, R.; Kovacikova, K.; Moesslacher, J.; Donckers, K.; Battisti, V.; Leyssen, P.; Langer, T.; Puerstinger, G.; Quérat, G.; Li, C.; Decroly, E.; Tas, A.; Marchand, A.; Chaltin, P.; Coutard, B.; van Hemert, M.; Neyts, J.; Delang, L. Novel Class of Chikungunya Virus Small Molecule Inhibitors That Targets the Viral Capping Machinery. *Antimicrob. Agents Chemother.* **2020**, *64* (7).
28. Ahmed, S. K.; Haese, N. N.; Cowan, J. T.; Pathak, V.; Moukha-Chafiq, O.; Smith, V. J.; Rodzinak, K. J.; Ahmad, F.; Zhang, S.; Bonin, K. M.; Streblow, A. D.; Streblow, C. E.; Kreklywich, C. N.; Morrison, C.; Sarkar, S.; Moorman, N.; Sander, W.; Allen, R.; DeFilippis, V.; Tekwani, B. L.; Wu, M.; Hirsch, A. J.; Smith, J. L.; Tower, N. A.; Rasmussen, L.; Bostwick, R.; Maddry, J. A.; Ananthan, S.; Gerdes, J. M.; Augelli-Szafran, C. E.; Suto, M. J.; Morrison, T. E.; Heise, M. T.; Streblow, D. N.; Pathak, A. K. Targeting Chikungunya Virus Replication by Benzoannulene Inhibitors. *J. Med. Chem.* **2021**, *64* (8), 4762–4786.
29. Gigante, A.; Gómez-SanJuan, A.; Delang, L.; Li, C.; Bueno, O.; Gamo, A. M.; Priego, E. M.; Camarasa, M. J.; Jochmans, D.; Leyssen, P.; Decroly, E.; Coutard, B.; Querat, G.; Neyts, J.; Pérez-Pérez, M. J. Antiviral Activity of [1,2,3]triazolo[4,5-d] pyrimidin-7(6H)-ones Against Chikungunya Virus Targeting the Viral Capping nsP1. *Antiviral Res.* **2017**, *144*, 216–222.
30. Gigante, A.; Canela, M. D.; Delang, L.; Priego, E. M.; Camarasa, M. J.; Querat, G.; Neyts, J.; Leyssen, P.; Pérez-Pérez, M. J. Identification of [1,2,3]triazolo[4,5-d] pyrimidin-7(6H)-ones as Novel Inhibitors of Chikungunya Virus Replication. *J. Med. Chem.* **2014**, *57* (10), 4000–4008.
31. Kovacikova, K.; Gorostiola González, M.; Jones, R.; Reguera, J.; Gigante, A.; Pérez-Pérez, M. J.; Pürstinger, G.; Moesslacher, J.; Langer, T.; Jeong, L. S.; Delang, L.; Neyts, J.; Snijder, E. J.; van Westen, G. J. P.; van Hemert, M. J. Structural Insights Into the Mechanisms of Action of Functionally Distinct Classes of Chikungunya Virus Nonstructural Protein 1 Inhibitors. *Antimicrob. Agents Chemother.* **2021**, *65* (7), e0256620.
32. Gómez-SanJuan, A.; Gamo, A. M.; Delang, L.; Pérez-Sánchez, A.; Amrun, S. N.; Abdelnabi, R.; Jacobs, S.; Priego, E. M.; Camarasa, M. J.; Jochmans, D.; Leyssen, P.; Ng, L. F. P.; Querat, G.; Neyts, J.; Pérez-Pérez, M. J. Inhibition of the Replication of Different Strains of Chikungunya Virus by 3-Aryl-[1,2,3]triazolo[4,5-d]pyrimidin-7(6 H)-ones. *ACS Infect. Dis.* **2018**, *4* (4), 605–619.
33. Kovacikova, K.; Morren, B. M.; Tas, A.; Albulescu, I. C.; van Rijswijk, R.; Jarhad, D. B.; Shin, Y. S.; Jang, M. H.; Kim, G.; Lee, H. W.; Jeong, L. S.; Snijder, E. J.; van Hemert, M. J. 6'-β-Fluoro-Homoaristeromycin and 6'-Fluoro-Homoneplanocin A Are Potent Inhibitors of Chikungunya Virus Replication Through Their Direct Effect on Viral Nonstructural Protein 1. *Antimicrob. Agents Chemother.* **2020**, *64* (4).

34. Jonsson, C. B.; Cao, X.; Lee, J.; Gabbard, J. D.; Chu, Y. K.; Fitzpatrick, E. A.; Julander, J.; Chung, D. H.; Stabenow, J.; Golden, J. E. Efficacy of a ML336 Derivative Against Venezuelan and Eastern Equine Encephalitis Viruses. *Antiviral Res.* **2019**, *167*, 25–34.
35. Schroeder, C. E.; Yao, T.; Sotsky, J.; Smith, R. A.; Roy, S.; Chu, Y. K.; Guo, H.; Tower, N. A.; Noah, J. W.; McKellip, S.; Sosa, M.; Rasmussen, L.; Smith, L. H.; White, E. L.; Aube, J.; Jonsson, C. B.; Chung, D.; Golden, J. E. Development of (E)-2-((1,4-Dimethylpiperazin-2-ylidene)amino)-5-nitro-N-phenylbenzamide, ML336: Novel 2-Amidinophenylbenzamides as Potent Inhibitors of Venezuelan Equine Encephalitis Virus. *J. Med. Chem.* **2014**, *57* (20), 8608–8621.
36. Skidmore, A. M.; Adcock, R. S.; Jonsson, C. B.; Golden, J. E.; Chung, D. H. Benzamidine ML336 Inhibits Plus and Minus Strand RNA Synthesis of Venezuelan Equine Encephalitis Virus Without Affecting Host RNA Production. *Antiviral Res.* **2020**, *174*, 104674.
37. Haese, N. N.; May, N. A.; Taft-Benz, S.; et al. Identification of Quinolinones as Antivirals Against Venezuelan Equine Encephalitis Virus. *Antimicrob. Agents Chemother.* **2021**, *65* (9), e0024421.
38. Bassetto, M.; De Burghgraeve, T.; Delang, L.; Massarotti, A.; Coluccia, A.; Zonta, N.; Gatti, V.; Colombano, G.; Sorba, G.; Silvestri, R.; Tron, G. C.; Neyts, J.; Leyssen, P.; Brancale, A. Computer-Aided Identification, Design and Synthesis of a Novel Series of Compounds With Selective Antiviral Activity Against Chikungunya Virus. *Antiviral Res.* **2013**, *98* (1), 12–18.
39. Giancotti, G.; Cancellieri, M.; Balboni, A.; Giustiniano, M.; Novellino, E.; Delang, L.; Neyts, J.; Leyssen, P.; Brancale, A.; Bassetto, M. Rational Modifications on a Benzylidene-Acrylohydrazide Antiviral Scaffold, Synthesis and Evaluation of Bioactivity Against Chikungunya Virus. *Eur. J. Med. Chem.* **2018**, *149*, 56–68.
40. Tardugno, R.; Giancotti, G.; De Burghgraeve, T.; Delang, L.; Neyts, J.; Leyssen, P.; Brancale, A.; Bassetto, M. Design, Synthesis and Evaluation Against Chikungunya Virus of Novel Small-Molecule Antiviral Agents. *Bioorg. Med. Chem.* **2018**, *26* (4), 869–874.
41. Ivanova, L.; Rausalu, K.; Ošeka, M.; Kananovich, D. G.; Žusinaite, E.; Tammiku-Taul, J.; Lopp, M.; Merits, A.; Karelson, M. Novel Analogues of the Chikungunya Virus Protease Inhibitor: Molecular Design, Synthesis, and Biological Evaluation. *ACS Omega* **2021**, *6* (16), 10884–10896.
42. Ivanova, L.; Rausalu, K.; Žusinaite, E.; Tammiku-Taul, J.; Merits, A.; Karelson, M. 1,3-Thiazolbenzamide Derivatives as Chikungunya Virus nsP2 Protease Inhibitors. *ACS Omega* **2021**, *6* (8), 5786–5794.
43. Battini, L.; Fidalgo, D. M.; Álvarez, D. E.; Bollini, M. Discovery of a Potent and Selective Chikungunya Virus Envelope Protein Inhibitor Through Computer-Aided Drug Design. *ACS Infect. Dis.* **2021**, 7 (6), 1503–1518.
44. Shechter, S.; Thomas, D. R.; Lundberg, L.; Pinkham, C.; Lin, S. C.; Wagstaff, K. M.; Debono, A.; Kehn-Hall, K.; Jans, D. A. Novel Inhibitors Targeting Venezuelan Equine Encephalitis Virus Capsid Protein Identified Using In Silico Structure-Based-Drug-Design. *Sci. Rep.* **2017**, 7 (1), 17705.
45. Zhang, S.; Garzan, A.; Haese, N.; Bostwick, R.; Martinez-Gzegozewska, Y.; Rasmussen, L.; Streblow, D. N.; Haise, M. T.; Pathak, A. K.; Augelli-Szafran, C. E.; Wu, M. Pyrimidone Inhibitors Targeting Chikungunya Virus nsP3 Macrodomain by Fragment-Based Drug Design. *PLoS One* **2021**, *16* (1), e0245013.
46. Martins, D. O. S.; Santos, I. A.; de Oliveira, D. M.; Grosche, V. R.; Jardim, A. C. G. Antivirals Against Chikungunya Virus: Is the Solution in Nature? *Viruses* **2020**, *12* (3).
47. Ledoux, A.; Cao, M.; Jansen, O.; Mamede, L.; Campos, P. E.; Payet, B.; Clerc, P.; Grondin, I.; Girard-Valenciennes, E.; Hermann, T.; Litaudon, M.; Vanderheydt, C.;

Delang, L.; Neyts, J.; Leyssen, P.; Frédérich, M.; Smadja, J. Antiplasmodial, Anti-chikungunya Virus and Antioxidant Activities of 64 Endemic Plants from the Mascarene Islands. *Int. J. Antimicrob. Agents* **2018**, *52* (5), 622–628.
48. Camini, F. C.; da Silva, T. F.; da Silva Caetano, C. C.; Almeida, L. T.; Ferraz, A. C.; Alves Vitoreti, V. M.; de Mello Silva, B.; de Queiroz Silva, S.; de Magalhães, J. C.; de Brito Magalhães, C. L. Antiviral Activity of Silymarin Against Mayaro Virus and Protective Effect in Virus-Induced Oxidative Stress. *Antiviral Res.* **2018**, *158*, 8–12.
49. Ferraz, A. C.; Moraes, T. F. S.; Nizer, W.; Santos, M. D.; Tótola, A. H.; Ferreira, J. M. S.; Vieira-Filho, S. A.; Rodrigues, V. G.; Duarte, L. P.; de Brito Magalhães, C. L.; de Magalhães, J. C. Virucidal Activity of Proanthocyanidin Against Mayaro Virus. *Antiviral Res.* **2019**, *168*, 76–81.
50. Campos, D.; Navarro, S.; Llamas-González, Y. Y.; Sugasti, M.; González-Santamaría, J. Broad Antiviral Activity of Ginkgolic Acid Against Chikungunya, Mayaro, Una, and Zika Viruses. *Viruses* **2020**, *12* (4).
51. Septembre-Malaterre, A.; Bedoui, Y.; Giry, C.; Gasque, P.; Guiraud, P.; Sélambarom, J. Quercetin Can Reduce Viral RNA Level of O'nyong-nyong Virus and Resulting Innate Immune Cytokine Responses in Cultured Human Synovial Fibroblasts. *Sci. Rep.* **2021**, *11* (1), 6369.
52. Henss, L.; Scholz, T.; Grünweller, A.; Schnierle, B. S. Silvestrol Inhibits Chikungunya Virus Replication. *Viruses* **2018**, *10* (11), 592.
53. Hwang, J.; Jiang, A.; Fikrig, E. A Potent Prolyl tRNA Synthetase Inhibitor Antagonizes Chikungunya and Dengue Viruses. *Antiviral Res.* **2019**, *161*, 163–168.
54. Feibelman, K. M.; Fuller, B. P.; Li, L.; LaBarbera, D. V.; Geiss, B. J. Identification of Small Molecule Inhibitors of the Chikungunya Virus nsP1 RNA Capping Enzyme. *Antiviral Res* **2018**, *154*, 124–131.
55. Nam, S.; Ga, Y. J.; Lee, J. Y.; Hwang, W. Y.; Jung, E.; Shin, J. S.; Chen, W.; Choi, G.; Zhou, B.; Yeh, J. Y.; Go, Y. Y. Radicicol Inhibits Chikungunya Virus Replication by Targeting Nonstructural Protein 2. *Antimicrob. Agents Chemother.* **2021**, *65* (7), e0013521.
56. Renaud, J. P.; Chari, A.; Ciferri, C.; Liu, W. T.; Remigy, H. W.; Stark, H.; Wiesmann, C. Cryo-EM in Drug Discovery: Achievements, Limitations and Prospects. *Nat. Rev. Drug Discov.* **2018**, *17* (7), 471–492.
57. De Rycker, M.; Baragana, B.; Duce, S. L.; Gilbert, I. H. Challenges and Recent Progress in Drug Discovery for Tropical Diseases. *Nature* **2018**, *559* (7715), 498–506.

CHAPTER FIVE

Inhibitors of Ebolavirus targeting innate immune evasion

Anna Laura Paulis[†], Angela Corona[†], and Enzo Tramontano[*,†]
Department of Life and Environmental Sciences, University of Cagliari, Monserrato, Italy
*Corresponding author: e-mail address: tramon@unica.it

Contents

1. Introduction

Ebola virus (EBOV) belongs the order *Mononegavirales*, *Filoviridae* family, genus Ebolavirus, which includes six species of enveloped, filamentous, linear non-segmented negative-sense RNA viruses (NNSVs): (EBOV), Sudan, Reston, Tai Forest, Bundibugyo, and Bombali viruses.[1–3] EBOV ssRNA- of approximately 19 kb that encodes for three structural proteins and seven non-structural proteins.[4,5] The viral glycoprotein (GP) mediates virion attachment and entry into the cell, the nucleoprotein forms the nucleocapsid with three other viral proteins (VP) VP30, VP35 and the large (L) RNA dependent RNA polymerase (RdRp), while VP24 and VP40 form the viral matrix. Other non-structural proteins are the secreted GP (sGP), the super small sGP (ssGP) and the D-peptide. These three proteins

† All the authors are contributed equally.

Annual Reports in Medicinal Chemistry, Volume 57
ISSN 0065-7743
https://doi.org/10.1016/bs.armc.2021.09.003
Copyright © 2021 Elsevier Inc.
All rights reserved.

are encoded by the virus but not present in the virion: sGP and ssGP are involved in the viral entry, while D-peptide is involved in the proteolysis process. Although their functions seem to be identified, their mechanism of action has yet to be fully explained.[5–8]

EBOV is the etiological agent of Ebola Virus Disease (EVD), a severe, most often fatal hemorrhagic fever that affects humans (HP) and non-humans primates (NHP). During the last decade, a major outbreak in West Africa spread between 2014 and 2016, that caused more than 11,000 deaths, was followed by recurring minor outbreaks, especially in Democratic Republic of the Congo (DRC). The last EVD epidemic event started in 2018, it is still ongoing and up to date has caused more than three thousand deaths only in DRC, and since 2021 it is also present in Guinea.[9] Due to the results of animal surveillance, both snakes and fruit bats are supposed to be natural reservoirs of this virus.[10] EBOV transmission follows a specific chain of events: first it is probably common an animal to animal infection, then the transmission occurs from animal to human and lastly from human to human. Several mechanisms are involved in this: consumption of potentially contaminated food, direct contact with body fluids (blood, saliva, sweat), vertical transmission (in utero and during delivery), contact with contaminated surfaces and materials.[11,12]

Once into the organism, EBOV preferentially infects macrophages and dendritic cells (DCs) following their differentiation from monocytes, causing high levels of cytokines and chemokines such as IL-1β, IL-6, IL-8, IL-10, MCP-1, MIP-1α, MIP-1β and TNFα that increase the overall endothelial permeability.[7] On the other hand, coagulation abnormalities are initiated early during EBOV infection followed by over-activation of the extrinsic pathway of coagulation that may promote disseminated intravascular coagulation.[13]

During the major outbreak in 2014–2016 the vaccine Ervebo, a recombinant vesicular-stomatitis virus-based vaccine, conferred immunogenicity against EBOV. Due to this result the European Medicine Agency (EMA) allowed its use during the 2018 outbreak in DRC as a possible transmission prevention tool available and for investigation. Erbevo was approved by FDA in 2020[14] together with Inmazeb, a monoclonal antibody cocktail that targets the EBOV glycoprotein (FDA, 2020). However, the road to the global protection against EVD is still long and would need a more systematic chronic treatment, economically sustainable, due to the persistence of residual virus in survivors that leads to long-term sequelae.[15]

The EBOV lethality is related to its ability to evade host innate immune response, especially during the early stages of infection.[16] It has been demonstrated that immune evasion functions of the EBOV VP35 protein are critical for virulence in NHPs.[17,18] In fact, a VP35 mutant EBOV does not cause lethal disease and instead elicits adaptive immune responses that can protect animals from wild-type EBOV challenge.[17] So as VP35, VP24 plays an important role on the evasion of innate immune response.[19] In fact, infection with a VP24 mutant EBOV results in marginal effects on the innate immune response pathways.[19] Moreover, pharmacological amplification of the innate immune response has been shown to overcome EBOV replication.[20] This proves that the IFN-beta pathway restoration by the block of viral components involved in immune evasion could lead to an effective antiviral approach.

2. Interferon activation

The innate immune response is the first, general, weapon developed against viral infections. It is structured in different levels of response due to the recognition of particular motifs termed pathogen o damage-associated molecular patterns (PAMPs and DAMPs), mediated by the so-called pattern recognition receptors (PRRs). These receptors recognize classical PAMPS and DAMPs such as pathogen cell wall components, including lipopolysaccharide (LPS), lipoteichoic acid, peptidoglycan, and glycans like β-1,3-glucan, conserved bacterial proteins such as flagellin and the type III secretion system (T3SS), and ribonucleic acid (RNA) and deoxyribonucleic acid (DNA).[21,22] Specifically, among nucleic acids, PRR recognizes the 5' triphosphate RNA, uncommon in host cytoplasmic RNA. Hence, once the pathogen succeeds in overcoming the physical barriers of the host defense, it is detected by PRRs and, depending on the PAMP or DAMP identified, these receptors trigger the production of interferons (IFNs) and cytokines through different pathways.

RNA virus infection leads to the activation of RLRs, a family of proteins characterized by the ability to bind specific RNA motifs, and establish an antiviral state through an ATP dependent mechanism.[23] The RLRs family includes RIG-I, MDA5 and laboratory of Genetics and Physiology 2 (LGP2). RIG-I and MDA5 present two caspase activation and recruitment domain (CARD) domains, a helicase domain, and a C-terminal domain (CTD), while LGP2 lacks the CARD domains[24] The CTD has a pocket binding specifically to 5′-PPP or 5′-PP groups and also contacts the unmethylated 2′-O group of the first nucleotide. Once the RNA is bound and detected, RIG-I and MDA5 recruit the mitochondrial anchored adaptor mitochondrial antiviral-signaling protein (MAVS) through the CARD domain, then trigger the activation of TBK1 and IKKε to activate

NF-κB and IRF family members, resulting in the expression of type I and III IFNs as well as pro-inflammatory cytokines and chemokines.[24,25] The type I IFNs, mostly produced by macrophages and DCs, induce an antiviral state in the neighboring cells determining the production of hundreds of IFN stimulated genes (ISGs).[26]

3. Interferon signaling

The IFN released binds to IFNα/β receptors (IFNAR1/2) on the surface of neighboring cells, these receptors are permanently associated with kinases belonging to the family Janus (JAK). The JAK/STAT pathway comprehends four members, JAK 1, 2, 3 and tyrosine kinase (Tyk) and seven Signal Transducer and Activation of Transcription (STAT 1–4, 5a, 5b and 6).[27] The activation of IFNARs determines the phosphorylation of tyrosine residues on the receptor tail,[28] this causes the recruitment of STAT1 and STAT2 through phosphorylated Tyk2 and JAK1.[29] Upon binding with JAK1, STATs are phosphorylated by JAKs and are released as heterodimer of phosphorylated STAT1 and STAT2.[30] The STAT1-STAT2 heterodimer interacts in the cytoplasm with the IRF9 protein forming a complex called IFN-stimulated gene factor 3 (ISGFR3) STAT1-STAT2 that is translocated into the nucleus due to the activity of some of the karyopherin alpha (KPNA) proteins, namely, the NPI-1 KPNAs (KPNA1, KPNA5 and KPNA6). The complex binds a specific sequence in the DNA called IFN Stimulated Responsive Element (ISRE) leading to transcription of ISGs.[31,32] Such ISGs can be divided into two categories due to their role, either antiviral "effectors" response or "regulators" of the response. The first ISGs improve the viral recognition and target the viral life cycle to stop viral infection and recruit immune cells, while second ISGs are essential in returning to normal balance cellular homeostasis post-infection, modulating the anti-viral response that could potently damage the host.[30]

Viruses have developed mechanisms to evade the host detection through several mechanisms, among which are (i) proteins that bind viral RNA reducing its exposition to RLRs and (ii) inhibitors of the IFN signaling. EBOV encodes two proteins, VP35 and VP24, that serve as potent suppressors of the RLRs mediated IFN production and the IFN signaling pathway. In particular, VP35 blocks the IFN production pathway triggered by viral RNA detection[33] while VP24 inhibits the IFN signaling sequestering karyopherin α1 and blocking the STAT1 system. In addition, as recently discovered, VP24 is also able to antagonize STAT3 both in its homodimeric and heterodimeric form.[34,35]

Given the key role of VP35 and VP24 in EBOV virulence,[36–39] a number of functional studies involving amino acidic substitutions coupled with crystallographic characterizations have been performed, defining the residues principally involved in the function of IFN antagonism for both proteins. Both EBOV proteins are considered validated drug targets.

4. VP35

VP35 is a 340 amino acid protein that contains three main functional domains: (i) the N-terminus, that possesses an NP-chaperoning domain[40] (PDB 4YPI), (ii) a homo-multimerization domain (PDB 6GBP), able to form various multimers, tetramers and trimers in solution[41–43] and (iii) the C-terminus domain presenting the IFN Inhibitory domain that binds dsRNA (RBD/IID) and is required to inhibit the RLRs signaling (PDB 3L25)[36,37,44,45] (Fig. 1). Similarly to its orthologous counterpart in the *Mononegavirales* order, VP35 is a key component of the viral RdRp complex.[46] VP35, bridging NP and RdRP, interacts with the NP-chaperoning domain, to prevent NP non-specifical associations with RNA, keeping it in the monomeric form and mediates the delivery of the NP-encapsidated

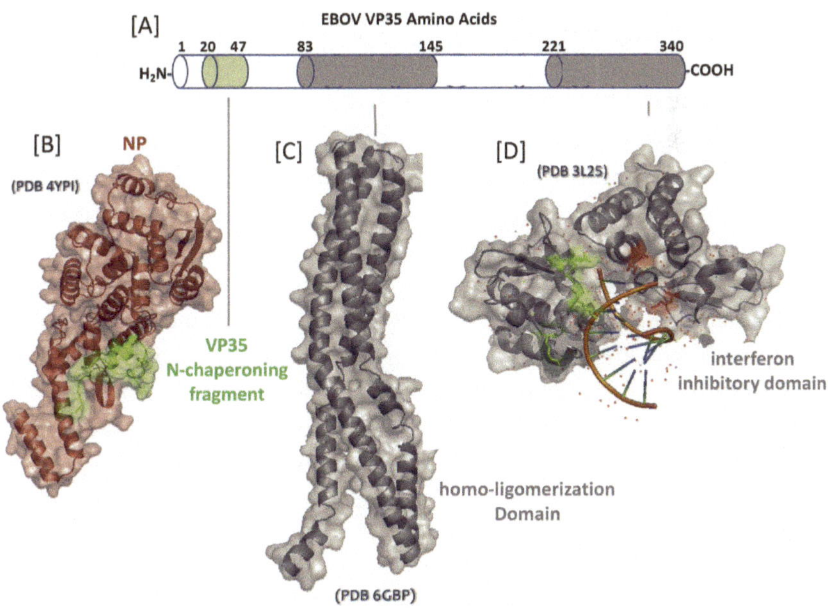

Fig. 1 VP35 protein structure. (A) VP35 subdomains; (B) structure of Ebola virus nucleoprotein N-terminal fragment (in orange) bound VP35 N-chaperoning fragment; (C) oligomerization domain of VP35; (D) VP35 interferon inhibitory domain with "backbone capping residues" (green) and "end capping residues" (red), in complex with dsRNA.

template to viral polymerase complex for viral RNA synthesis. In addition, VP35 has been demonstrated to interact with the N-terminal region of the viral L protein but this molecular function remains to be characterized.[47] In this contest, VP35 is also involved in viral genome packaging and nucleocapsid formation[48]Recently discovered VP35-associated NTPase and helicase-like activities suggest its potential involvement also in viral RNA remodeling.[49] Of note, its homo-multimerization domain is responsible for binding the viral polymerase L[50,51] and it is also required for full IFN antagonism.

In the VP35 C-terminus domain the preservation of the IID structure is critical for EBOV. Crystallographic analyzes that solved the structures of EBOV IID in complex with dsRNA[52] and Reston IID dsRNA[53] revealed that VP35 uses a bimodal strategy to bind the dsRNA. The protein acts as a functional dimer where one monomer binds the terminal nucleotides and the proximal phosphate backbone ("backbone-capping") while a second monomer binds the sugar-phosphate backbone of both dsRNA strands ("end-capping") (Fig. 1D). IID mutations have been shown to suppress the ability of VP35 to bind to dsRNA and further resulted in loss of its IFN antagonistic function.[37] In the "backbone-capping" monomer, R305, K309, R312 and K339 amino acid residues directly interact with the dsRNA phosphate backbone, while in the "end-capping", amino acid residues K339, Q274 and I278 bind dsRNA terminal nucleotides.[52,53] Mutations in residues R305, K309, R312, K319, R322, F239 and K339 strongly affect suppression of IFN-β.[37–39,52,54–57] Interestingly, VP35 double mutant R319A/K322A (KRA) EBOV has been shown to be completely avirulent in guinea pigs,[38] and the simultaneous presence of F239A, K319A, and R322A mutations (VP35m), in EBOV decreased the viral growth rate and attenuated the virulence in animal models.[17] Moreover, the interaction with the host protein 8 kDa dynein light chain (DLC8 or LC8) enhances VP35 stability during viral replication, while the VP35 IID binds also the dsRNA-binding protein, 76 KD (DRBP76; also referred to as NFIL3, NF90 and other names)[58]

The VP35 suppression of the IFN-b production is exerted at several levels of the RLR cascade (Fig. 2), by preventing the RLR activation by sequestering viral dsRNA,[37] formed as viral replication intermediate, and also inhibiting the RIG-I cascade at several levels[7,59] interacting (i) with RNA-activated protein kinase (PKR) activator (PACT) to disable its interaction with RIG-I[60,61]; (ii) with TBK-1 and IKKε[62] to block their interactions with their substrate IRF-3; (iii) with both IRF-3 and IRF-7 to promote their SUMOylation,[63] hence blocking type I IFN production mediated by tripartite motif-containing protein 6 (TRIM6),[64–67] and (iv) repressing the stress granule assembly.[68,69]

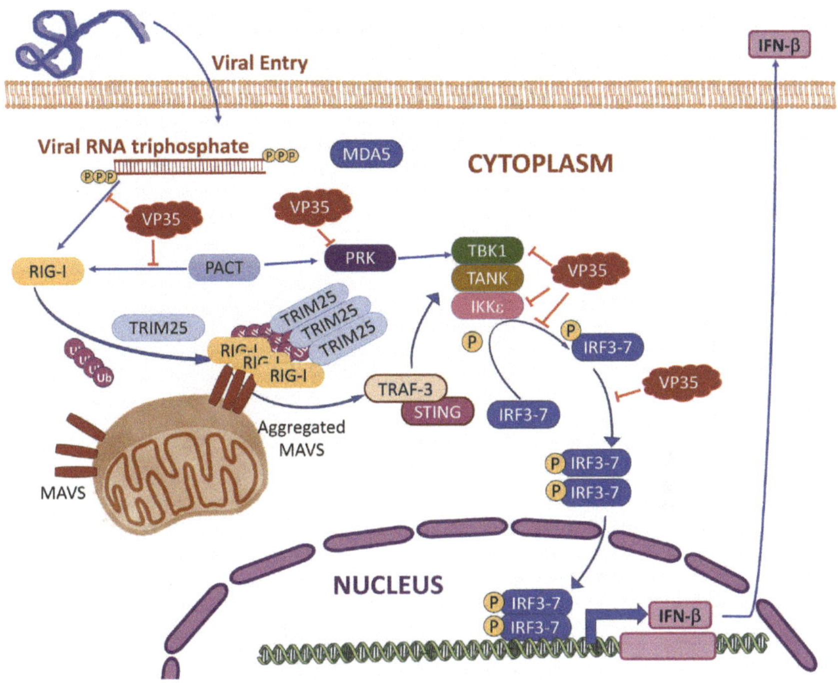

Fig. 2 Scheme of VP35 inhibition of IFN-α/β production.

5. VP24

VP24 is a 251 amino acid protein required for the nucleocapsid formation of EBOV virions in the cytoplasm of infected cells together with NP, VP30, VP35, and the L protein.[70] VP24 facilitates the condensation of the viral nucleocapsid, interacting with NP protein through the amino acids V170 and N171 (Fig. 3), thereby ensuring its efficient packaging and the production of infectious virions. Indeed only NP, VP35 and VP24 have been shown to be required to produce structures that are morphologically indistinguishable from nucleocapsids observed during EBOV infection.[71] VP24 has been also found to be present in the nucleus, suggesting that it undergoes nucleocytoplasmic trafficking regulating its activity. In fact, VP24 contains a nuclear export sequences (NES) signal, required for the export of the protein from the nucleus to the cytoplasm where is required for the viral particles formation.[72] The secondary, but relevant, activity of VP24 is the inhibition of the cellular IFN signaling cascade through the interaction with the NPI-1 subfamily of karyopherin-α proteins (KPNα1, KPNα5, and KPNα6) and by blocking the translocation of PySTAT-1 to

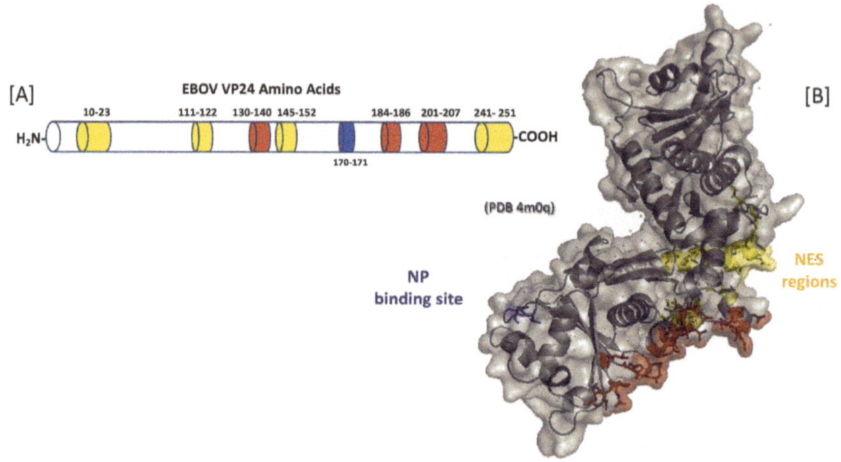

Fig. 3 VP24 structure (gray) with residues critical for karyopherin-α interaction (red), NES signals (yellow), NP binding site (blue).

the nucleus.[31,73,74] The interaction with KPNαhas been characterized by crystal structure analysis[75] and involves an hydrophobic portion including amino acid residues L115, L121, D124, W125, T128, T129, N130, T131, N135, R137, T138, R140, Q184, N185, H186, L201, E203, P204, D205 and S207 (Fig. 3).[75] However, also amino acid residues 142–146 were reported to be important for the ISG54 inhibition, since mutations at residues 142–146 were able to increase the expression of ISG54 reporter up to 90%, thus drastically reducing VP24 anti-IFN activity.[74]

As demonstrated by Vidal et al, VP24 undergoes SUMOylation.[76] This modification is essential for the stability of the protein, and in fact VP24 mutated in the putative SUMOylation site has a significantly reduced ability to interact with KPNA5 and to block the IFN activity, it exhibits decreased stability, and shows a reduced capacity to inhibit IFN-induced nuclear translocation of phosphorylated STAT1.[76] In a recently published work, VP24 has been shown to interact with STAT3, inhibiting its activity, in infected cells; the two proteins co-localize in cytoplasmatic viral replication/ inclusion bodies, suggesting that STAT3 is included into these virions. The interaction was also confirmed in a cell line defective for IFN, Vero cells, excluding possible impact by virus-induced type-I IFN. The interaction does not affect STAT3 ability to bind DNA, and is moreover STAT1-independent. Further studies are required to determine how this interaction can affect innate immune response in viral infected cells[35] (Fig. 4).

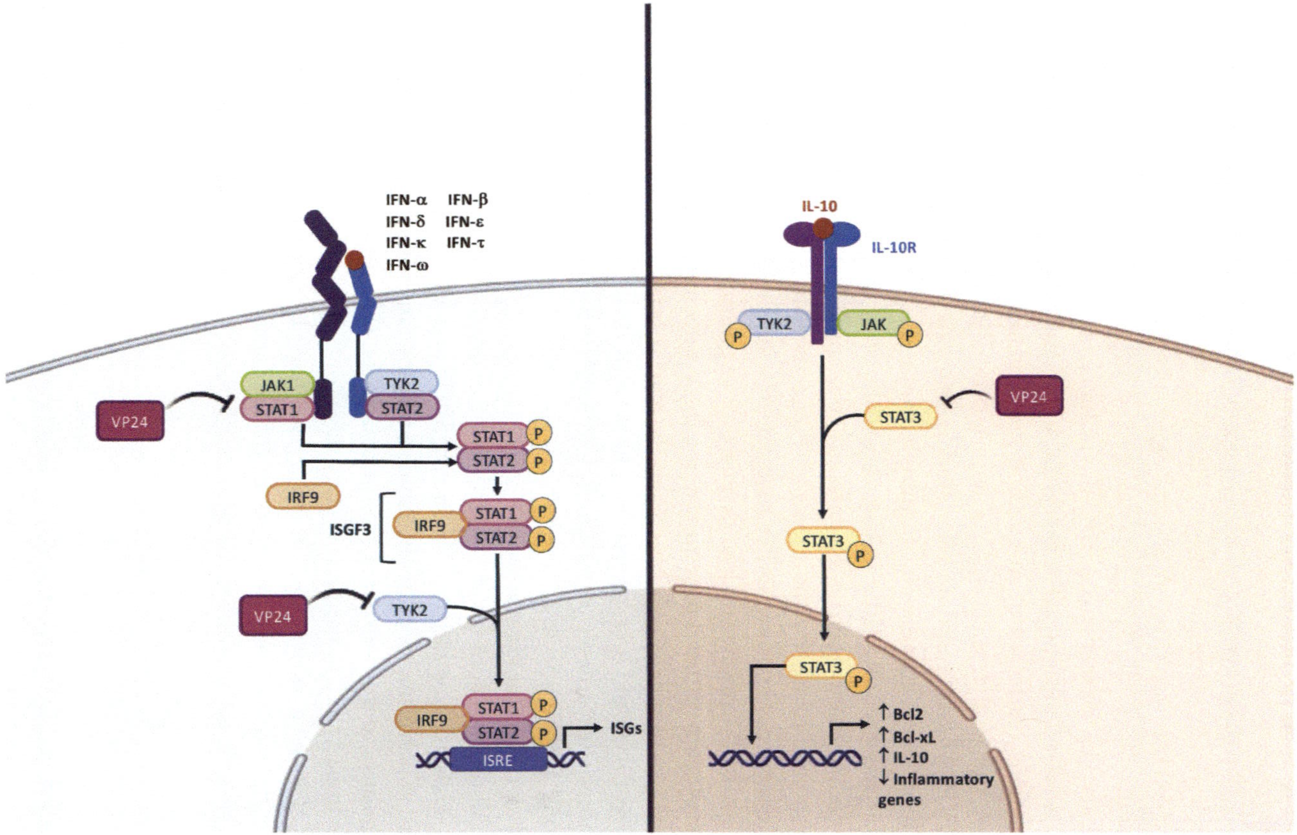

Fig. 4 Scheme of EBOV VP24 inhibition of IFN-α/β signaling.

6. EBOV inhibitors targeting innate immune evasion

The suppression of innate immune response is critical during EBOV infections. As mentioned above, EBOV has evolved two principal strategies to avoid the interruption of viral replication cycle: VP35 and VP24 viral proteins block the IFN production and the IFN signaling, respectively, and are key elements for EBOV virulence. Several mutational studies at the level of binding sites for both VP35 and VP24 have demonstrated how their contribution in evading host innate immunity is important to allow the viral replication. As reported by Ilinykh et al. disabling VP35 and VP24 innate response antagonizing domains (IRAD) result in the modification of the expression of multiple genes involved in maturation of DC and the induction of the innate and adaptive immune responses in a time dependent manner and with different consequences for the two proteins.[19] On the one hand, the abolition of VP35 activity increases the expression of multiple genes involved in DC maturation,[77] on the other hand disabling the VP24 IRAD resulted in the decrease of expression of many genes whose level was upregulated in response to EBOV infection. The combined effect of the two mutated IRAD increases the levels of transcripts of almost all subtypes of IFNs in DC compared to the wt EBOV infection[19] The characterization of the differences of binding affinity of the complex VP24- KPNA among EBOV species demonstrated that a higher affinity corresponds to higher IFN inhibitory activity.[78]

All of these considerations lead to the conclusion that these two proteins are validated drug targets.[36–39] In fact, VP35 and VP24 are currently investigated as promising drug targets, since a therapy targeting their IFN inhibitory functions and the consequent restoration of the IFN response will have a positive impact in the treatment of early stages of EVD. In the last decade a number of *in silico, in vitro* and *in vivo studies* have been performed, identifying potential active compounds among synthetic and naturally derived molecules.

6.1 VP35 inhibitors

Since the determination of the first VP35 IID crystal structure,[45] the purification of the first soluble His-tagged full length rVP35[79] and the availability of effective pull-down assays,[37,55] *in silico* and *in vitro* studies have been carried out to identify ligands possibly able to inhibit VP35 RdRp cofactor activity and revert itsIFN inhibitory effect.

At first, a screening of FDA approved drugs docked on the VP35 structure (PDB ID: 4IBI),[80] lead out to a high LibDock score for *amodiaquine* and *cloroquine* (Fig. 5). Interestingly those molecules previously demonstrated anti-EBOV activity.[81] Later, the compassionate treatment of EVD patients with a combination of amodiaquine-artesunate resulted in an increased survival,[82] while no increase of the survival benefit was registered for treatment of animals or EVD patients with cloroquine.[83–85]

Later, Brown and colleagues docked a total of 5.4 million drug-like small molecules from the ZINC library to the VP35 IID (PDB ID: 3FKE), and identified and co-crystallized *N*-aryl pyrrole-containing compounds bound to the VP35 IID, showing that they bind near a pocket formed by residues from the alpha-helical and beta-sheet subdomains, including amino acid residues A221, R225, Q241, L242, K248, K251, P293, I295, I297, D302, and F328, and proved they are be able to disrupt the interaction between VP35 and NP, so affecting VP35 polymerase cofactor function[86] and that compound *GA017* (Fig. 5) is able to block over 90% of viral replication at 125 μM concentration. Unfortunately, the impact of such compound on VP35 IFN antagonist function was not investigated. Later, a sensitive in vitro high-throughput screen was developed to identify inhibitors targeting the VP35 – NP interface[87]

Differently, Glanzer and colleagues, applied an ensemble-flexible strategy reducing the search space for docking to CBP of VP35 IID (PDB ID: 3FKE) and in particular the molecular space around the amino acid residues R305, K309, R312, K319, R322 and K339, known to be involved in the dsRNA binding, among which residues R305, K309, R322 and K339 were designated as flexible. Among the 80,000 compounds, ZINC05328460 (Fig. 5) showed the highest predicted affinity showed the greatest inhibition of VP35-dsRNA binding ($IC_{50} = 4$ μM).[88]

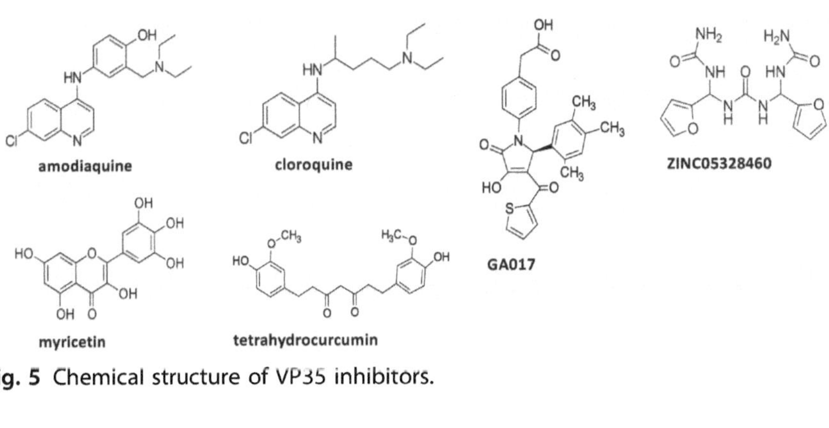

Fig. 5 Chemical structure of VP35 inhibitors.

Also, VP35 IID (PDB ID: 3FKE) was used for a pharmacophore based virtual screening predicting a potential high binding affinity for FDA-approved deslanoside and digoxin as well as the deslanoside analogue ZINC77291634, with predicted binding energies of −12.9 kcal/mol, −12.2 kcal/mol, and −10.0 kcal/mol, respectively.[89] A further pharmacophore model (HypoA of 4-acetyl-3-hydroxy-1-phenyl-1H-pyrrol-2 (5H)-one derivates) was used to identify seven VP35 potential inhibitors (cpd 1–7) with 3D QSAR prediction and molecular docking method.[90]

Later, Daino and colleagues that utilizing a fluorescent dsRNA oligomer ligand, found that *epigallocatechin 3-O gallate* (EGCG) and the flavonol *myricetin* (Fig. 5) are able to suppress the VP35-dsRNA binding with an IC_{50} values of 43.5 μM and 2.7 μM, respectively.[91] Of note, among natural products, also the curcuminoid metabolite *tetrahydrocurcumin* (Fig. 5) was predicted by molecular dynamics simulations to potentially bind EBOV VP35 (PDB ID: 4IBB).[92]

The effect of the VP35 inhibition of the RIG-I pathway was investigated by development of gene-reporter assays,[93] used to screen natural extracts identifying some activity form the extract from the leaves of *Asphodelus microcarpus* Salzm.et Vivi against the viral protein. The extract was able to counteract the IFN VP35 inhibitory effect, partially restoring IFN production.[93,94]

The possibility to Inhibit VP35 was investigated also by cell-penetrating peptides (CPPs). Serving as "intracellular blockers." These molecules were developed to the aim of facilitating the entry of active compounds into cells.[95,96] Differently from conventional antibodies that are unable to target viral proteins localized within the cytoplasm, these molecules take advantage of electrostatic interactions between the negatively charged membrane components and their positive charged portions, such as CPP the nona-arginine (R9), displaying a lack of cytotoxicity that would make transbodies-based therapy a promising approach compared to the current antibodies treatments.[97,98] Seesuay and colleagues used R9 to mediate the transport of human single chain variable antibodies fragments (HuscFvs) or transbodies directed against VP35 that specifically recognize the IID function.[97]

HepG2 cells transiently expressing EBOV VP35 were treated with R9-HuscFv3 and R9-HuscFv8 and gene expression of IFN-β promoter and EIF2AK2, encoding for PKR, were evaluated, showing a restoration of IFN-β promoter and EIF2AK2 expression compared to the expression of IFNB1 and EIF2AK2 in not treated VP35-transducted cells, leading to the host innate immunity rescue.[97]

R9-HuscFv3 was predicted to interact with IID residues of the CBP (R305, K309, R312, R322, and K339), the end-cap motif (F235, F239, and I340), other residues known to interact with the dsRNA (P233, T237, Q274, C275, I278, A306, and S310), and also with D271.[97] R9-HuscFv8 was shown to interact with the CBP residues R322 and K339, the end-cap residue I340 and with amino acid residues S272, C275, I278, and Q279, important for the VP35-dsRNA binding, as well as other residues (E269, Q270, and D271).[97] In addition, the lack of cytotoxicity renders transbodies-based therapy a promising approach compare to the current antibodies treatments.[97,98] Similarly, Flego and colleagues targeted VP35 in the cytoplasm by as single chain fragment variable (scFv) identified by Phage display technology, all scFvs were expressed in cell cytoplasm as intrabodies; a luciferase reporter gene inhibition assay performed in A549 cells showed that two of the scFvs can significantly hamper the inhibition of the IFN-β-induced RIG-I signaling cascade mediated by EBOV VP35.[99] All the selected scFvs were able to interact with VP35 in ELISA assys, one of the scFvs being also able to react in Western Blot assay (WB).

6.2 VP24 inhibitors

With the aim of developing VP24 inhibitors, several studies have been performed leading to the identification of a number of promising small molecules (Fig. 6).

The investigation of an FDA-approved database of antivirals, identified the HIV protease inhibitor, *indinavir* (Fig. 6), as potential VP24 interactor due to the putative three hydrogen-bonding interactions between: (i) the O2 atom of Indinavir and the nitrogen atom from the sidechain of residue Gln94 in VP24, (ii) the atom N4 of Indinavir and the oxygen atom in the sidechain of VP24's Gln94, and (iii) the O4 atom of Indinavir and the oxygen atom from the main chain of Gln94 of VP24.[100]

An *in silico* study have designed antisense phosphorodiamidate morpholino oligomers (PMOs) to target Ebola EBOV virus genes.[101] AVI-7537 is a PMO*plus* that binds directly to the viral VP24 transcript RNA blocking the protein production and showing 61% of protection from lethal challenges., Of note, meanwhile 100% of protection was reached with AVI-6003, a PMO combination of AVI-7287 (targeted to VP24) and AVI-7288 (targeted to NP).[101]

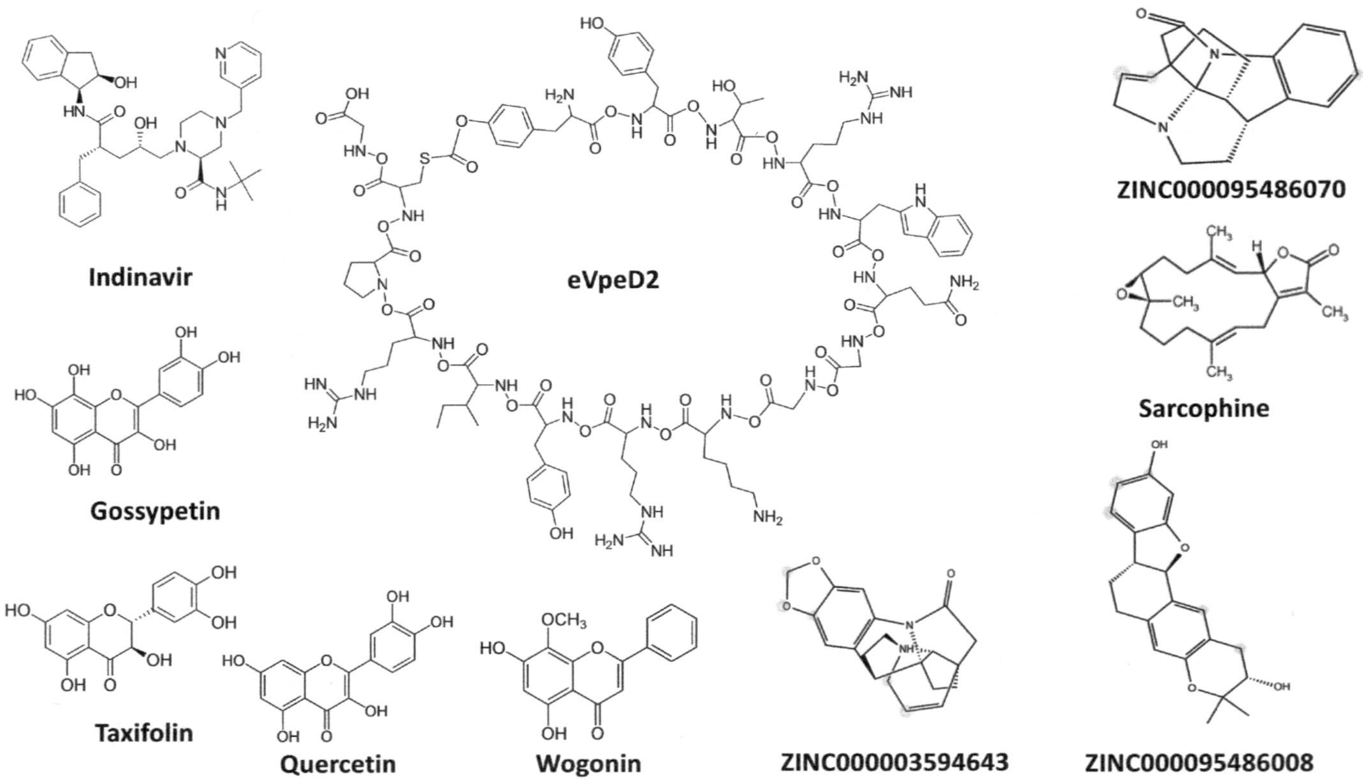

Fig. 6 Chemical structure of VP24 inhibitors.

Some macrocyclic peptides, generated with the by RaPID (Random non-standard Peptides Integrated Discovery) system are have been shown to be able to disrupt the interaction between eVP24 and KPNA5, demonstrating that this interaction is druggable and could lead to the production of unique anti-viral drugs.[102] The most promising one peptide, *eVpeD2* (Fig. 6), was able to inhibit this interaction in a dose-dependent manner, with an IC_{50} value of approximately 9 μM.[102]

On the basis of an *in silico* study that identified as potential EBOV inhibitors the natural compounds *Gossypetin* (Fig. 6), isolated from the flowers and the calyx of *Hibiscus sabdariffa*, and Taxifolin, present in *Taxus chinensis*, *Larix sibirica*, *Pinus roxburghii* and *Cedrus deodara*,[103] a library of flavonoids have been screened to identify molecules that shared a similar structure with the previously identified molecules. *Quercetin* and *Wogonin* were selected to be tested together with *Gossypetin*, *Taxifolin* (Fig. 6) and Tricetin in a luciferase assay to evaluate the inhibition of VP24 activity. The most interesting molecule was Quercetin, that showed a significant efficacy in restoring IFN stimulation in a dose-dependent manner. To further characterize its mechanism of action, Quercetin ability to block the VP24 inhibition of P-STAT1 translocation into the nucleus was evaluated performing an immunofluorescent assay, observing that it is able to block VP24 function and induce the recovery of P-STAT1 translocation. Since VP24 is known to interact with KPNA5, to determine whether Quercetin was able to inhibit this interaction, a co-immunoprecipitation was performed confirming a reduction in VP24/KPNA5 co-immunoprecipitation and indicating that Quercetin interferes with this interaction. Quercetin was then finally tested on EBOV replication in a IFN-susceptible cell-line model.[104] Of note, when with the derivative form Quercetin 3-β-O-D-glucoside (Q3G) derivative was tested on EBOV replication it was less efficient than the non-glucosylated form suggesting that Quercetin indeed mainly targets the VP24 anti-IFN function while Q3C probably mainly targets EBOV entry.[104]

Kwofie et al. with the purpose of identify putative lead compounds from the African flora with the potential to inhibiting the activity VP24 functions screened an integrated library of >7600 African natural products obtained from the AfroDB and NANPDB databases, using pharmaco informatics and molecular docking methodologies. For 19 of these compounds, it was characterized the physicochemical and pharmacological profiles, using ADMET Predictor™, SwissADME and DataWarrior, and the VP24 putative binding mechanism using LigPlot+, predicting the activity spectra to determine their anti-EBOV activity. Four natural products, namely *ZINC000095486070*, *ZINC000003594643*, *ZINC000095486008* and *sarcophine* (Fig. 6), were found to be potential VP24 inhibitors and will deserve further biological testing to verify their ability as potential EBOV biotherapeutic agents[105]

7. Conclusions

EVD is an endemic disease with a high fatality rate and long-term sequelae, that would require a systematic chronic treatment, enlightening the need to explore all the possible targets and mechanisms for the development of new versatile and cost-effective anti-EBOV therapeutics. The IFN-beta pathway is critical for counteracting viral infections and, EVD fatal outcome is strongly related to the viral ability to hijack the innate immune response. EBOV VP35 and VP24 proteins are key factors for viral replication and IFN-beta suppression, and also validated drug targets. Several small molecules were identified as able to block these viral components involved in immune evasion, showing that an antiviral approach targeted at the restoration of the innate immune response is a feasible and promising goal.

Acknowledgments

A.C. and E.T. FDS2019 EBOV CUPF72F20000270007.

References

1. Goldstein, T.; Anthony, S. J.; Gbakima, A.; et al. The Discovery of Bombali Virus Adds Further Support for Bats as Hosts of Ebolaviruses. *Nat. Microbiol.* **2018**, *3* (10). https://doi.org/10.1038/s41564-018-0227-2.
2. Kuhn, J. H. Guide to the Correct Use of Filoviral Nomenclature. *Curr. Top. Microbiol. Immunol.* **2017**, *411*, 447–460. https://doi.org/10.1007/82_2017_7.
3. Geisbert, T. W. Marburg and ebola hemorrhagic fevers (filoviruses). In *Mandell, Douglas, and Bennett's Principles and Practice of Infectious Diseases*; Elsevier, 2015. https://doi.org/10.1016/B978-1-4557-4801-3.00166-1.
4. Wan, W.; Kolesnikova, L.; Clarke, M.; et al. Structure and Assembly of the Ebola Virus Nucleocapsid. *Nature* **2017**, *551* (7680). https://doi.org/10.1038/nature24490.
5. Banerjee, A.; Pal, A.; Pal, D.; Mitra, P. Ebolavirus Interferon Antagonists-Protein Interaction Perspectives to Combat Pathogenesis. *Brief. Funct. Genom.* **2018**, *17* (6), 392–401. https://doi.org/10.1093/bfgp/elx034.
6. Kühl, A.; Pöhlmann, S. How Ebola Virus Counters the Interferon System. *Zoo. Publ. Health* **2012**, *59*. https://doi.org/10.1111/j.1863-2378.2012.01454.x.
7. Messaoudi, I.; Amarasinghe, G. K.; Basler, C. F. Filovirus Pathogenesis and Immune Evasion: Insights From Ebola Virus and Marburg Virus. *Nat. Rev. Microbiol.* **2015**, *13* (11). https://doi.org/10.1038/nrmicro3524.

8. Volchkova, V. A.; Klenk, H.-D.; Volchkov, V. E. Delta-Peptide Is the Carboxy-Terminal Cleavage Fragment of the Nonstructural Small Glycoprotein sGP of Ebola Virus. *Virology* **1999**, *265* (1). https://doi.org/10.1006/viro.1999.0034.
9. Centers for Disease Control and Prevention. *Centers for Disease Control and Prevention, National Center for Emerging and Zoonotic Infectious Diseases (NCEZID), Division of High-Consequence Pathogens and Pathology (DHCPP), Viral Special Pathogens Branch (VSPB)*; 2021.
10. Adekanmbi, O.; Ilesanmi, O.; Lakoh, S. Ebola: A Review and Focus on Neurologic Manifestations. *J. Neurol. Sci.* **2021**, *421*. https://doi.org/10.1016/j.jns.2021.117311.
11. Bebell, L. M.; Oduyebo, T.; Riley, L. E. Ebola Virus Disease and Pregnancy: A Review of the Current Knowledge of Ebola Virus Pathogenesis, Maternal, and Neonatal Outcomes. *Birth Def. Res.* **2017**, *109* (5). https://doi.org/10.1002/bdra.23558.
12. Williams, J.; Mai, C. T.; Mulinare, J.; Isenburg, J.; Flood, T. J.; Ethen, M.; Frohnert, B.; Kirby, R. S. Centers for Disease Control and Prevention. Updated estimates of neural tube defects prevented by mandatory folic Acid fortification - United States, 1995-2011. MMWR Morb Mortal Wkly Rep. 2015 Jan 16, *64*(1), 1-5. PMID: 25590678; PMCID: PMC4584791.
13. Geisbert, T. W.; Young, H. A.; Jahrling, P. B.; Davis, K. J.; Kagan, E.; Hensley, L. E. Mechanisms Underlying Coagulation Abnormalities in Ebola Hemorrhagic Fever: Overexpression of Tissue Factor in Primate Monocytes/Macrophages Is a Key Event. *J. Infect. Dis.* **2003**, *188* (11). https://doi.org/10.1086/379724.
14. Ollmann, S. E. A Vaccine Against Ebola Virus. *Cell* **2020**, *181* (1), 6. https://doi.org/10.1016/j.cell.2020.03.011.
15. Malvy, D.; McElroy, A. K.; de Clerck, H.; Günther, S.; van Griensven, J. Ebola Virus Disease. *The Lancet* **2019**, *393* (10174), 936–948. https://doi.org/10.1016/S0140-6736(18)33132-5.
16. Mahanty, S.; Bray, M. Pathogenesis of Filoviral Haemorrhagic Fevers. *Lancet Infect. Dis.* **2004**, *4* (8), 487–498. https://doi.org/10.1016/S1473-3099(04)01103-X.
17. Woolsey, C.; Menicucci, A. R.; Cross, R. W.; et al. A VP35 Mutant Ebola Virus Lacks Virulence but Can Elicit Protective Immunity to Wild-Type Virus Challenge. *Cell Rep.* **2019**, *28* (12), 3032–3046.e6. https://doi.org/10.1016/j.celrep.2019.08.047.
18. Zinzula, L.; Tramontano, E. Strategies of Highly Pathogenic RNA Viruses to Block dsRNA Detection by RIG-I-Like RECEPTORS: HIDE, Mask, Hit. *Antivir. Res.* **2013**, *100* (3). https://doi.org/10.1016/j.antiviral.2013.10.002.
19. Ilinykh, P. A.; Lubaki, N. M.; Widen, S. G.; et al. Different Temporal Effects of Ebola Virus VP35 and VP24 Proteins on Global Gene Expression in Human Dendritic Cells. *J. Virol.* **2015**, *89* (15). https://doi.org/10.1128/JVI.00924-15.
20. Jasenosky, L. D.; Cadena, C.; Mire, C. E.; et al. The FDA-Approved Oral Drug Nitazoxanide Amplifies Host Antiviral Responses and Inhibits Ebola Virus. *iScience* **2019**, *19*, 1279–1290. https://doi.org/10.1016/j.isci.2019.07.003.
21. Takeuchi, O.; Akira, S. Pattern Recognition Receptors and Inflammation. *Cell* **2010**, *140* (6), 805–820. https://doi.org/10.1016/j.cell.2010.01.022.
22. Briard, B.; Place, D. E.; Kanneganti, T.-D. DNA Sensing in the Innate Immune Response. *Physiology* **2020**, *35* (2). https://doi.org/10.1152/physiol.00022.2019.
23. Bruns, A. M.; Horvath, C. M. Antiviral RNA Recognition and Assembly by RLR Family Innate Immune Sensors. *Cytokine Growth Fact. Rev.* **2014**, *25* (5). https://doi.org/10.1016/j.cytogfr.2014.07.006.
24. Rehwinkel, J.; Gack, M. U. RIG-I-Like Receptors: Their Regulation and Roles in RNA Sensing. *Nat. Rev. Immunol.* **2020**, *20* (9). https://doi.org/10.1038/s41577-020-0288-3.

25. Wu, B.; Hur, S. How RIG-I Like Receptors Activate MAVS. *Curr. Opin. Virol.* **2015**, *12*. https://doi.org/10.1016/j.coviro.2015.04.004.
26. Pothlichet, J.; Meunier, I.; Davis, B. K.; et al. Type I IFN Triggers RIG-I/TLR3/NLRP3-Dependent Inflammasome Activation in Influenza A Virus Infected Cells. *PLoS Pathog.* **2013**, *9* (4). https://doi.org/10.1371/journal.ppat.1003256.
27. Aaronson, D. S. A Road Map for Those Who Don't Know JAK-STAT. *Science* **2002**, *296* (5573). https://doi.org/10.1126/science.1071545.
28. Brooks, A. J.; Dai, W.; O'Mara, M. L.; et al. Mechanism of Activation of Protein Kinase JAK2 by the Growth Hormone Receptor. *Science* **2014**, *344* (6185). https://doi.org/10.1126/science.1249783.
29. Decker, T.; Müller, M.; Stockinger, S. The Yin and Yang of Type I Interferon Activity in Bacterial Infection. *Nat. Rev. Immunol.* **2005**, *5* (9). https://doi.org/10.1038/nri1684.
30. Raftery, N.; Stevenson, N. J. Advances in Anti-Viral Immune Defence: Revealing the Importance of the IFN JAK/STAT Pathway. *Cell. Mol. Life Sci.* **2017**, *74* (14). https://doi.org/10.1007/s00018-017-2520-2.
31. Reid, S. P.; Leung, L. W.; Hartman, A. L.; et al. Ebola Virus VP24 Binds Karyopherin α1 and Blocks STAT1 Nuclear Accumulation. *J. Virol.* **2006**, *80* (11). https://doi.org/10.1128/JVI.02349-05.
32. Lukhele, S.; Boukhaled, G. M.; Brooks, D. G. Type I Interferon Signaling, Regulation and Gene Stimulation in Chronic Virus Infection. *Semin. Immunol.* **2019**, *43*. https://doi.org/10.1016/j.smim.2019.05.001.
33. Basler, C. F.; Wang, X.; Muhlberger, E.; et al. The Ebola Virus VP35 Protein Functions as a Type I IFN Antagonist. *Proc. Natl. Acad. Sci. U. S. A.* **2000**, *97* (22). https://doi.org/10.1073/pnas.220398297.
34. Shelemba, A. A.; Lushnikova, E. L.; Kolesnikov, S. I.; Nepomnyashchikh, L. M.; Chepurnov, A. A. Role of Ebola Virus vp24 Protein in Inhibition of Interferonogenesis. *Bull. Exp. Biol. Med.* **2016**, *160* (3), 350–352. https://doi.org/10.1007/s10517-016-3168-x.
35. Harrison, A. R.; Todd, S.; Dearnley, M.; et al. Antagonism of STAT3 Signalling by Ebola Virus. *PLoS Pathogens* **2021**, *17* (6). https://doi.org/10.1371/journal.ppat.1009636.
36. Leung, D. W.; Prins, K. C.; Borek, D. M.; et al. Structural Basis for dsRNA Recognition and Interferon Antagonism by Ebola VP35. *Nat. Struct. Mol. Biol.* **2010**, *17* (2). https://doi.org/10.1038/nsmb.1765.
37. Cárdenas, W. B.; Loo, Y.-M.; Gale, M.; et al. Ebola Virus VP35 Protein Binds Double-Stranded RNA and Inhibits Alpha/Beta Interferon Production Induced by RIG-I Signaling. *J. Virol.* **2006**, *80* (11). https://doi.org/10.1128/JVI.02199-05.
38. Prins, K. C.; Delpeut, S.; Leung, D. W.; et al. Mutations Abrogating VP35 Interaction With Double-Stranded RNA Render Ebola Virus Avirulent in Guinea Pigs. *J. Virol.* **2010**, *84* (6). https://doi.org/10.1128/JVI.02459-09.
39. Hartman, A. L.; Bird, B. H.; Towner, J. S.; Antoniadou, Z.-A.; Zaki, S. R.; Nichol, S. T. Inhibition of IRF-3 Activation by VP35 Is Critical for the High Level of Virulence of Ebola Virus. *J. Virol.* **2008**, *82* (6). https://doi.org/10.1128/JVI.02344-07.
40. Kirchdoerfer, R. N. N.; Abelson, D. M. M.; Li, S.; Wood, M. R. R.; Saphire, E. O. O. Assembly of the Ebola Virus Nucleoprotein from a Chaperoned VP35 Complex. *Cell Rep.* **2015**, *12* (1), 140–149. https://doi.org/10.1016/j.celrep.2015.06.003.
41. Zinzula, L.; Nagy, I.; Orsini, M.; Weyher-Stingl, E.; Bracher, A.; Baumeister, W. Structures of Ebola and Reston Virus VP35 Oligomerization Domains and Comparative Biophysical Characterization in All Ebolavirus Species. *Structure* **2019**, *27* (1). https://doi.org/10.1016/j.str.2018.09.009.
42. Ramaswamy, V. K.; Di Palma, F.; Vargiu, A.; et al. Insights Into the Homo-Oligomerization Properties of N-Terminal Coiled-Coil Domain of Ebola Virus VP35 Protein. *Virus Res.* **2018**, *247* (6), 61–70. https://doi.org/10.1016/j.virusres.2018.02.003.

43. Di Palma, F.; Daino, G. L.; Ramaswamy, V. K.; et al. Relevance of Ebola Virus VP35 Homo-Dimerization on the Type I Interferon Cascade Inhibition. *Antiviral Chem. Chemother.* **2019**, *27*, 1–13. https://doi.org/10.1177/2040206619889220.
44. Groseth, A.; Charton, J. E.; Sauerborn, M.; et al. The Ebola Virus Ribonucleoprotein Complex: A Novel VP30-L Interaction Identified. *Virus Res.* **2009**, *140* (1-2), 8–14. https://doi.org/10.1016/j.virusres.2008.10.017.
45. Leung, D. W.; Ginder, N. D.; Fulton, D. B.; et al. Structure of the Ebola VP35 Interferon Inhibitory Domain. *Proc. Natl. Acad. Sci. U. S. A.* **2009**, *106* (2), 411–416. https://doi.org/10.1073/pnas.0807854106.
46. Mühlberger, E.; Weik, M.; Volchkov, V. E.; et al. Comparison of the Transcription and Replication Strategies of Marburg Virus and Ebola Virus by Using Artificial Replication Systems. *J. Virol.* **1999**, *73* (3), 2333–2342.
47. Leung, D. W.; Borek, D.; Luthra, P.; et al. An Intrinsically Disordered Peptide From Ebola Virus VP35 Controls Viral RNA Synthesis by Modulating Nucleoprotein-RNA Interactions. *Cell Rep.* **2015**, *11* (3), 376–389. https://doi.org/10.1016/j.celrep.2015.03.034.
48. Takamatsu, Y.; Kolesnikova, L.; Becker, S. Ebola Virus Proteins NP, VP35, and VP24 are Essential and Sufficient to Mediate Nucleocapsid Transport. *Proc. Natl. Acad. Sci. U. S. A.* **2018**, *115* (5), 1075–1080. https://doi.org/10.1073/pnas.1712263115.
49. Shu, T.; Gan, T.; Bai, P.; et al. Ebola Virus VP35 has Novel NTPase and Helicase-Like Activities. *Nucleic Acids Res.* **2019**, *1*, 1–15. https://doi.org/10.1093/nar/gkz340.
50. Moller, P.; Pariente, N.; Klenk, H.-D. H.-D. H.; et al. Homo-Oligomerization of Marburgvirus VP35 is Essential for Its Function in Replication and Transcription. *J. Virol.* **2005**, *79* (23), 14876–14886. https://doi.org/10.1128/JVI.79.23.14876-14886.2005.
51. Reid, S. P.; Cárdenas, W. B.; Basler, C. F. Homo-Oligomerization Facilitates the Interferon-Antagonist Activity of the Ebolavirus VP35 Protein. *Virology* **2005**, *341* (2), 179–189. https://doi.org/10.1016/j.virol.2005.06.044.
52. Leung, D. W.; Borek, D.; Farahbakhsh, M.; et al. Crystallization and Preliminary X-ray Analysis of Ebola VP35 Interferon Inhibitory Domain Mutant Proteins. *Acta Crystallograph. Sect. F Struct. Biol. Crystal. Commun.* **2010**, *66* (6), 689–692. https://doi.org/10.1107/S1744309110013266.
53. Kimberlin, C. R.; Bornholdt, Z. A.; Li, S.; Woods, V. L.; MacRae, I. J.; Saphire, E. O. Ebolavirus VP35 Uses a Bimodal Strategy to Bind dsRNA for Innate Immune Suppression. *Proc. Natl. Acad. Sci. U. S. A.* **2010**, *107* (1), 314–319. https://doi.org/10.1073/pnas.0910547107.
54. Hartman, A. L.; Ling, L.; Nichol, S. T.; Hibberd, M. L. Whole-Genome Expression Profiling Reveals That Inhibition of Host Innate Immune Response Pathways by Ebola Virus Can Be Reversed by a Single Amino Acid Change in the VP35 Protein. *J. Virol.* **2008**, *82* (11), 5348–5358. https://doi.org/10.1128/JVI.00215-08.
55. Zinzula, L.; Esposito, F.; Pala, D.; Tramontano, E. DsRNA Binding Characterization of Full Length Recombinant Wild Type and Mutants Zaire Ebolavirus VP35. *Antivir. Res.* **2012**, *93* (3), 354–363. https://doi.org/10.1016/j.antiviral.2012.01.005.
56. Hartman, A. L.; Towner, J. S.; Nichol, S. T. A C-Terminal Basic Amino Acid Motif of Zaire Ebolavirus VP35 is Essential for Type I Interferon Antagonism and Displays High Identity With the RNA-Binding Domain of Another Interferon Antagonist, the NS1 Protein of Influenza a Virus. *Virology* **2004**, *328* (2), 177–184. https://doi.org/10.1016/j.virol.2004.07.006.
57. Hartman, A. L.; Dover, J. E.; Towner, J. S.; Nichol, S. T. Reverse Genetic Generation of Recombinant Zaire Ebola Viruses Containing Disrupted IRF-3 Inhibitory Domains Results in Attenuated Virus Growth In Vitro and Higher Levels of IRF-3 Activation Without Inhibiting Viral Transcription or Replication. *J. Virol.* **2006**, *80* (13), 6430–6440. https://doi.org/10.1128/JVI.00044-06.

58. Basler, C. F.; Krogan, N. J.; Leung, D. W.; Amarasinghe, G. K. Virus and Host Interactions Critical for Filoviral RNA Synthesis as Therapeutic Targets. *Antivir. Res.* **2019**, *162*. https://doi.org/10.1016/j.antiviral.2018.12.006.
59. Fanunza, E.; Frau, A.; Corona, A.; Tramontano, E. Insights Into Ebola Virus VP35 and VP24 Interferon Inhibitory Functions and Their Initial Exploitation as Drug Targets Elisa. *Infect. Disord. Drug Target.* **2018**, *19*, 1–12. https://doi.org/10.2174/1871526519666181123145540.
60. Luthra, P.; Ramanan, P.; Mire, C. E.; et al. Mutual Antagonism Between the ebola Virus VP35 Protein and the RIG-I Activator PACT Determines Infection Outcome. *Cell Host Microbe.* **2013**, *14* (1), 74–84. https://doi.org/10.1016/j.chom.2013.06.010.
61. Leung, D. W.; Prins, K. C.; Basler, C. F.; Amarasinghe, G. K. Ebolavirus VP35 is a Multifunctional Virulence Factor. *Virulence* **2010**, *1* (6), 526–531. https://doi.org/10.4161/viru.1.6.12984.
62. Prins, K. C.; Cárdenas, W. B.; Basler, C. F. Ebola Virus Protein VP35 Impairs the Function of Interferon Regulatory Factor-Activating Kinases IKKepsilon and TBK-1. *J. Virol.* **2009**, *83* (7), 3069–3077. https://doi.org/10.1128/JVI.01875-08.
63. Chang, T.-H.; Kubota, T.; Matsuoka, M.; et al. Ebola Zaire Virus Blocks Type I Interferon Production by Exploiting the Host SUMO Modification Machinery. Basler CF, ed. *PLoS Pathog.* **2009**, *5* (6), e1000493. https://doi.org/10.1371/journal.ppat.1000493.
64. Bharaj, P.; Atkins, C.; Luthra, P.; et al. The Host E3-Ubiquitin Ligase TRIM6 Ubiquitinates the Ebola Virus VP35 Protein and Promotes Virus Replication. Lyles DS, ed. *J. Virol.* **2017**, *91* (18), e00833-17. https://doi.org/10.1128/jvi.00833-17.
65. Chen, Z. J.; Sun, L. J. Nonproteolytic Functions of Ubiquitin in Cell Signaling. *Mol. Cell.* **2009**, *33* (3), 275–286. https://doi.org/10.1016/j.molcel.2009.01.014.
66. Oudshoorn, D.; Versteeg, G. A.; Kikkert, M. Regulation of the Innate Immune System by Ubiquitin and Ubiquitin-Like Modifiers. *Cytokine Growth Fact. Rev.* **2012**, *23* (6), 273–282. https://doi.org/10.1016/j.cytogfr.2012.08.003.
67. Rajsbaum, R.; García-Sastre, A. Viral Evasion Mechanisms of Early Antiviral Responses Involving Regulation of Ubiquitin Pathways. *Trends Microbiol.* **2013**, *21* (8), 421–429. https://doi.org/10.1016/j.tim.2013.06.006.
68. Le Sage, V.; Cinti, A.; McCarthy, S.; et al. Ebola Virus VP35 Blocks Stress Granule Assembly. *Virology* **2017**, *502*, 73–83. https://doi.org/10.1016/j.virol.2016.12.012.
69. Nelson, E. V.; Schmidt, K. M.; Dog, S.; et al. *Ebola Virus Does Not Induce Stress Granule Formation During* **2016**, *90* (16), 7268–7284. https://doi.org/10.1128/JVI.00459-16. Editor.
70. Huang, Y.; Xu, L.; Sun, Y.; Nabel, G. J. The Assembly of Ebola Virus Nucleocapsid Requires Virion-Associated Proteins 35 and 24 and Posttranslational Modification of Nucleoprotein. *Mol. Cell* **2002**, *10* (2). https://doi.org/10.1016/S1097-2765(02)00588-9.
71. Banadyga, L.; Hoenen, T.; Ambroggio, X.; Dunham, E.; Groseth, A.; Ebihara, H. Ebola Virus VP24 Interacts With NP to Facilitate Nucleocapsid Assembly and Genome Packaging. *Sci. Rep.* **2017**, *7* (1). https://doi.org/10.1038/s41598-017-08167-8.
72. Harrison, A. R.; David, C. T.; Rawlinson, S. M.; Moseley, G. W. The Ebola Virus Interferon Antagonist VP24 Undergoes Active Nucleocytoplasmic Trafficking. *Viruses* **2021**, *13* (8). https://doi.org/10.3390/v13081650.
73. Reid, S. P.; Valmas, C.; Martinez, O.; Sanchez, F. M.; Basler, C. F. Ebola Virus VP24 Proteins Inhibit the Interaction of NPI-1 Subfamily Karyopherin α Proteins With Activated STAT1. *Journal of Virology* **2007**, *81* (24). https://doi.org/10.1128/JVI.01097-07.
74. Mateo, M.; Reid, S. P.; Leung, L. W.; Basler, C. F.; Volchkov, V. E. Ebolavirus VP24 Binding to Karyopherins Is Required for Inhibition of Interferon Signaling. *J. Virol.* **2010**, *84* (2). https://doi.org/10.1128/JVI.01372-09.

75. Xu, W.; Edwards, M. R.; Borek, D. M.; et al. Ebola Virus VP24 Targets a Unique NLS Binding Site on Karyopherin Alpha 5 to Selectively Compete With Nuclear Import of Phosphorylated STAT1. **2015**, *16* (2), 187–200. https://doi.org/10.1016/j.chom.2014.07.008.Ebola.
76. Vidal, S.; el Motiam, A.; Seoane, R.; et al. Regulation of the Ebola Virus VP24 Protein by SUMO. *J. Virol.* **2019**, *94* (1). https://doi.org/10.1128/JVI.01687-19.
77. Lubaki, N. M.; Ilinykh, P.; Pietzsch, C.; et al. The Lack of Maturation of Ebola Virus-Infected Dendritic Cells Results From the Cooperative Effect of at Least Two Viral Domains. *J. Virol.* **2013**, *87* (13). https://doi.org/10.1128/JVI.03316-12.
78. Schwarz, T. M.; Edwards, M. R.; Diederichs, A.; et al. VP24-Karyopherin Alpha Binding Affinities Differ Between Ebolavirus Species, Influencing Interferon Inhibition and VP24 Stability. *J. Virol.* **2017**, *91* (4). https://doi.org/10.1128/JVI.01715-16.
79. Zinzula, L.; Esposito, F.; Mühlberger, E.; et al. Purification and Functional Characterization of the Full Length Recombinant Ebola Virus VP35 Protein Expressed in E. coli. *Prot. Exp. Purif.* **2009**, *66* (1), 113–119. https://doi.org/10.1016/j.pep.2009.02.008.
80. Ekins, S.; Freundlich, J. S.; Coffee, M. A Common Feature Pharmacophore for FDA-Approved Drugs Inhibiting the Ebola Virus. *F1000Research* **2014**, *3* (0), 277. https://doi.org/10.12688/f1000research.5741.1.
81. Madrid, P. B.; Chopra, S.; Manger, I. D.; et al. A Systematic Screen of FDA-Approved Drugs for Inhibitors of Biological Threat Agents. *PLoS One* **2013**, *8* (4). https://doi.org/10.1371/journal.pone.0060579, e60579.
82. Gignoux, E.; Azman, A. S.; de Smet, M.; et al. Effect of Artesunate–Amodiaquine on Mortality Related to Ebola Virus Disease. *New Engl. J. Med.* **2016**, *374* (1), 23–32. https://doi.org/10.1056/NEJMoa1504605.
83. Bishop, B. M. Potential and Emerging Treatment Options for Ebola Virus Disease. *Ann. Pharmacother.* **2015**, *49* (2), 196–206. https://doi.org/10.1177/1060028014561227.
84. Dowall, S. D.; Bosworth, A.; Watson, R.; et al. Chloroquine Inhibited Ebola Virus Replication In Vitro but Failed to Protect Against Infection and Disease in the In Vivo Guinea Pig Model. *J. Gen. Virol.* **2015**, *96* (12), 3484–3492. https://doi.org/10.1099/jgv.0.000309.
85. Falzarano, D.; Safronetz, D.; Prescott, J.; Marzi, A.; Feldmann, F.; Feldmann, H. Lack of Protection Against Ebola Virus From Chloroquine in Mice and Hamsters. *Emerg. Infect. Dis.* **2015**, *21* (6), 1065–1067. https://doi.org/10.3201/eid2106.150176.
86. Brown, C. S.; Lee, M. S.; Leung, D. W.; et al. In Silico Derived Small Molecules Bind the Filovirus VP35 Protein and Inhibit Its Polymerase Cofactor Activity. *J. Mol. Biol.* **2014**, *426* (10), 2045–2058. https://doi.org/10.1016/j.jmb.2014.01.010.
87. Liu, G.; Nash, P. J.; Johnson, B.; et al. A Sensitive in Vitro High-Throughput Screen to Identify Pan-filoviral Replication Inhibitors Targeting the VP35-NP Interface. *ACS Infect. Dis.* **2017**, *3* (3), 190–198. https://doi.org/10.1021/acsinfecdis.6b00209.
88. Glanzer, J. G.; Byrne, B. M.; McCoy, A. M.; James, B. J.; Frank, J. D.; Oakley, G. G. In Silico and In Vitro Methods to Identify Ebola Virus VP35-dsRNA Inhibitors. *Bioorg. Med. Chem.* **2016**, *24* (21), 5388–5392. https://doi.org/10.1016/j.bmc.2016.08.065.
89. Shah, R.; Panda, P. K.; Patel, P.; Mumbai, N.; Farm, A.; Road, G. D. Pharmacophore Based Virtual Screening and Molecular Docking Studies of Inherited Compounds. *World J. Pharmacy Pharm. Sci.* **2015**, *4* (05), 1268–1282.
90. Ren, J. X.; Zhang, R. T.; Zhang, H.; Cao, X. S.; Liu, L. K.; Xie, Y. Identification of Novel VP35 Inhibitors: Virtual Screening Driven New Scaffolds. *Biomed. Pharmacother.* **2016**, *84*, 199–207. https://doi.org/10.1016/j.biopha.2016.09.034.
91. Daino, G. L.; Frau, A.; Sanna, C.; et al. Identification of Myricetin as an Ebola Virus VP35-Double-Stranded RNA Interaction Inhibitor Through a Novel Fluorescence-Based Assay. *Biochemistry* **2018**, *57* (44), 6367–6378. https://doi.org/10.1021/acs.biochem.8b00892.

92. Baikerikar, S. Curcumin and Natural Derivatives Inhibit Ebola Viral Proteins: An In Silico Approach. *Pharmacog. Res.* **2017**, *9* (5), 15. https://doi.org/10.4103/pr.pr_30_17.
93. Cannas, V.; Daino, G. L.; Corona, A.; Esposito, F.; Tramontano, E. A Luciferase Reporter Gene Assay to Measure Ebola Virus Viral Protein 35-Associated Inhibition of Double-Stranded RNA-Stimulated, Retinoic Acid-Inducible Gene 1-Mediated Induction of Interferon β. *J. Infect. Dis.* **2015**, *212* (suppl 2), S277–S281. https://doi.org/10.1093/infdis/jiv214.
94. Di Petrillo, A.; Fais, A.; Pintus, F.; et al. Broad-Range Potential of Asphodelus microcarpus Leaves Extract for Drug Development. *BMC Microbiol.* **2017**, *17* (1), 159. https://doi.org/10.1186/s12866-017-1068-5.
95. Kristensen, M.; Birch, D.; Nielsen, H. M. Applications and Challenges for Use of Cell-Penetrating Peptides as Delivery Vectors for Peptide and Protein Cargos. *Int. J. Mol. Sci.* **2016**, *17* (2), 185. https://doi.org/10.3390/ijms17020185.
96. Liu, J.; Gaj, T.; Patterson, J. T.; Sirk, S. J.; Barbas, C. F. Cell-Penetrating Peptide-Mediated Delivery of TALEN Proteins via Bioconjugation for Genome Engineering. *PLoS One* **2014**, *9* (1). https://doi.org/10.1371/journal.pone.0085755, e85755.
97. Seesuay, W.; Jittavisutthikul, S.; Sae-Lim, N.; Sookrung, N.; Sakolvaree, Y.; Chaicumpa, W. Human Transbodies That Interfere With the Functions of Ebola Virus VP35 Protein in Genome Replication and Transcription and Innate Immune Antagonism Article. *Emerg. Microbes Infect.* **2018**, 7 (1). https://doi.org/10.1038/s41426-018-0031-3.
98. Corti, D.; Misasi, J.; Mulangu, S.; et al. Protective Monotherapy Against Lethal Ebola Virus Infection by a Potently Neutralizing Antibody. *Science* **2016**, *351* (6279), 1339–1342. https://doi.org/10.1126/science.aad5224.
99. Flego, M.; Frau, A.; Accardi, L.; et al. Intracellular Human Antibody Fragments Recognizing the VP35 Protein of Zaire Ebola Filovirus Inhibit the Protein Activity. *BMC Biotechnol.* **2019**, *19* (1), 64. https://doi.org/10.1186/s12896-019-0554-2.
100. Zhao, Z.; Martin, C.; Fan, R.; Bourne, P. E.; Xie, L. Drug Repurposing to Target Ebola Virus Replication and Virulence Using Structural Systems Pharmacology. *BMC Bioinform.* **2016**, *17* (1). https://doi.org/10.1186/s12859-016-0941-9.
101. Iversen, P.; Warren, T.; Wells, J.; et al. Discovery and Early Development of AVI-7537 and AVI-7288 for the Treatment of Ebola Virus and Marburg Virus Infections. *Viruses* **2012**, *4* (11). https://doi.org/10.3390/v4112806.
102. Song, X.; Lu, L.; Passioura, T.; Suga, H. Macrocyclic Peptide Inhibitors for the Protein–Protein Interaction of Zaire Ebola Virus Protein 24 and Karyopherin Alpha 5. *Org. Biomol. Chem.* **2017**, *15* (24). https://doi.org/10.1039/C7OB00012J.
103. Raj, U.; Varadwaj, P. K. Flavonoids as Multi-Target Inhibitors for Proteins Associated With Ebola Virus: In Silico Discovery Using Virtual Screening and Molecular Docking Studies. *Interdiscip. Sci. Comput. Life Sci.* **2016**, *8* (2). https://doi.org/10.1007/s12539-015-0109-8.
104. Fanunza, E.; Iampietro, M.; Distinto, S.; et al. Quercetin Blocks Ebola Virus Infection by Counteracting the VP24 Interferon-Inhibitory Function. *Antimicrobial Agents and Chemotherapy* **2020**, *64* (7). https://doi.org/10.1128/AAC.00530-20.
105. Kwofie, S. K.; Broni, E.; Teye, J.; et al. Pharmacoinformatics-based identification of potential bioactive compounds against Ebola virus protein VP24. *Computers in Biology and Medicine.* **2019**, *113*. https://doi.org/10.1016/j.compbiomed.2019.103414.

CHAPTER SIX

Uracil derivatives as non-nucleoside inhibitors of viral infections

Anastasia Khandazhinskaya[a,*], Elena Matyugina[a], and Mikhail Novikov[b]
[a]Engelhardt Institute of Molecular Biology, Russian Academy of Science, Moscow, Russian Federation
[b]Department of Pharmaceutical & Toxicological Chemistry, Volgograd State Medical University, Volgograd, Russian Federation
*Corresponding author: e-mail address: khandazhinskaya@bk.ru

Contents

Annual Reports in Medicinal Chemistry, Volume 57
ISSN 0065-7743
https://doi.org/10.1016/bs.armc.2021.08.001
Copyright © 2021 Elsevier Inc.
All rights reserved.

1. Introduction

Heterocyclic bases as a part of nucleosides and nucleotides are important components of DNA and RNA, and in addition, they take part in a huge number of key metabolic processes, as cofactors or regulators of hundreds of reactions of various types.[1] In this regard, any modification of the nucleic base can have a significant effect on the recognition and inhibition of the corresponding enzymes, and, thus, on the spectrum of its activity.[1–3] At the moment, analogs and derivatives of nucleic acid components (NA) are important elements of anticancer,[1,2,4] antibacterial,[5,6] antifungal[7,8] and antiviral therapies.[1–3,9,10]

The antiviral activity of pyrimidine base derivatives has been known since 1963, when it was shown that 2′-deoxy-5-iodocytidine inhibits the development of herpesvirus infection.[11] Pyrimidine-containing nucleoside analogs became the first drugs for the treatment of infection caused by human immunodeficiency virus type 1 (HIV-1) and are still used in therapy. Zidovudine,[12] stavudine,[13] lamivudine[14,15] and emtricitabine[16] are nucleoside inhibitors of HIV-1 reverse transcriptase (NRTI HIV-1). Pyrimidine-containing analogs of NA components are widely used to treat infections caused by viruses of the herpes group, in particular: brivudine against varicella-zoster virus[17] and cidofovir against cytomegalovirus.[18] The main mechanism of the antiviral activity of NRTI is intracellular phosphorylation to the corresponding triphosphorylated nucleoside analogs, which then act as chain terminator substrates of viral polymerases.[1]

It was later found that uracil derivatives are capable of inhibiting HIV-1 RT by a non-competitive (non-nucleoside) mechanism. HIV-1 non-nucleoside RT inhibitors (HIV-1 NNRTIs) do not require preliminary metabolic transformations, since they bind to the viral enzyme in the hydrophobic RT pocket (approximately 10 Å from the RT active site) and disrupt RT conformation required for polymerizing activity.[19,20] More than fifteen hundred uracil-containing HIV NNRTIs have been synthesized over the past three decades and more than one hundred and fifty papers have been published. Each group of scientists optimizes their own type of inhibitors, thus the data obtained are rather scattered and often contradictory. The most well-known types of uracil-containing HIV NNRTIs are mentioned in reviews on various aspects of research on different HIV NNRTIs.[19,21–24]

It has become clear over time that in addition to inhibiting HIV RT by nucleoside and non-nucleoside mechanisms, uracil derivatives can inhibit the RNase H and Integrase (IN) activities of HIV. Besides uracil derivatives proved to be effective NNIs of other viruses. Some of them, such as viruses of the herpes family (HCMV and HSV) and hepatitis (HCV and HBV), are widespread HIV coinfections and are well investigated. Other viruses, such as respiratory syncytial virus (RSV), human adenovirus (HAdV), tick-borne encephalitis virus (TBEV) and SARS-CoV-2, are less studied and there are no highly effective and selective drugs against them.

This review is devoted to the systematization of data on the evolution of various types of uracil-containing NNIs over 30 years, discussion of successful approaches to the design and synthesis, and generalization of the found patterns in the effect of structure on antiviral activity against HIV-1 and other viral infections. Separate sections describe nucleoside analogs that act as HIV-1 NNRTI and dimeric molecules containing active components aimed at inhibiting various viral targets.

2. 6-Substituted uracils derivatives

2.1 Derivatives of 6-arylsulfanyl- and 6-arylselenyluracil

2.1.1 Evolution of group

1-(2-Hydroxyethoxymethyl)-6-(phenylthio)thymine **1** (HEPT; Fig. 1) was synthesized by Hiromishi Tanaka and Tadashi Miyasaka from Showa University (Tokyo, Japan). In 1987, HEPT was sent to the Rega Institute (Leuven, Belgium) to evaluate its activity against HSV. This choice was prompted by its structural similarity to a well-known anti-HSV drug acyclovir, which also had a 2-hydroxyethoxymethyl side chain. There was no activity of HEPT against HSV. However, Masanori Baba (from Fukushima Medical College in Japan) found that HEPT is active against HIV-1 in MT-4 cell culture, but does not inhibit HIV-2.[25] The mechanism of action study showed that, unlike other nucleoside analogs, HEPT triphosphate does not interact with HIV-1 reverse transcriptase (HIV-1 RT) and does not affect HIV-1 replication.[26]

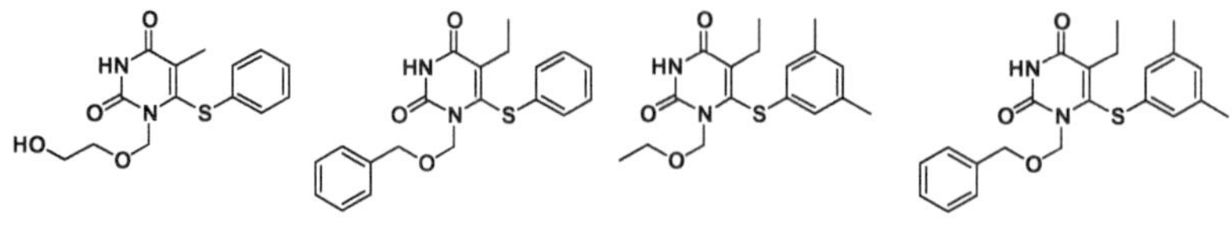

1 HEPT	2 E-BPU	3 E-EPU-dM	4 E-BPU-dM
EC_{50} 7.0 μM	EC_{50} 0.0059 μM	EC_{50} 0.0054 μM	EC_{50} 0.0032 μM
CC_{50} 740 μM	CC_{50} 30 μM	CC_{50} >100 μM	CC_{50} >20 μM
SI 106	SI 6122	SI >16100	SI >6250

Fig. 1 Structures and properties of compounds **1–4.**

The mechanism of action of HEPT became clear after synthesis and evaluation of more potent HEPT derivatives, such as 1-benzyloxymethyl-5-ethyl-6-phenylthiouracil **2** (E-BPU),[27] 1-ethyl-1-ethoxymethyl-6-(3,5-dimethylphenylthio)uracil **3** (E-EPU-dM) and 1-benzyloxymethyl-5-ethyl-6-(3,5-dimethylphenylthio)uracil **4** (E-BPU-dM).[28,29] Compounds **2–4** in the nanomolar concentrations inhibited HIV-1 replication of both the wild strain and its numerous mutant variants, including resistant to AZT, in the MT-4 cell culture (Fig. 1). The compounds have been shown to inhibit RT of HIV-1 but not of HIV-2.[30]

The study of HIV-1 mutant strains resistant to HEPT showed that they had one from mutations 188-Tyr > His, 181-Tyr > Cys or 106-Val > Ala in HIV-1 reverse transcriptase. These virus strains showed sufficiently lower sensitivity to HEPT derivatives. Moreover, HEPT-resistant viral strains were also cross-resistant to all other HIV-1 NNRTIs including TIBO, nevirapine and TSAO.[31] Thus, it was shown that HEPT derivatives bind to the allosteric site of the reverse transcriptase located close to the active site of the enzyme and functionally associated with it.

The evolution of HEPT derivatives described above has taken several directions. The synthesis and antiviral properties of analogs with modification of aryl fragment at uracil 6 position are described. Thus, the analog of compound **2**—1-benzyloxymethyl-5-ethyl-6-(α-pyridylthio)uracil **5** (Fig. 2) inhibited HIV-1 replication in the CEM-IW cell culture at subnanomolar concentration and had a selectivity index (SI) of more than 1100.[32] 6-Cyclohexylthio-1-ethoxymethyl-5-isopropyluracil **6** (TNK-6123, Fig. 2) inhibited viral replication at low nanomolar concentration but was less toxic and had a SI of almost 25,000.[33]

6-(2-Naphthylthio)-1-ethoxymethyl-5-isopropyl uracil **7** (Fig. 2) and its analogs with different substituents in the naphthyl fragment and modifications at 1 and 5 positions were significantly inferior in antiviral activity.[34]

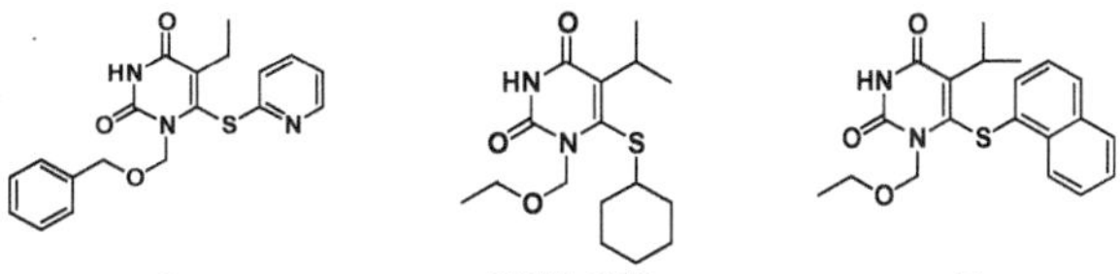

5
EC_{50} 0.0009 μM
-
SI >1100

6 TNK-6123
EC_{50} 0.003 μM
CC_{50} 74.8 μM
SI 24933

7
EC_{50} 0.057 μM
CC_{50} 38.33 μM
SI 579

Fig. 2 Structures and properties of compounds **5–7**.

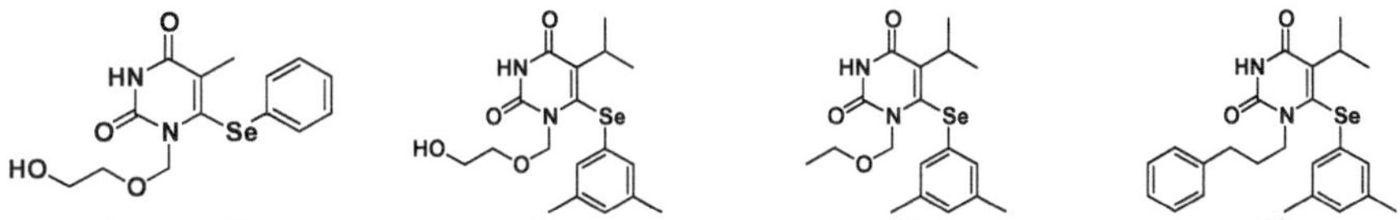

8 Se-HEPT
EC_{50} 0.96 μM (HIV-1)
CC_{50} >200 μM
SI >208

9
EC_{50} 0.019 μM
CC_{50} 113 μM
SI 5950

10
EC_{50} 0.0047 μM
CC_{50} >200 μM
SI >42600

11
EC_{50} 0.0006 μM
-
SI 44 833

Fig. 3 Structures and properties of compounds **8–11**.

The replacement of the sulfur atom in HEPT with selenium gave rise to new group of NNRTIs.[35] In that regard, 1-[(2-hydroxyethoxy)methyl]-6-(phenylselenyl)thymine **8** (Fig. 3) in 0.96 μM concentration inhibited the cytopathic effect of HIV-1 (strain LAV) in PBM cell culture by 50% with SI more than 200. In the same cell culture, HEPT had an EC_{50} 5.30 μM. Additionally, in contrast to 6-phenylthio analogs, 6-(phenylselenyl)uracil derivatives exhibited a pronounced antiviral effect against HIV-2. Thus, the EC_{50} of compound **8** in PBM cell culture infected with HIV-2 (strain ROD-2) was 25.6 μM, and the selectivity index was 7. The mechanism of action was studied using poly(rA)-oligo (dT) as a template primer. The study showed that unlike HEPT **1** and its 6-(phenylthio) analogs **2–4**, Se-HEPT **8** has a weak effect on the viral enzyme at IC_{50} 90.3 μM. Se-HEPT **8** and its analogs do not show any cytotoxicity in PBM cell culture, but some of them demonstrated noticeable cytotoxicity in Vero cell culture.[35] This suggests that 6-phenylselenyl uracil derivatives may be toxic in rapidly dividing cells, such as human bone marrow cells, and have a negative effect on the process of hematopoiesis.[36]

The introduction of isopropyl substituent at 5 position and methyl groups at the meta-positions of the phenyl ring enhance the antiviral properties. Thus, 6-[(3,5-dimethylphenyl)selenyl]-1-[(2-hydroxyethoxy)methyl]-5-isopropyluracil **9** (Fig. 3) at 19 nM concentration inhibited the cytopathic effect of HIV-1 in the CEM-SS cell culture by 50% with SI 5950. 6-[(3,5-Dimethylphenyl)selenyl]-1-ethoxymethyl-5-isopropyluracil **10** (Fig. 3), devoid hydroxyl group in the side chain, was more active (EC_{50} 0.0047 μM in CEM-SS cells infected by HIV-1) and not cytotoxic, so its SI exceeded 42,600.[37]

The rapid development of viral resistance (HIV-1 strain IIIB) to 6-arylselenyl uracil derivatives in vitro was shown in MT-2 cell culture. DNA sequence analysis of mutant HIV-1 strains showed that the cause of resistance was a single nucleotide change (TAT→TGT) in the RT gene leading to the substitution of Tyr for Cys either at 181 position or at 188 position of HIV-1 RT.[38]

Kim et al.[39] described the synthesis and inhibitory properties of 6-arylselenyl-uracil derivatives containing a butyl or 3-phenylpropyl substituent at the N1 position. It was found that 5-isopropyl uracil derivatives more effectively suppress HIV-1 reproduction than 5-ethyl analogs. The most active compound of this series was 1-(3-phenylpropyl)-5-isopropyl-6-(3,5-dimethylphenylselenyl)uracil **11** (Fig. 3), which protected the cells by 50% from the cytopathic effect of HIV-1 in 0.0006 μM concentration with SI 44833.

2.1.2 Chemical synthesis of 6-arylsulfanyl- and 6-arylselenyluracil derivatives

The methods of synthesis of 6-thio- and 6-selenium uracil derivatives described in this section are mostly similar. There are three main synthetic strategies:

- Introduction of the arylsulfanyl or arylselenyl fragment via the dilithium salt of uracil;
- The use of 2-thiouracil as a substrate for the introduction of arylsulfanyl and arylselenyl fragments;
- Cyclization of the pyrimidine ring.

The key stage in the synthesis of 6-thio- and 6-selenium uracil derivatives in the first approach is the treatment of the corresponding uracil derivative with an excess of lithium diisopropylamide (LDA) at a reduced temperature, followed by reaction with diaryldisulfide or diaryldiselenide.

One of the methods of HEPT **1** synthesis includes four steps (Scheme 1). The first one is the reaction of 1-[(2-hydroxyethoxy)methyl]pyrimidine **13** with *tert*-butyldimethylsilyl chloride (TBDMSCl) in DMF in the presence of imidazole. The resulting silyl ether **14** is treated with a 2.5-fold excess of

LDA in THF at −70 °C, which leads to the dilithium salt **15**. The interaction of **15** with excess of diphenyl disulfide leads to the formation of 6-(phenylthio)-derivative **16** with 90% yield. Subsequent acid hydrolysis of **16** gives the target HEPT **1**. Compound **12** was obtained from the corresponding starting materials in the same way.[25,40] 6-(Arylselenyl)uracil derivatives were synthesized similarly. Thus, to obtain 1-[(2-hydroxyethoxy) methyl]-6-(phenylselenyl)thymine **8**, diphenyl selenide was added to dilithium intermediate **15**, which led to the formation of 6-(phenylselenyl)-derivative **16** in 82% yield. Removal of the protective group gives the target product **8** with 55% total yield.[36]

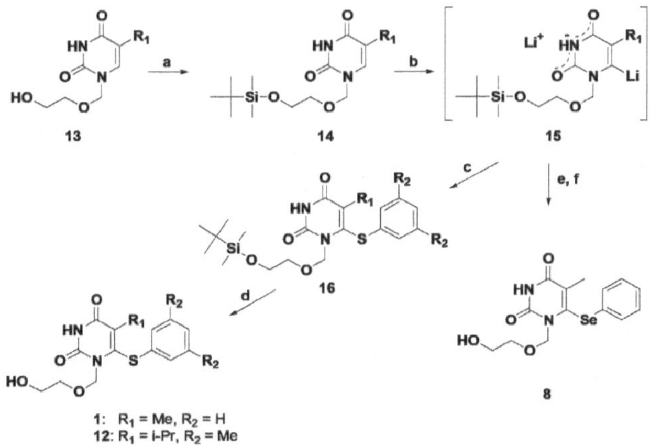

Scheme 1 Reagents and conditions: (a) TBDMSCl, DMF, imidazole, rt., 14 h, 90%; (b) LDA, THF, −70 °C, 1 h; (c) ArSSAr, THF, −70 °C, 1 h; (d) THF, aqueous acetic acid; (e) PhSeSePh, THF, −78 °C, 30 min, 82%; (f) Tetra-*n*-butylammonium fluoride, THF, rt., 30 min, 75%.

In the case of the synthesis of 1-(alkoxymethyl)- and 1-(benzyloxymethyl) derivatives **2** and **3** the scheme can be simplified (Scheme 2),[29] since there are no stages of introducing and removal of the protective TBDMSCl-group. 1-(Benzyloxymethyl)-6-(phenylsulfanyl)thymine (R_1=Ph, R_2=CH_3, R_3=H) was obtained by lithiation of the starting 1-(benzyloxymethyl) thymine followed by the introduction of a phenylsulfanol fragment with the yield of 58%. However, under the conditions of this method, compounds **2**, **3** and **4** with ethyl substituent at position 5, as well as 5-isopropyl containing compounds, are formed in low yields.[29]

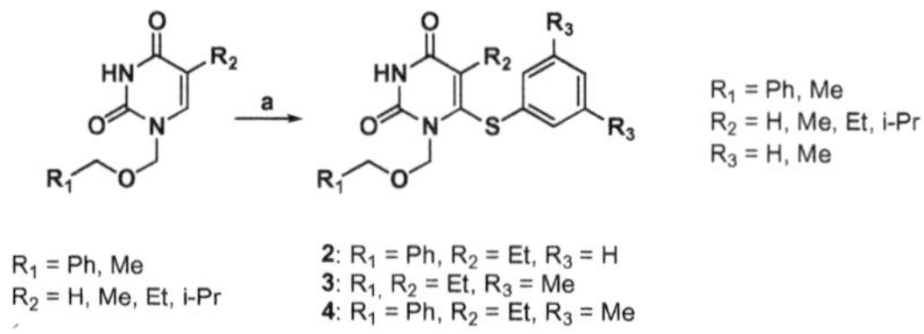

Scheme 2 Reagents and conditions: (a) LDA, THF, −70 °C, 1 h, then ArSSAr, THF, −70 °C, 1 h.

Treatment of 1-(benzyloxymethyl)-5-ethyluracil with α,α′-dipyridyl sulfide under the same conditions led to the formation of compound **5** in 43% yield (Scheme 3), and α,α′-dipyridyl selenide to selenium-containing derivative **17** in 40% yield.[32]

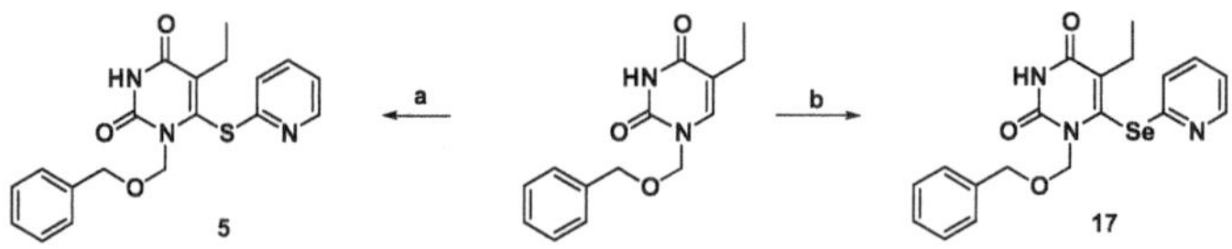

Scheme 3 Reagents and conditions: (a) LDA, THF, −78 °C, 1 h, α,α′-dipyridyl disulfide, −78 °C, 1 h, 43%; (b) LDA, THF, −70 °C, 1 h, α,α′-dipyridyl diselenide, −78 °C, 1 h, 40%.

The yield of 6-substituted products is noticeably affected by the presence of a substituent at 5 position of the pyrimidine ring. Alkyl substituents significantly reduce the yield, apparently due to steric and electronic effects. However, the use of 5-substituted 2-thiouracils can increase the yield of both 5-alkyl-6-phenylsulfanyl-substituted products and the corresponding 6-arylselenic derivatives. Scheme 4 shows an example of the synthesis of 1-((2-hydroxyethoxy)methyl)-5-(isopropyl)-6-(3,5-dimethylphenylsulfanyl)

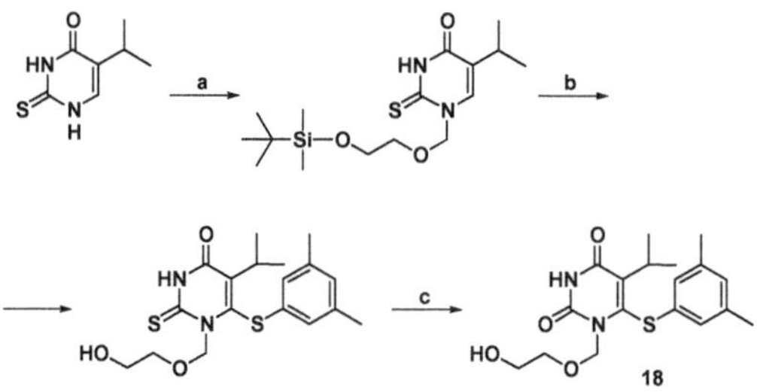

Scheme 4 Reagents and conditions: (a) (i) HMDS, $(NH_4)_2SO_4$, reflux, 15 h; (ii) (2-acetoxyethoxy)methyl acetate, CsI, reflux, 2 h; (iii) 1N NaOH, MeOH, rt., 3 h; (iv) TBDMSCl, MeCN, imidazole, rt., 14 h, 38%; (b) (i) LDA, THF, −70 °C, 1 h; (ii) ArSSAr, THF, −70 °C, 1 h; (iii) THF, conc. HCl, rt., 61%; (c) 1N NaOH, 35% H_2O_2, 1 h, 84%.

uracil **18** in 20% total yield.[41] 1-(Benzyloxymethyl)-5-ethyl-(phenylsulfanyl) uracil **2**, 1-(ethoxymethyl)-5-ethyl-6-(3,5-dimethylphenylsulfanyl)uracil **3**, 1-(benzyloxymethyl)-5-ethyl-6-(3,5-dimethylphenylsulfanyl)uracil **4** and 1-(ethoxymethyl)-5-isopropyl-6-(3,5-dimethylphenylselenyl)uracil **10** were obtained in a similar way; the yields for the starting 5-alkyl-2-thiouracils did not excide 20%.[29,37]

Scheme 5 shows an example of the synthesis of 1-alkyl- and aralkyl-substituted 6-arylselenyl-5-alkyluracils by the construction of pyrimidine ring.[39] The scheme includes the alkaline hydrolysis of the starting ethyl ester of 3,3-di(methylthio)acrylic acid **19** in a water-ethanol solution. The resulting carboxylic acid **20** was converted into the acid chloride **21** by oxalyl chloride, which was treated with silver isocyanate in benzene. The resulting isocyanate **22** reacted with 3-phenylpropylamine, which led to the corresponding ureide **23**. Its cyclization in acetic acid in the presence of a catalytic amount of methanesulfonic acid gave 1-(3-phenylpropyl)-5-isopropyl-6-methylsulfanyluracil **24**, which was oxidized by m-CPBA to the corresponding sulfone **25**. Its treatment with 3,5-dimethylphenylselenol and NaOH in ethanol led to the target compound **11**.[39]

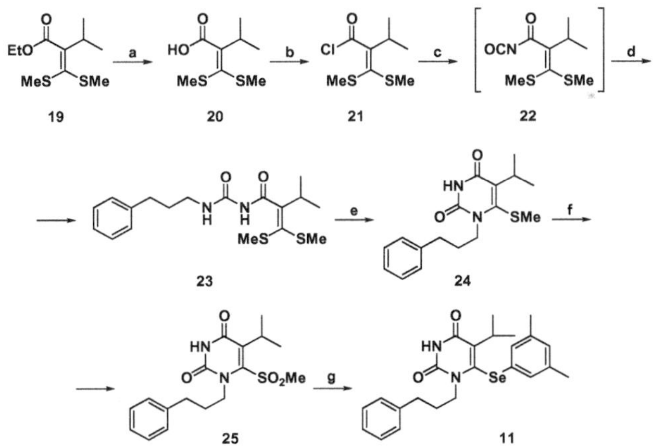

Scheme 5 Reagents and conditions: (a) 2 N KOH, EtOH, reflux, 3 h, 82%; (b) $(COCl)_2$, DMF (cat.), benzene, N_2, rt., 3 h, 91%; (c) AgOCN, benzene, reflux, N_2, 30 min; (d) $Ph(CH_2)_3NH_2$, −78 °C to rt., 1 h, 88%; (e) $MeSO_3H$, AcOH, 80 °C, 1 h, 86%; (f) m-CPBA, benzene, reflux, 16 h; (g) 3,5-dimethylselenol, NaOH, EtOH, rt., N_2, 2 h, 98%.

The synthesis of 6-arylsulfanyl- and 6-arylselenyl-substituted uracils is described in more detail in the review.[42]

2.2 6-Benzyl derivatives of HEPT

2.2.1 Evolution of group

Subsequent structure–activity relationship (SAR) studies led to 6-benzyl-1-ethoxymethyl-5-isopropyluracil **26** (MKC-442, Fig. 4).[43] MKC-442 has reached Phase III extended clinical trials as HIV-1 NNRTI, but its development was stopped without clarification of why.

1-Ethoxymethyl-5-isopropyl-6-(1-naphthylmethyl)uracil **27** (Fig. 4) showed worse properties than MKC-442 **26**. The compound inhibited the replication of HIV-1 in the MT4 cell culture at EC_{50} 0.017 μM and had an SI of 2229, in the CEM cell culture—EC_{50} 0.042 μM and SI 480.[44]

6-(3,5-Dimethylbenzyl)-1-(ethoxymethyl)-5-isopropyluracil **28** (I-EPU-dM, GCA-186, Fig. 4) in MT-4 cell culture showed higher activity against HIV-1 than compound **26** (MKC-442) and suppressed viral replication at a concentration of less than 1 nM.[45]

Further development of HEPT analogs consisted in side chain modifications at the N1 position of uracil. As a result 6-(3,5-dimethylbenzyl)-5-isopropyl-1-(3-methylbut-2-enyloxymethyl)uracil **29** (Fig. 4) was developed, which inhibited HIV-1 replication in MT-4 cell culture in concentration less than 1 nM and had a SI more than 667,000. However, like all representatives of the HEPT series, the compound had a low genetic barrier. The inhibitory activity of compound **29** decreased more than 3500 times because of mutations in the hydrophobic pocket (Y181C, K103N, or Y181C + K103N).[46]

The introduction of a phenyl substituent into the side chain of compound **29** also gave a highly active compound, although inferior in protective effect. 1-((E)-Cinnamyloxymethyl)-6-(3,5-dimethylbenzyl)-5-ethyluracil **30** in 0.003 μM concentration protected MT-4 cell culture by 50% from the cytopathic effect of HIV-1 and had a selectivity index 10,667.[47] Similar properties were demonstrated by uracil derivatives with (2-phenoxyethyl) oxymethyl substituents at N1 position and a 6-(3,5-dimethoxybenzyl) fragment.[48]

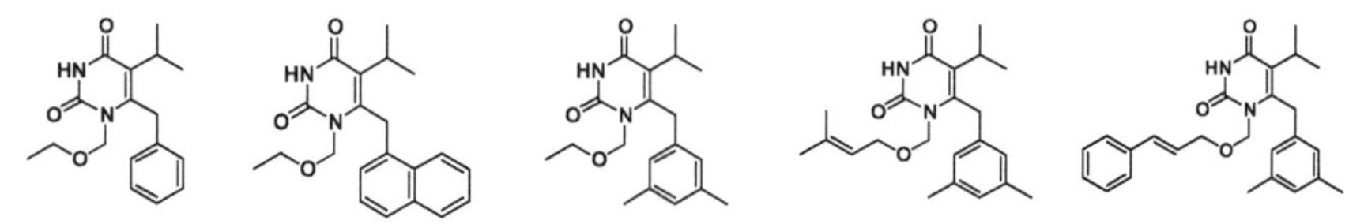

26 (MKC-442)	**27**	**28**	**29**	**30**
EC_{50} 0.015 μM	EC_{50} 0.017 μM	EC_{50} 0.0006 μM	EC_{50} 0.00015 μM	EC_{50} 0.003 μM
CC_{50} 102 μM	CC_{50} 38.33 μM	CC_{50} 43 μM	CC_{50} >100 μM	CC_{50} 32 μM
SI 68000	SI 2229	SI 72 000	SI >667000	SI 10667

Fig. 4 Structures and properties of compounds **26–30**.

Modifications of the linkage between aryl moiety and uracil residue, in particular, the introduction of a 1-arylvinyl moiety at 6 position of uracil, did not lead to an increase in activity. The most effective inhibitor in this group was 6-[1-(3,5-dimethylphenyl)vinyl]-1-ethoxymethyl-5-ethyluracil **31** (Fig. 5) in MT-4 cells infected with HIV-1 (EC_{50} 0.035 μM and SI > 2857). It was shown that the compound **31** suppressed mutant HIV-1 strains Y181C and Y181C + K103N at IC_{50} 17 and 29 μM, respectively.[49]

Loksha et al.[50] described the synthesis and anti-HIV-1 properties of 6-benzyl-5-ethyl uracil derivatives containing a fluorine atom in the methylene fragment. It was found that 3-[(3-(ethoxymethyl)-5-ethyl-1,2,3,6-tetrahydro-2,6-dioxopyrimidin-4-yl)-fluoromethyl]benzonitrile **32** (Fig. 5) exhibited inhibitory properties against the wild strain of HIV-1 in MT-4 cell culture (EC_{50} 0.07 μM). However, its activity against mutant HIV-1 strains Y181C and Y181C + K103N was 2-3 orders lower. A good genetic barrier was demonstrated by 6-[difluoro(3-methyl-5-cyanophenyl)methyl]-1-ethyl-5-isopropyluracil **33** (Fig. 5), which suppressed the replication of HIV-1 in both the wild strain and its mutant variants (Y181C and K103N + Y181C) at nanomolar concentrations: EC_{50} 4.2, 4.3, and 9.0 nM, respectively.[51]

Dimers containing two 5-ethyl uracil residues linked at 6 positions by aromatic fragment for example 5-ethyl-1-ethoxymethyl-6-[3-[(5-ethyl-1-ethoxymethyl-2,6-dioxopyrimidin-4(1*H*,3*H*)-yl)methyl]-benzyl] pyrimidine-2,4(1*H*, 3*H*)-dione (**34**) were synthesized by Ali et al.[52] The study of anti-HIV-1 properties in MT-4 cell culture revealed weak activity against the wild strain of HIV-1 and no activity against mutant strains (Y181C and K103N + Y181C).[52] The introduction of a 1-naphthoyl substituent at 6 position of the uracil residue led to a significant decrease in antiviral activity.[53,54] HEPT analogs with 6-alkenyl or 6-cyclohexenyl substituents were also inactive.[55]

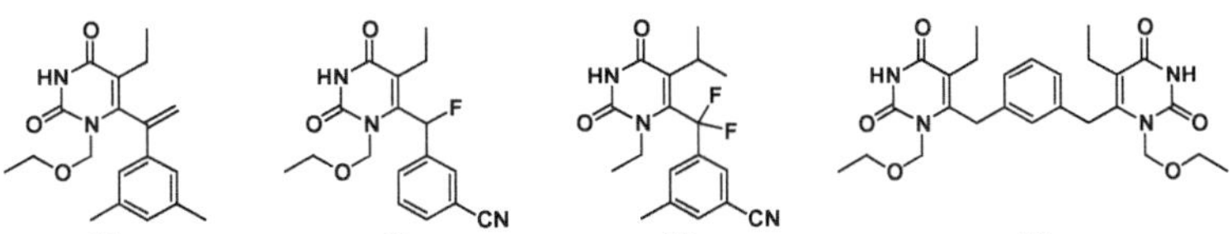

31	32	33	34
EC_{50} 0.035 μM	EC_{50} 0.07 μM	EC_{50} 0.0042 μM	EC_{50} 47 μM
CC_{50} >100 μM	CC_{50} >100 μM	-	CC_{50} >100 μM
SI >2857	SI >1428	-	SI >2

Fig. 5 Structures and properties of compounds **31–34**.

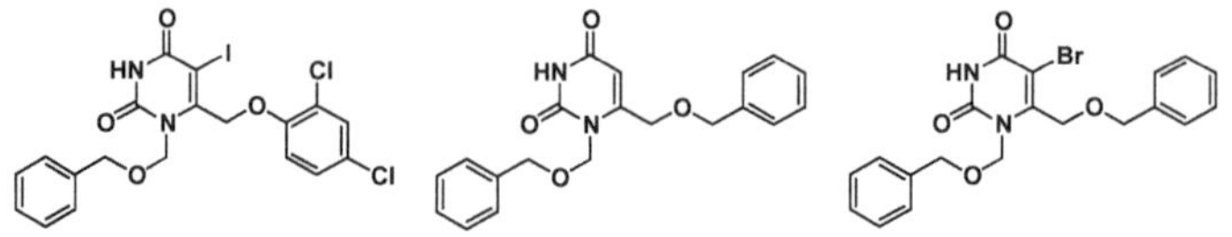

35
EC_{50} 0.16 μM
-
-

36
EC_{50} 0.31 μM (HIV-1)
EC_{50} 32 μM (Influenza H1N1)
CC_{50} 69 μM

37
EC_{50} 7 μM (HIV-1)
EC_{50} 0.3 μM (Influenza H1N1)
CC_{50} 29 μM

Fig. 6 Structures and properties of compounds **35–37**.

Numerous 1-substituted 6-arylamino and 6-aryloxy uracil derivatives have been described. However, their anti-HIV-1 activity was lower than the activity of compounds described above.[40,56,57] Among 1-substituted 6-[(2,4-dichlorophenoxy)methyl]-derivatives of uracil, 1-(benzyloxymethyl)-6-(2,4-dichlorophenoxymethyl)-5-iodouracil **35** (Fig. 6), should be mentioned which inhibited viral replication in the culture of CEM/0 cells with EC_{50} 0.16 μM, an order of magnitude better than HEPT.[58] Its homologs—6-benzyloxymethyl uracil derivatives **36** and **37** (Fig. 6) were active against HIV-1 (**36** with EC_{50} 0.31 μM, IC_{50} (HIV RT) 0.74 μM; **37** with EC_{50} 7.0 μM) in MT-4 cell culture, and for the H1N1 influenza virus (**36** with EC_{50} 32 μM; **37** with EC_{50} 0.3 μM) in MDCK cell culture.[59]

2.2.2 Details of the chemical synthesis of 6-benzyl derivatives of HEPT

The synthesis of compound **29**, presented in Scheme 6, included the condensation of 2,4-bis(trimethylsilyloxy)pyrimidine **38** with acetal-bis-(2-methyl-2-propenyloxy)methane[60] in acetonitrile in the presence of TMS triflate, which led to the target 6-(3,5-dimethylbenzyl)-5-ethyl-1-(3-methylbut-2-enyloxymethyl)uracil with 29% yield.[46] Compounds **35–37** were synthesized under similar conditions.[40,59]

Introducing a benzyl moiety into the 6 position of the uracil residue is also described. The starting 1-substituted 5-alkyluracil **39** was treated with an excess of 2,2,6,6-tetramethylpiperidide lithium[61] or LDA[45] and the resulting Li-salt was condensed with benzaldehyde. Next, 6-phenylhydroxymethyl

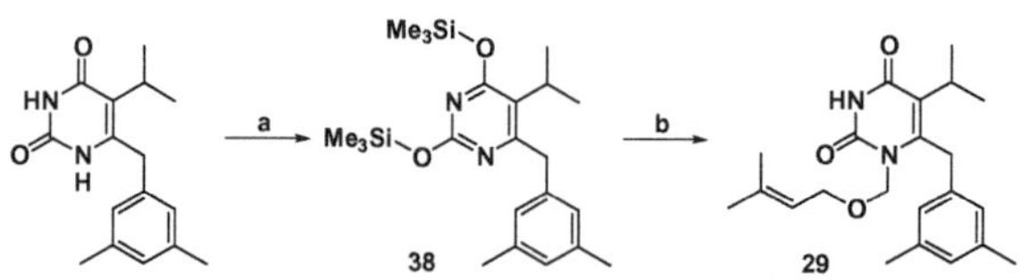

Scheme 6 Reagents and conditions: (a) HMDS, $(NH_4)_2SO_4$, reflux, overnight; (b) acetonitrile, bis-(2-methyl-2-propenyloxy)methane, TMS triflate, −40 °C, 72%.

derivative **40** was acetylated with acetic anhydride and subjected to hydrogenolysis in ethanol in the presence of Pd/C (Scheme 7) to give compound **28** or corresponding analog with different substituents R_1 and R_2.

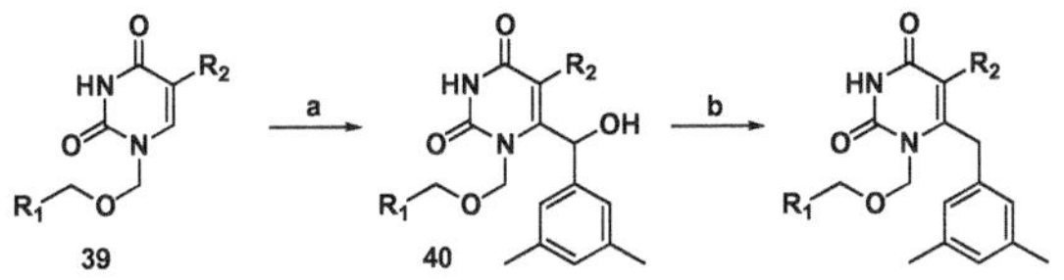

$R_1 = CH_3$, Ph; R_2 = Me, Et, i-Pr.

28: $R_1 = CH_3$, R_2 = i-Pr
and other analogues

Scheme 7 Reagents and conditions: (a) LDA, 3,5-dimethylbenzaldehyde, THF, −70°C, 1 h, 60%–84%; (b) (i) Ac_2O, Py, rt., overnight; (ii) H_2, Pd/C, EtOH, 24 h, rt., 48%–85%.

Another method for introducing a benzyl fragment into the pyrimidine ring is based on the use of Grignard reagents. Compound **41** was treated with isopropyl magnesium chloride in the presence of LiCl in THF, and then benzyl bromide was added to give 6-benzyl derivative **42**. Its hydrolysis led to 6-benzyl-5-isopropyl uracil **43** and after alkylation to compound **26** (MKC-442) (Scheme 8).[62]

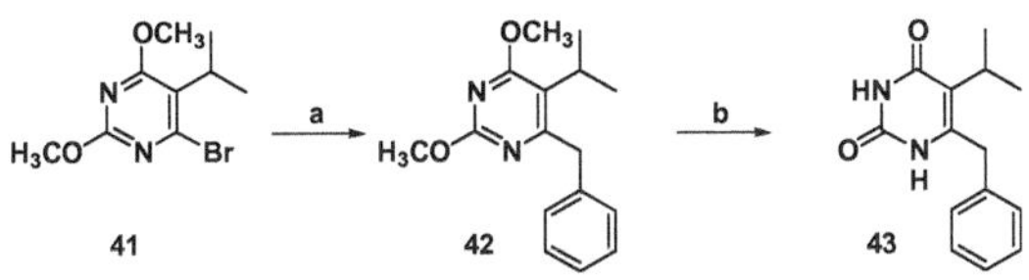

Scheme 8 Reagents and conditions: (a) (i) *i*-PrMgCl·LiCl, THF, rt., 5 h; (ii) $PhCH_2Br$, 20 h, 87%; (b) HCl, MeOH, Δ, 5 h, 92%.

An alternative method for the synthesis of 1-substituted 6-benzyluracil derivatives includes the use of starting 4-aryl-3-oxo-butyric acid esters **44**, which are condensed with thiourea in the presence of alkaline metal alcoholates. As a result, 2-thiouracils **45** were obtained in good yields (Scheme 9). Their treatment with a 10% aqueous solution of 2-chloroacetic acid gave 6-benzyluracils **46**—starting material for synthesis of compounds **26–30**.[63,64]

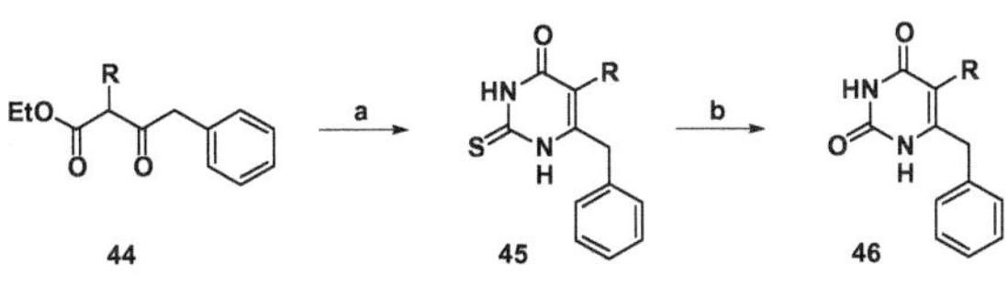

R = Me, Et, i-Pr.

Scheme 9 Reagents and conditions: (a) thiourea, NaOEt, EtOH, Δ; (b) 10%aq. $ClCH_2COOH$, Δ.

2.3 6-Benzoyl containing HEPT analogs

2.3.1 Evolution of group

Son et al. described a new set of HEPT **1** analogs where the sulfur atom was replaced by a carbonyl group which bonds pyrimidine to a 3,5-dimethylphenyl fragment.[56] It resulted to a number of highly active 6-(3,5-dimethylbenzoyl)-5-ethyl- and 5-isopropyl uracils containing an alkenyl or alkynyl substituent in 1 position. The most interesting among of synthesized compounds was 1-(4-acetoxy-trans-2-butenyl)-5-isopropyl-6-(3,5-dimethylbenzoyl)uracil **47** (Fig. 7), which inhibited HIV-1 replication at concentration less than 9 nM with SI more than 6487.

The introduction of pyridylmethyl substituents at 1 position of the 6-benzoyluracil residue led to a number of compounds with a significantly higher genetic barrier. So, 1-[(2-amino-6-fluoropyridin-4-yl)methyl]-5-isopropyl-6-(3-methyl-5-cyanobenzoyl)pyrimidine-2,4(1*H*, 3*H*)-dione **48** (Fig. 7) in MT-2 cells showed similar inhibitory properties against wild-type HIV-1 and strains carrying Y181C or K103N mutations.[65] Substitution of a cyano group for a fluoromethyl group gave a highly active 6-[3-(fluoromethyl)-5-methylbenzoyl]-5-isopropyl-1-(pyridin-4-ylmethyl)uracil **49** (Fig. 7), which suppressed HIV-1 replication at EC_{50} 2 nM and had SI > 48,000.

Buckheit's group studied the effect of different substituents at 1 position of uracil, in particular cyclopropylmethyl, cyclobutylmethyl, cyclohexylmethyl, benzyl, 1-cyclopenten-1-ylmethyl and 3-cyclopenten-1-ylmethyl, on the activity of 6-(3,5-dimethylbenzoyl)-5-alkyluracils.[66–68] 1-(Cyclopent-3-en-1-ylmethyl)-6-(3,5-dimethylbenzoyl)-5-ethyluracil **50** and 1-(cyclopropylmethyl)-6-(3,5-dimethylbenzoyl)-5-isopropyluracil **51** (Fig. 8) showed the best antiviral properties. The study of antiviral efficacy of **50** and **51** against laboratory and clinical isolates of HIV-1, HIV-2 and simian immunodeficiency virus (SIV) was carried out in various cell lines:

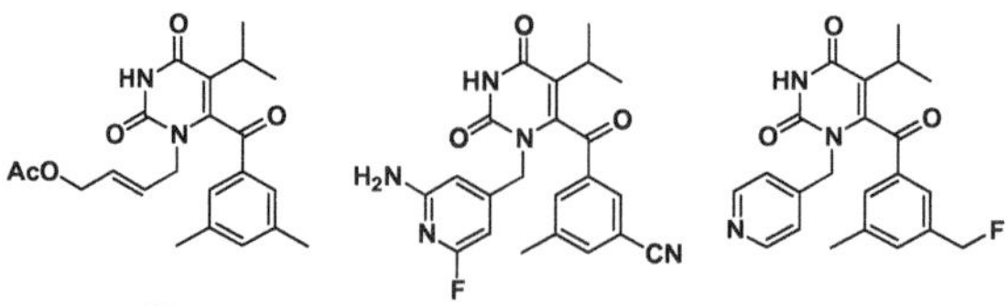

47
EC_{50} <0.009 μM
CC_{50} 58.39 μM
SI >6487

48
EC_{50} 0.0034 μM (wt)
EC_{50} 0.0067 μM (Y181C)
EC_{50} 0.0016 μM (K103N)

49
EC_{50} 0.002 μM
CC_{50} 92.8 μM
SI 48305

Fig. 7 Structures and properties of compounds **47–49**.

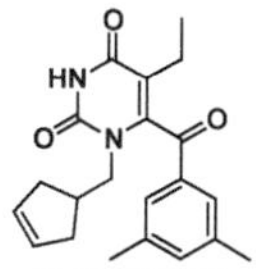
50 (IQP-0410)
EC_{50} 0.2 nM (HIV-1), SI 2,777,778
EC_{50} = 16 μM (HIV-2), SI > 71
EC_{50} = 0.18 μM (HIV-1 entry inhibition),
SI = 15714; IC_{50} = 0,07 μM (RT)

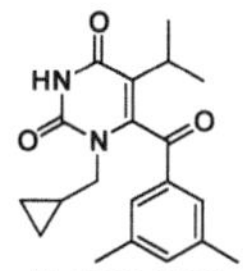
51 (IQP-0528)
EC_{50} = 0.2 nM (HIV-1), SI > 2,000,000
EC_{50} = 0.1 μM (HIV-2) SI > 13030
EC_{50} = 12 nM (HIV-1 entry inhibition),
SI = 100778; IC_{50} = 1,97 μM (RT)

Fig. 8 Structures and properties of compounds **50** and **51**.

T cells (CEM-SS, H9 and MT2), B cells (AA5), monocytes (U937), and a hybrid T/B cell line (174 × CEM). The compounds exhibited high activity against HIV-1 in different cell lines with IC_{50} from 0.0009 to 0.01 μM, without toxic effects in all cell lines up to a concentration of 1000 μM, above which precipitation of substances was observed. SI for **50** and **51** in CEM-SS-cells infected with IIIB-strain of HIV-1 exceeded 2,000,000. Inhibitory effect of compounds **50** and **51** was found against ROD-strain of HIV-2, although 50–200 times less than in the case of HIV-1 inhibition. The selectivity index of compounds **50** and **51** was more than 71 and more than 13,030, respectively. The mechanism of action study showed that compounds **50** and **51** inhibit HIV-1 RT both in the case of homopolymeric poly(rC)-oligo(dG) and in the case of heteropolymeric rRNA template primer of the assay system, but are inactive against HIV-2 RT. Additional study of the mechanism of action revealed that compounds **50** and **51** do not inhibit the activity of integrase, protease, RNase H at concentrations up to 100 μM. Studies in HeLa–LTR cells infected with HIV-1 or HIV-2 have shown that these compounds effectively suppress the attachment of the virus to the cell, and the inhibitory concentrations of virus attachment are close to the inhibitory concentrations obtained for HIV-2 in cell culture.

Inhibition of HIV-1 and HIV-2 entry occurs due to the recognition of a conformational binding site formed during the interaction of the virus with target cells but does not include direct inhibition of gp120-CD4 binding. To further study the modes of action of compound **50**, resistant strains of HIV-1 and HIV-2 were selected and characterized. In the case of HIV-1, the emergence of resistance was in three phases: the emergence of individual amino acid changes in the NNRTI binding pocket; changes in envelope glycoproteins gp120 and gp41; subsequent multiple additional changes to RT. In the case of HIV-2, resistance to inhibition of penetration was achieved without changes in RT. Several mutations were identified in envelope glycoproteins, which confirm the mechanism of inhibition of penetration of compound **50** against HIV-2.[69] Compound **50** was tested against HIV-1 strains resistant to NNRTIs. Viruses with amino acid mutations L100I, T139I, and M184V remained completely sensitive to the action of the compound. A slight decrease in efficiency was observed against viruses with K101E (18 times) and V108I (18 times) mutations. A higher level of resistance was found for viruses with mutations A98G + V108I (70 times), P236L (33 times), Y188H (55 times), K103N (>50 times) and Y181C (10–375 times). Thus, compound **50** has a better resistance profile than nevirapine.[70] Dual mechanism of action, high potency, minimal toxicity to vaginal cells and natural flora, and broad spectrum of activity against clinical wild-type and drug-resistant viruses make these molecules attractive candidates for microbicide development.[71] In the absence of an effective HIV vaccine, the discovery of microbicide candidates for preventing sexual transmission of HIV is considered an important approach to preventing further spread of infection. For this reason, the applicability of compound **51** was evaluated as a vaginal gel[72,73] and polymer films.[74,75] In 2019, ImQuest BioSciences conducted clinical trials of a microbicide gel containing compound **51** (IQP-0528).[76]

2.3.2 Details of the chemical synthesis of 6-benzoyl derivatives of HEPT

The synthesis of 6-benzoyl uracil derivatives **53** presented in Scheme 10. It includes the condensation of 6-chloropyrimidine **52** with phenylacetonitrile in a DMF solution in the presence of NaH. Resulting 6-(α-cyanobenzyl) pyrimidine derivatives without isolation from the reaction mixture were subjected to oxidation with pure oxygen for 1 h. Subsequent acid hydrolysis gave the corresponding 6-benzoyluracils **53** with 86% yield.[77,78]

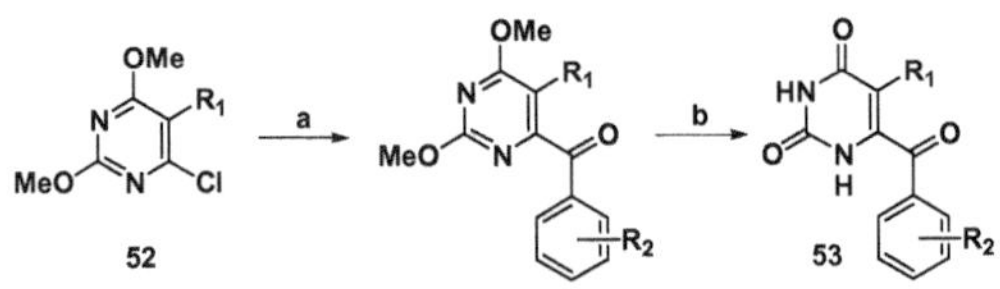

Scheme 10 Reagents and conditions: (a) (i) $ArCH_2CN$, NaH, DMF; (ii) O_2, 1 h; (b) HCl, MeOH, Δ.

Another method was proposed by Bardagí and Rossi.[79] It does not involve oxidation and is based on the Stille reaction. 2,4-Dimethoxy-6-(trimethylstannyl)pyrimidine **54** was obtained with 76% yield by treating 2,4-dimethoxy-6-chloropyrimidine with sodium trimethylstannyl. The following condensation of **54** with aroyl chloride in the presence of

$Pd(PPh_3)_2Cl_2$ in toluene and further hydrolysis led to 6-benzoyluracil **55** in 54% yield (Scheme 11).

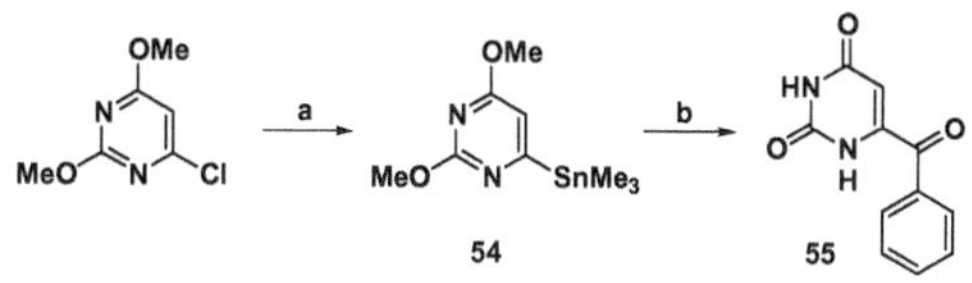

Scheme 11 Reagents and conditions: (a) $NaSnMe_3$; (b) (i) PhCOCl, toluene, $Pd(PPh_3)_2Cl_2$; (ii) HCl, MeOH, Δ.

Compound **48** was synthesized starting from 3,5-dichloro-2,4, 6-trifluoropyridine **56**, which was then condensed with dibenzyl malonate and subjected to mono-decarboxylation. Hydrogenolysis of benzyl ester **57** gave the corresponding acid, following decarboxylation and bromination led to bromide **58**. Replacement of bromide with sodium benzoate, removal of chlorine atoms by hydrogenolysis and cleavage of benzoate gave 2,6-difluoro-4-hydroxymethylpyridine **59**. Substitution of fluoride with *p*-methoxybenzylamine led to the formation of mesylate **60**. Condensation of **60** with 6-(3-cyano-5-methylbenzoyl)-5-isopropyl uracil **61** gave the uracil derivative **62**. Final deprotection resulted in the target compound **48** (Scheme 12).[51]

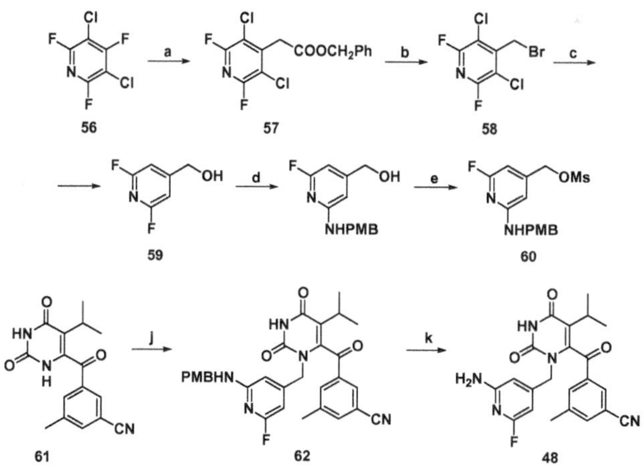

Scheme 12 Reagents and conditions: (a) (i) dibenzyl malonate, NaH, DMF; (ii) DMSO/H_2O, 130 °C, 75%; (b) (i) H_2, Pd/C, THF, 78%–89%; (ii) HgO, Br_2, chlorobenzene, 145 °C, 75%–80%; (c) (i) NaOBz, DMF, 66%–81%; (ii) H_2 (50 psi), Pd/C, EtOH, 61%; (iii) NaOMe, MeOH, 95%–98%; (d) 4-methoxybenzylamine, 125 °C, 90%–94%; (e) methanesulfonyl chloride, triethylamine, $CHCl_3$; (j) mesylate **60**, K_2CO_3, LiI, DMF, 32%–55%; (k) CAN, ACN/H_2O, 77%.

3. N1-Substituted uracil derivatives

3.1 Evolution of group

In 2009, Anderson's group described the anti-HIV-1 properties of 1-[2-(2-benzyl-4-methylphenoxy)ethyl]uracil **63** (Fig. 9). Despite its moderate inhibitory properties, compound **63** became a start point for the design of new branch of highly active anti-HIV-1 agents.[80] The introduction of chlorine atoms into the diphenylmethyl fragment resulted to better antiviral activity. Thus, 1-(2-(4-chloro-2-(2,5-dichlorobenzyl)phenoxy)ethyl)uracil **64** (Fig. 9) showed EC_{50} 0.38 μM in the MT-2 cell culture. However, the presence of halogens in the molecule led to increased toxicity (CC_{50} 15 μM). A significant improvement in inhibitory properties was observed when the CH_2 group was replaced with an oxygen atom and one of the chlorine atoms with CN group. In particular, (E)-3-(3-chloro-5-(4-chloro-2-(2-(2,4-dioxo-3,4-dihydropyrimidin-1(2*H*)-yl)ethoxy)phenoxy) phenyl)acrylonitrile **65** (Fig. 9) suppressed HIV-1 replication in extremely low concentration—EC_{50} 0.055 nM. Even though compound **65** was quite toxic (CC_{50} 10 μM), its selectivity index was 181818. It was shown that mutations (Y181C and K103N/Y181C) in the hydrophobic pocket of RT lead to a decrease in the activity of compounds **64** and **65** by several orders.[81]

Benzophenone-uracil derivatives have been synthesized and studied as analogs of benzophenone NNRTIs described by GlaxoSmithKline.[82,83] The most active among them was 1-[2-[4-chloro-2-(3,5-dimethylbenzoyl)-phenoxy]ethyl]uracil **66** (Fig. 10) with EC_{50} 0.016 μM in CEM cell culture infected by HIV-1, its SI was more than 19,500. Compound **66** had lower activity against the double mutant HIV-1 (K103N/Y181C).

The compound **66** showed marked activity against wild-type HIV-1 RT (K*i* 2.4 μM) and retained active against most mutant forms of HIV-1 RT (L100I, K103N, V106I, Y181C, Y188L, G190A and K103N/Y181C)

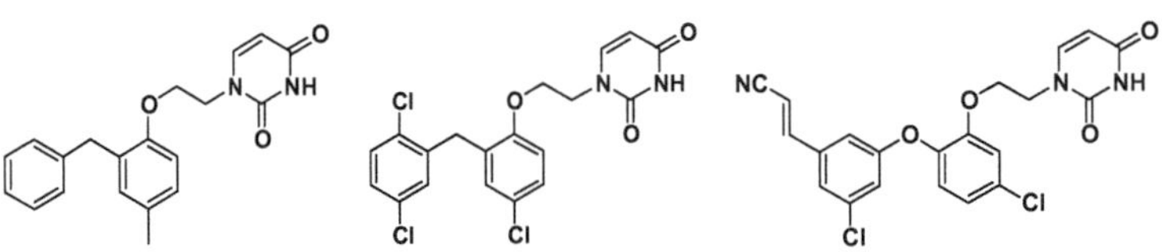

63
EC_{50} 4.8 μM
CC_{50} 72 μM
SI 15

64
EC_{50} 0.38 μM
CC_{50} 15 μM
SI 39

65
EC_{50} 0.000055 μM
CC_{50} 10 μM
SI 181818

Fig. 9 Structures and properties of compounds **63–65**.

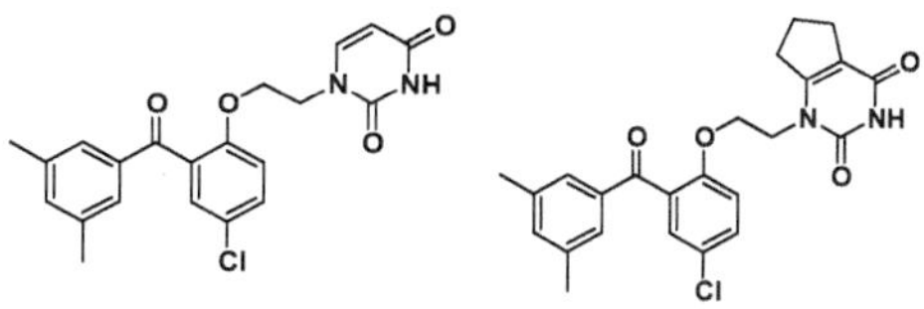

66
EC_{50} 0.016 μM
CC_{50} >313 μM
SI >19588

67
EC_{50} 0.0075 μM
CC_{50} >50 μM
SI >6666

Fig. 10 Structures and properties of compounds **66** and **67**.

except for V106A. The activity profile of compound **66**, retaining inhibitory properties towards most mutant RT forms, like one of efavirenz, although the absolute K*i* values were closer to that of nevirapine.[83]

The next stage in the development of benzophenone-uracil derivatives was the synthesis and study of the anti-HIV-1 properties of 1-[2-[2-(3, 5-dimethylbenzoyl)-4-chlorophenoxy]ethyl]-6,7-dihydro-1*H*-cyclopenta[*d*] pyrimidine-2,4-(3H,5H)-dione **67** (Fig. 10). The compound **67** inhibited viral replication with EC_{50} 7.5 nM in MAGI-CCR5 cell culture and its selectivity index exceeded 6666.[84]

The introduction of a cyanvinyl fragment into an NNRTI molecule, as a rule, leads to an increase in the anti-HIV-1 effect. This was shown for rilpivirine[85] or compound **65**.[81] However, the cytotoxicity of the compounds also increases: the cyanvinyl fragment is easily oxidized in the cell and can produce toxic metabolites. For this reason, the cyanvinyl residue was incorporated into the second cycle of the bicyclic NNRTI moiety. Thus, by probing a number of heterocyclic systems, 8-(2-(2-(2,4-dioxo-3,4-dihydropyrimidin-1(2*H*)-yl)ethoxy)phenoxy)indolizine-2-carbonitrile **68** was synthesized (Fig. 11). The EC_{50} of **68** was 0.38 nM, which is weaker than the EC_{50} of compound **65**. However, the cytotoxicity of compound **68** was lower than that of **65** and SI exceeded 263157.[86] The disadvantage of compound **68** was a drop in activity against mutant Y181C and double mutant K103N/Y181C.[86]

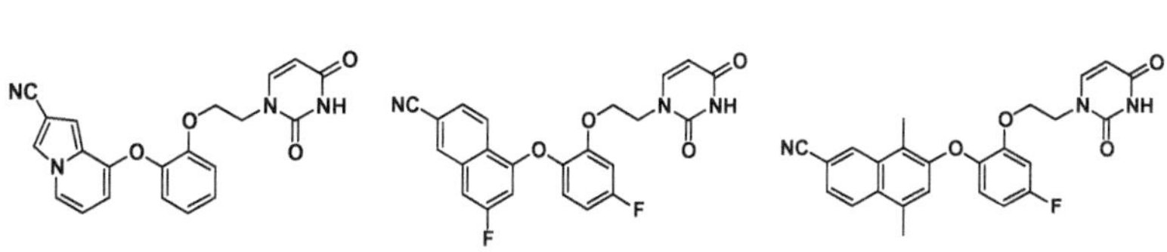

68
EC_{50} 0.00038 μM
CC_{50} >100 μM
SI >263157

69
EC_{50} 0.0019 μM
CC_{50} >100 μM
SI >52631

70
EC_{50} 0.0019 μM
CC_{50} >100 μM
SI >52631

Fig. 11 Structures and properties of compounds **68–70**.

Several attempts have been made to develop compounds with a higher genetic barrier. Such compound turned out to be 5-(2-(2-(2,4-dioxo-3,4-dihydropyrimidin-1(2*H*)-yl)ethoxy)-4-fluorophenoxy)-7-fluoro-2-naphthonitrile **69** (Fig. 11), which inhibited HIV-1 replication in MT-2 cell culture with EC_{50} 1.9 nM (wild type), 5.6 nM (Y181C) and 21 nM (K103N/Y181C).[87]

Replacement of 1-naphthyl fragment in compound **69** by 2-naphthyl fragment led to a series of potent inhibitors. Among them 7-(2-(2-(2,4-dioxo-3,4-dihydropyrimidin-1(2*H*)-yl)ethoxy)-4-fluorophenoxy)-5,8-dimethyl-2-naphthonitrile **70** (Fig. 11) suppressed HIV-1 replication in MT-2 cell culture with EC_{50} 1.9 nM (wild type), 28 nM (Y181C), and 410 nM (K103N/Y181C).[88] When comparing the genetic barrier, compound **70** is noticeably inferior to **69**, which was chosen for preclinical trials along with **65**.[89]

The preclinical potential of compounds **65** and **69** was first estimated in BALB/c mice.[89] A single abdominal injection of compounds led to a prolonged serum residence time up to 48 h and a maximum concentration (C_{max}) 4000–15,000 times higher than their therapeutic/effective concentrations. Observed C_{max} values were 4–15 times lower than their cytotoxic concentrations in MT-2 cells. Compound **69** showed increased area under the curve and decreased plasma clearance compared to compound **65** and efavirenz (NNRTI standard). Therefore, the overall pharmacokinetic profile of compound **69** was better than for compound **65** and efavirenz. In addition, both compounds were well tolerated with no signs of acute toxicity in BALB mice. Thus, compounds **65** and **69** besides high anti-HIV-1 activity in vitro, demonstrated safety and increased circulation time in vivo, which makes them promising for further development as new HIV NNRTIs.[89] Further study of compound **69** revealed a synergistic antiviral activity with existing anti-HIV drugs (Emtricitabine, Tenofovir disoproxil fumarate, Elvitegravir, Festinavir). The efficacy of compound **69** was also demonstrated in humanized mice (Hu-PBL). The compound completely suppressed viral load and prevented the loss of human CD^{4+} T cells. A long-acting nano-formula of compound **69** (compound nanoparticle) in lactide-coglycolide polymer was developed and showed stable maintenance of drug concentrations in plasma and drug efficacy for almost 3 weeks after a single dose.[90]

Another class of NNRTIs containing diaryl ether is described. 5-Substituted uracils carrying a 3-phenoxybenzyl substituent at the N1 position showed significant activity against HIV-1 RT, and in the infected

MT-4 cells. The most promising compound was 3-[6-bromo-2-fluoro-3-(5-methyl-2,4-dioxo-3,4-dihydro-2H-pyrimidin-1-ylmethyl)phenoxy]-5-chlorobenzonitrile **71** (Fig. 12), which inhibited the activity of the wild-type viral enzyme and the double mutant K103N/Y181C with IC_{50} 3 nM. In cell culture, **71** inhibited HIV-1 with EC_{50} 1 nM (wild type) and 11 nM (K103N/Y181C). Also, **71** affected the activity of several RT variants containing the most clinically significant mutations in nanomolar concentrations.[91]

Another group of compounds was chimeric structures containing a diaryl fragment of Etravirine/Rilpivirine and a 1-ethyl uracil fragment, which is part of compounds **63–69**. The most active compound was (E)-3-(4-((5-chloro-2-((2-(2,4-dioxo-3,4-dihydropyrimidin-1(2*H*)-yl)ethyl) amino)pyrimidin-4-yl)oxy)-3,5-dimethylphenyl)acrylonitrile **72** (Fig. 12), which inhibited the cytopathic effect of HIV-1 in MT-4 cells by 50% with EC_{50} 0.0056 μM and SI more than 50,454. The compound **72** inhibited HIV-1 RT with IC_{50} 1.58 μM, which is comparable to the effect of Nevirapine and Etravirine. However, the study of its genetic barrier showed that **72** significantly loses the anti-HIV-1 effect in mutants (K103N, L100I, Y181C, E138K, and Y188L) and becomes almost completely inactive with double mutations (Y181C/K103N, F227L/V106A).[92]

3.2 Synthesis of N1-substituted uracil derivatives

The synthesis of all uracil derivatives **63–70** consisted of three main stages:

(A) The formation of the diaryl part;

(B) Synthesis of 2-bromoethyl intermediate;

(C) Synthesis of target compound by condensation of 2-bromoethyl intermediate with an appropriate nucleophile.

The synthesis of compounds **63**, **64**, **66** and **67** is presented in Scheme 13.

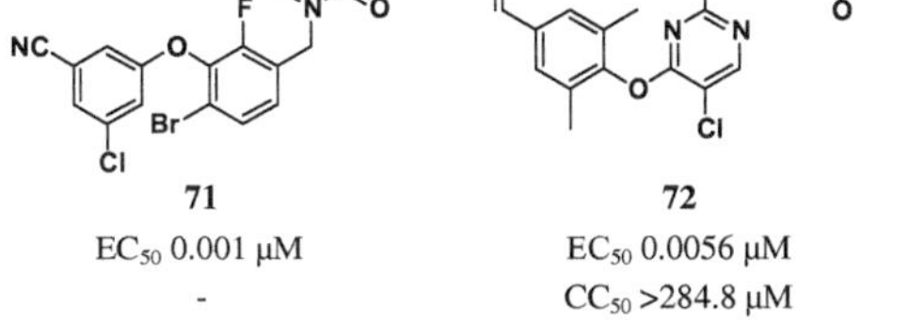

71	**72**
EC_{50} 0.001 μM	EC_{50} 0.0056 μM
-	CC_{50} >284.8 μM
-	SI >50454

Fig. 12 Structures and properties of compounds **71** and **72**.

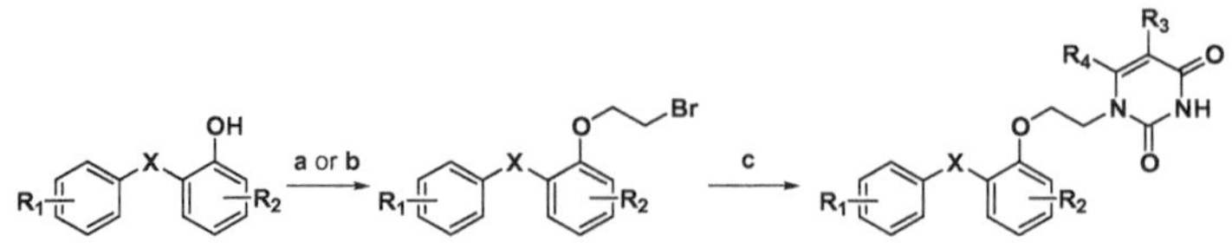

73 74 63, 64, 66, 67

Scheme 13 Reagents and conditions: (a) DIAD, 2-bromoethanol, Ph_3P, THF, rt., overnight; (b) 1,2-dibromoethane, acetone, Δ, 48 h; 68%–92%; (c) 2,4-bis (trimethylsilyloxy)pyrimidine, 170–180 °C, 1.5 h; 38%–90%.

(A) 2-Benzyl-4-methylphenol **73a** (R_1=H, R_2=4-Me, X=CH_2) and 4-chloro-2-(2,5-dichlorobenzyl)phenol **73b** (R_1=2,5-Cl_2, R_2=4-Cl, X=CH_2), which are part of 1-[2-(2-benzylphenoxy)ethyl]uracils **63** and **64**, were obtained by the Friedel-Crafts reaction from benzyl halides or benzyl alcohols with phenols in the presence of Lewis acids—$ZnCl_2$, $AlCl_3$ or the reaction of benzyl halides with potassium phenolates in an anhydrous non-polar solvent—benzene, toluene.[93,94] 4-Chloro-2-(3, 5-dimethylbenzoyl)phenol **73c** (R_1=3,5-Me_2, R_2=4-Cl, X=C(O)), which is a part of 1-[2-[4-chloro-2-(3,5-dimethylbenzoyl)phenoxy] ethyl]-derivatives of uracil **66** and **67**, was obtained by Fries rearrangement.

(B) 2-Bromoethyl esters **74** were obtained by two methods: (1) by the Mitsunobu reaction (DIAD, Ph_3P, 2-R-phenol and 2-bromoethanol)[81] or (2) by treating phenol **73** with an excess of 1,2-dibromoethane in acetone in the presence of K_2CO_3.[95]

(C) The synthesis of target uracil derivatives **63**, **64**, **66** and **67** was carried out by condensation of equimolar amounts of 2,4-bis(trimethylsilyloxy)-pyrimidine and bromides without solvent upon heating to 170–180 °C for 1.5 h.[96]

Synthesis of compound **65** (Scheme 14), containing a cyanvinyl fragment,[81] was different from the synthesis of compounds **63**, **64**, **66** and **67**. Despite the noticeable structural differences between these compounds, the synthesis also consisted of the three mentioned stages: A–C.

Stage A included several steps and began with the condensation of 1-fluoro-2-methoxy-4-nitrobenzene **75** and 3-chloro-5-iodophenol **76** under the conditions of the S_NAr reaction. The NO_2 group of resulting anisole **77** was reduced in the presence of Fe and NH_4Cl to give the corresponding amine **78**, which was converted to iodide **79** by treatment with concentrated HCl and CuCl under conditions of Sandmeyer reaction.[81] Heck reaction of iodide **79** with acrylonitrile in the presence of

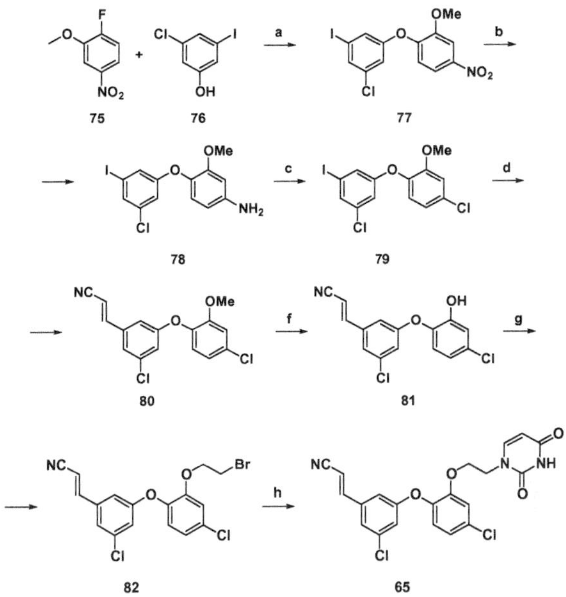

Scheme 14 Reagents and conditions: (a) K_2CO_3, DMSO, 60 °C, overnight; (b) Fe, NH_4Cl, EtOH, H_2O, 75 °C, 2 h, 100%; (c) $NaNO_2$, HCl conc., CuCl, 0 °C, 2 h, 40%; (d) acrylonitrile, $PdCl_2(PPh_3)_2$, Et_3N, DMF, 140 °C, 2 h, 21%–53%; (f) LiCl, DMF 160 °C, 16 h; (g) DIAD, 2-bromoethanol, Ph_3P, THF, rt., overnight; (h) 3-benzoyluracil, K_2CO_3, DMF, rt., overnight then NH_4OH, MeOH, rt.; 23%.

Et_3N and a catalytic amount of $PdCl_2(PPh_3)_2$ in DMF led to the formation of the (E)-cyanovinyl anisole derivative **80**. Hydrolysis of the methoxy group of anisole **80** was carried out by the action of lithium chloride in boiling DMF and gave the target (E)-cyanovinyl-phenol derivative **81**.[81]

Stage B of the synthesis included the introduction of a 2-bromoethyl on the phenolic hydroxyl of compound **81**. Bromide **82**, formed under the Mitsunobu reaction conditions (DIAD, Ph_3P, 2-R-phenol and 2-bromoethanol), was used in the next step to synthesize the target compound **65**.[81]

Stage C. The presence of an unsaturated fragment in bromide **82** made this compound labile to oxidation. So, the use of 2,4-bis(trimethylsilyloxy) pyrimidine for the condensation was not allowed since it proceeds under severe conditions (heating at 170–180 °C). To overcome this problem, N-benzoyluracil was alkylated by bromide **82**. Benzoyl group was hydrolyzed then in accordance with a previously developed method.[97]

Scheme 15 shows the synthesis of compound **68**, which differed from described above, but also has three stages.

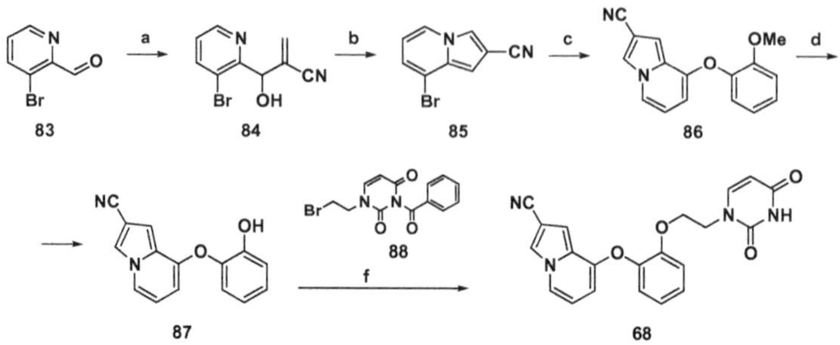

Scheme 15 Reagents and conditions: (a) acrylonitrile, DABCO, heat, 2 h; (b) Ac_2O, 100 °C, 2 h→140 °C, 16 h; (c) CuI, Cs_2CO_3, 2-methoxyphenol, 2,2,6,6-tetramethyl-3,5-heptanedione, dioxane, 120 °C, 18 h; (d) BBr_3, DCM, −78 °C→0 °C, 3 h; (f) **88**, K_2CO_3, DMF, 60 °C, 3 h then NH_4OH, dichloromethane, 16 h, 29%.

Stage A. Synthesis of the indolizine system containing a cyano group at 2 position is carried out in accordance with a previously developed method.[98] Treatment of 3-bromopyridine-2-carbaldehyde **83** with acrylonitrile in the presence of DABCO leads to the formation of 2-[(3-bromopyridin-2-yl)(hydroxy)methyl]prop-2-enenitrile **84**. Compound **84** is heated with acetic anhydride to afford 8-bromoindolizine-2-carbonitrile **85**. Reaction of nitrile **85** with 2-methoxyphenol in the presence of cesium carbonate, catalytic amounts of copper iodide I and 2,2,6,6-tetramethyl-3,5-heptanedione gave 8-(2-methoxyphenoxy)indolizine-2-carbonitrile **86**. Hydrolysis of the methoxy group of anisole **86** by action of BBr_3 provided 8-(2-hydroxyphenoxy)indolizine-2-carbonitrile **87**.

Stage B. 3-Benzoyl-1-(2-bromoethyl)pyrimidine-2,4(1*H*,3*H*)-dione **88** was obtained by alkylation of 3-benzoyluracil with an excess of 1,2-dibromoethane in the presence of potassium carbonate in DMF. The yield of bromide **88** was 37%.[86]

Stage C. The synthesis of the target compound **68** was carried out by the reaction of phenol **87** and a 2-bromoethyl uracil derivative **88** in the presence of potassium carbonate in DMF, followed by hydrolytic elimination of the benzoic acid residue.[86]

There were also three stages A–C in the synthesis of compound **69** (Scheme 16).[87]

Stage A. Synthesis of 7-fluoro-5-(4-fluoro-2-methoxyphenoxy)-2-naphthonitrile **89** by the condensation of 4-fluoro-2-methoxyphenol **90** and 5-bromo-7-fluoro-2-naphthonitrile **91** in the presence of cesium carbonate and catalytic amounts of CuI and 2,2,6,6-tetramethyl-3,5-heptanedione,[99,100] followed by hydrolysis of the methoxy group

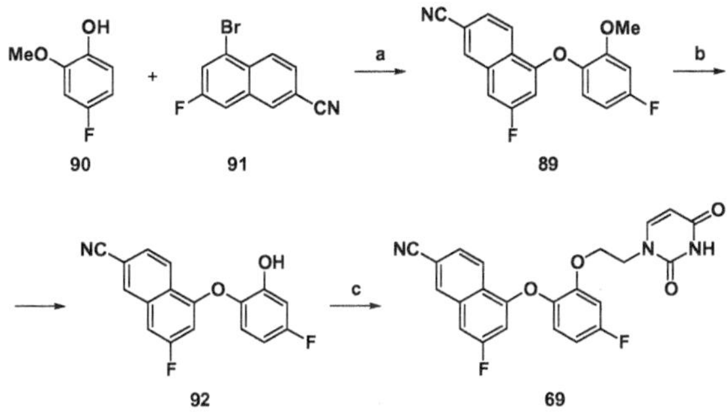

Scheme 16 Reagents and conditions: (a) CuI, Cs_2CO_3, 2-bromo-5-fluoroanisole, 2,2,6,6-tetramethyl-3,5-heptanedione, dioxane, 100 °C, 48 h; (b) BBr_3, DCM, −78 °C→0 °C, 3 h; (c) **88**, K_2CO_3, DMF, 60 °C, 3 h then NH_4OH, dichloromethane, 16 h, 28%.

of anisole **89** by treatment with BBr_3 gave 7-fluoro-5-(4-fluoro-2-hydroxyphenoxy)-2-naphthonitrile **92**.

Stage B. 3-Benzoyl-1-(2-bromoethyl)uracil **88** was obtained as described above.[86]

Stage C. The synthesis of the target compound **69** was carried out by the reaction of phenol **92** and 3-benzoyl-1-(2-bromoethyl)uracil **88** in the presence of potassium carbonate in a DMF, followed by hydrolytic elimination of the benzoic acid residue.[87]

The synthesis of compound **71** (Scheme 17) included the constructing a diaryl fragment (stage A), its conversion to bromide (stage B), and condensation with uracil (stage C).[91] The interaction of 3-chloro-5-hydroxybenzonitrile **93** with an excess of 1,4-dibromo-2,3-difluorobenzene **94** in the presence of NaH in DMF led to 3-chloro-5-(3,6-dibromo-2-fluorophenoxy)benzonitrile **95** in good yield. Then benzonitrile **95** was converted into aldehyde **96** by the action of i-PrMgCl and subsequent treatment with DMF. Next it was reduced to benzyl alcohol **97** with following conversion into benzyl bromide **98**. Final alkylation of thymine by bromide **98** led to the formation of a target **71**.

Compound **72** was synthesized by the reaction of 2,4,5-trichloropyrimidine **99** with 3-(4-hydroxy-3,5-dimethylphenyl)acrylonitrile in the presence of potassium carbonate in accordance with the previously described procedure (Scheme 18).[101,102] As a result, 2,5-dichloropyrimidine **100** was formed. It was condensed with 1-(2-aminoethyl)uracil[103] in *N*-methylpyrrolidone (NMP) solution, which led to the formation of the target **72**.

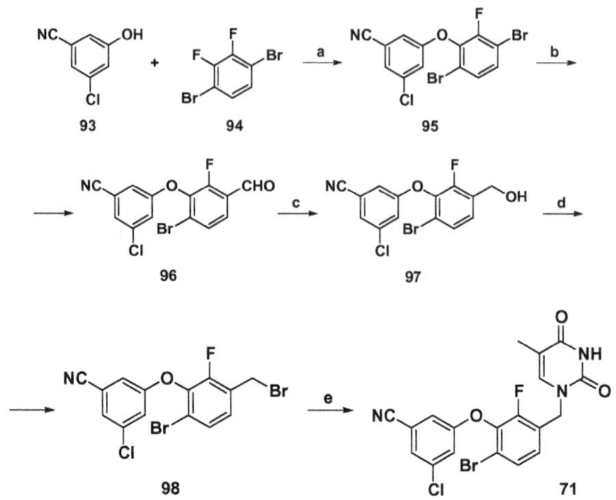

Scheme 17 Reagents and conditions: (a) NaH, DMF, 125 °C, 2 h; (b) iPrMgCl, CuCN·2LiCl, THF then DMF, −78 °C; (c) $NaBH_4$, THF/MeOH, 22 °C 24 h; (d) PBr_3, CH_2Cl_2, 22 °C or MsCl, NEt_3, CH_2Cl_2, 22 °C, then LiBr, THF, 80 °C; (e) thymine, KI, K_2CO_3, MeCN, 85 °C, 90%.

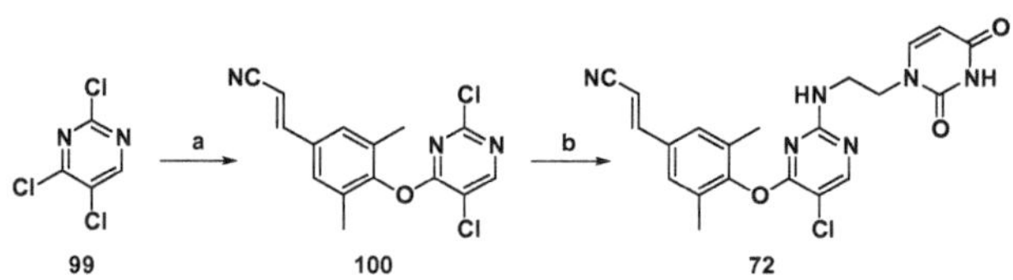

Scheme 18 Reagents and conditions: a) 3-(4-hydroxy-3,5-dimethylphenyl)acrylonitrile, K_2CO_3, DMF, rt.; b) 1-(2-aminoethyl)uracil, DIPEA, NMP, 100 °C, 60%.

4. 1,3-Disubstituted uracils

4.1 Evolution of group

Maruyama et al. described the synthesis and antiviral properties of many 3-benzyl uracil derivatives, which were active against both HIV-1 and CMV. The parent 1-cyanomethyl-3-benzyluracil had weak inhibitory properties. However, the introduction of methyl groups in the *meta* positions led to a noticeable increase in anti-HIV-1 activity. 1-Cyanomethyl-3-(3,5-dimethylbenzyl)uracil **101** (Fig. 13) had EC_{50} 0.59 μM, and SI more than 786. The introduction of a benzyl radical into uracil 1 position significantly increased the anti-HIV-1 activity: 1-benzyl-3-(3,5-dimethylbenzyl) uracil **102** (Fig. 13) inhibited the cytopathic effect of the virus by 50% at 0.234 μM, but the cytotoxicity also increased and as a result SI was 829.

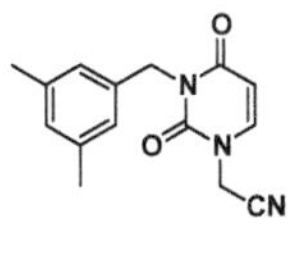

101
EC_{50} 0.59 μM
CC_{50} >464 μM
SI >786

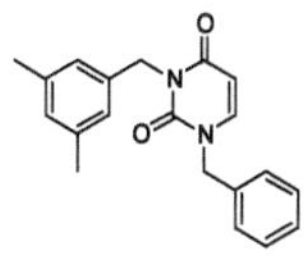

102
EC_{50} 0.234 μM
CC_{50} 194 μM
SI 829

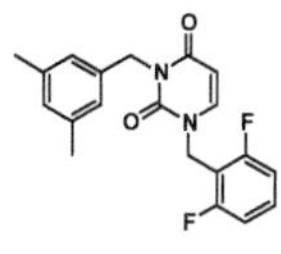

103
EC_{50} 0.115 μM
CC_{50} 326 μM
SI 2835

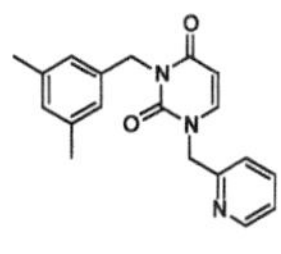

104
EC_{50} 0.04 μM
CC_{50} 145 μM
SI 3625

Fig. 13 Structures and properties of compounds **101–104**.

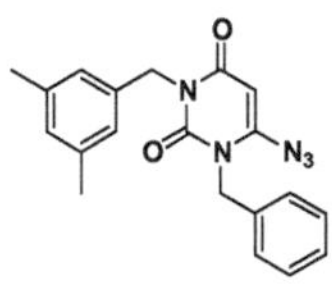

105
EC_{50} 0.088 μM
CC_{50} 40.5 μM
SI 460

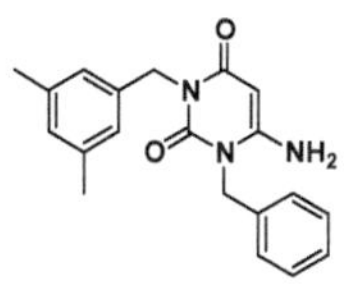

106
EC_{50} 0.06 μM
CC_{50} 50.1 μM
SI 835

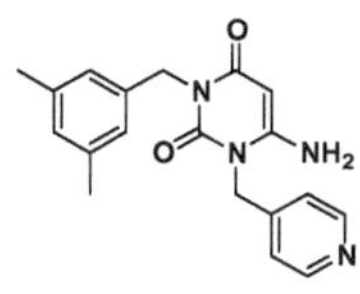

107
EC_{50} 0.03 μM
CC_{50} >100 μM
SI >2863

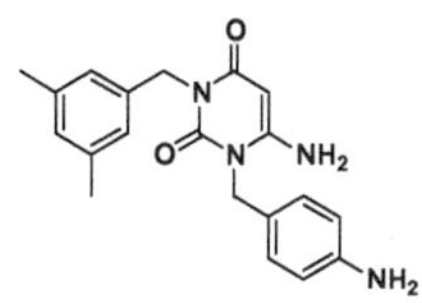

108
EC_{50} 0.01 μM
CC_{50} >20 μM
SI >2000

Fig. 14 Structures and properties of compounds **105–108**.

The introduction of fluorine atoms into the benzyl fragment (compound **103**, Fig. 13) improved anti-HIV-1 properties.[104]

Further studies have shown that the introduction of a heterocyclic fragment also helped to enhance antiviral activity. Thus, 3-(3,5-dimethylbenzyl)-1-(2-picolyl)uracil **104** (Fig. 13) inhibited HIV-1 replication with EC_{50} 0.04 μM and SI 3625.[105]

The antiviral properties of 6-substituted 1-benzyl-3-(3,5-dimethylbenzyl)uracil derivatives have been studied. 6-Azido-1-benzyl-3-(3,5-dimethylbenzyl)uracil **105** (Fig. 14) and 6-amino analogue **106** (Fig. 14) exhibited high anti-HIV-1 activity in MT-4 cells.[106] To determine the resistance profiles of these compounds, HIV-1-infected MT-4 cells were cultured with increasing concentrations of each compound. After serial passages of infected cells, virus variants turned out 500 times less sensitive to uracil derivatives **105** and **106** when compared to the wild type. Sequence analysis of the RT mutant HIV-1 variants showed the Y181C mutation in the RT polymerase domain for all viruses resistant to **105** and **106**. Docking studies using the RT crystal structure supported the necessities of amino acid residues Leu100, Val106, Tyr181 and Trp229 for the inhibitory effect of **106**. Four additional mutations have also been identified (K451R, R461K, T468P, and D471N) in the RT RNase H domain. However, their role in the formation of resistance is unclear.[107]

Further studies led to the discovery of a more active compound, 6-amino-3-(3,5-dimethylbenzyl)-1-(4-picolyl)uracil **107** (Fig. 14), which suppressed HIV-1 replication in MT-4 cells with EC_{50} 0.03 μM and SI of almost 3000. Docking using **107** as a ligand showed that the hydrogen of the 6-amino group forms a hydrogen bond with the amide group of the Lys101 residue (NH···O=C), and the 3,5-dimethylbenzyl fragment is located in the hydrophobic region, formed by residues Tyr181, Tyr188, Trp229 and Leu234 from HIV-1 RT. The nitrogen of the 4-picolyl fragment forms a hydrogen bond with the amide group of the Lys101 (N···H–N) residue.[108]

High anti-HIV-1 activity was found for 6-amino-3-(3,-5-dimethylbenzyl)-1-(4-aminobenzyl)uracil **108** (Fig. 14), which in MT-4 cell culture suppressed HIV-1 replication with EC_{50} 0.01 μM and had SI more than 2000. Docking using compound **108** as a ligand showed that the hydrogen of the 6-amino group forms a hydrogen bond with the amide group of the Lys101 residue (NH···O=C), the 3,5-dimethylbenzyl fragment is in the hydrophobic region and gives π-π stacking with aromatic rings of Tyr181 and Tyr188 residues. In addition, a CH–π interaction was observed between the methyl group of the 3,5-dimethylbenzyl fragment and the indole fragment of the Trp229 residue. The data obtained for **108** were the almost the same as for 1-(4-picolyl)uracil **107**. However, **108** with a 4-aminobenzyl group in N1 position had stronger inhibitory properties than **107**. It was assumed that the reason for this was better solubility of compounds **108** in water. The amino group of the 4-aminobenzyl fragment **108** can form hydrogen bonds with water molecules outside of HIV-1 RT, which can stabilize the ligand **108** in the hydrophobic pocket, thus enhancing the activity against HIV-1.[109]

A number of 1-[2-(aryloxy)ethyl]-3-benzyl uracil derivatives were synthesized and their antiviral properties were studied. It was found that the better anti-HIV-1 effect among compound of this group was shown by 1-[2-(4-methylphenoxy)ethyl]-3-(3,5-dimethylbenzyl)uracil **109** (Fig. 15).[110]

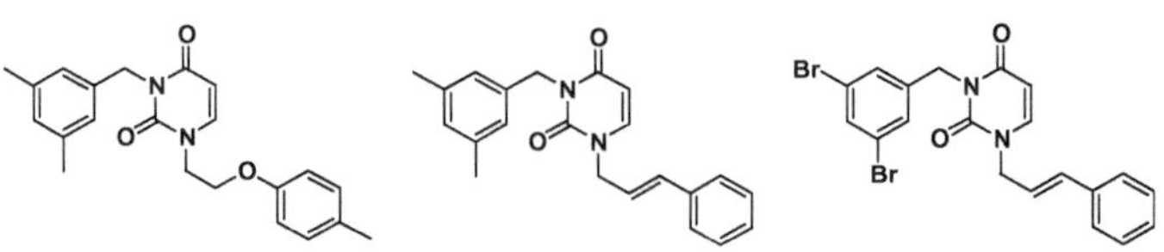

109	110	111
EC_{50} 0.27 μM	EC_{50} 1.27 μM	EC_{50} 1.72 μM
CC_{50} >343 μM	CC_{50} 257 μM	CC_{50} 231 μM
SI >1270	SI 202	SI 134

Fig. 15 Structures and properties of compounds **109–111**.

The compound inhibited the cytopathic effect of the virus by 50% in the CEM-SS cells with EC_{50} 0.27 μM and its SI exceeded 1270. Another group of compounds containing a cinnamyl substituent at 1 position was tested against HIV-1 in the CEM SS cells. The most active were 1-cinnamyl-3-(3,5-dimethylbenzyl)uracil **110** (Fig. 15) (EC_{50} 1.27 μM, CC_{50} 257 μM) and 1-cinnamyl-3-(3,5-dibromobenzyl)uracil **111** (Fig. 15) (EC_{50} 1.72 μM and CC_{50} 231 μM). The antiviral testing of **109**, **110**, and **111** against HIV-1 RT showed that the compounds inhibit the wild-type enzyme at IC_{50} 0.26, 0.55, and 0.31 μM, respectively. Compounds **109**, **110**, and **111** retained a pronounced activity against the L100I and G190A mutants, and inhibited RT with the substitution of V106A or Y181C at much higher concentrations, but were not active against the double mutant K103N/Y181C.[111] It was also found that the introduction of an amide fragment into the bridge significantly reduces the activity against HIV-1.[110]

4.2 Synthesis of 1,3-disubstituted uracil derivatives

1-Cyanomethyl-3- (3,5-dimethylbenzyl)uracil **101** was obtained in 65% yield by the Mitsunobu reaction: treatment of the starting 1-cyanomethyluracil **112**[112] with 3,5-dimethylbenzyl alcohol in the presence of excess PPh_3 and diisopropyl azodicarboxylate (DIAD) in THF (Scheme 19).

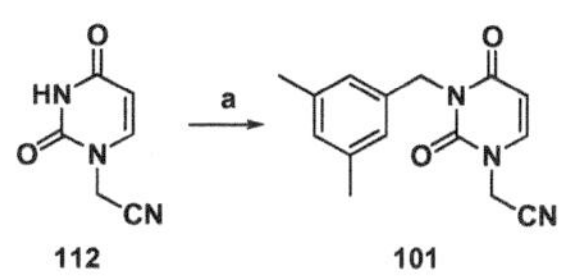

Scheme 19 Reagents and conditions: (a) THF, 3,5-dimethylbenzyl alcohol, PPh_3, DIAD, 50 °C, overnight; 65%.

For the synthesis of 3-(3,5-dimethylbenzyl)-1-(pyridyl-2-methyl)uracil **104**, Maruyama et al.[105] developed another approach based on the formation of 3-(3,5-dimethylbenzyl)uracil **113**. Its synthesis consisted in the treatment of 1-(tetrahydrofuran-2-yl)uracil **114**[113] with 3,5-dimethylbenzyl alcohol in the presence of PPh_3 excess and DIAD in THF. The yield of 3-(3,5-dimethylbenzyl)-1-(tetrahydrofuran-2-yl)uracil **115** was 68%. Its hydrolysis with 1M HCl gave uracil **113** in 69% yield. Treatment of compound **113** with 2-picolyl chloride hydrochloride in DMF in the presence of potassium carbonate led to the target compound **104** in 86% yield (Scheme 20).

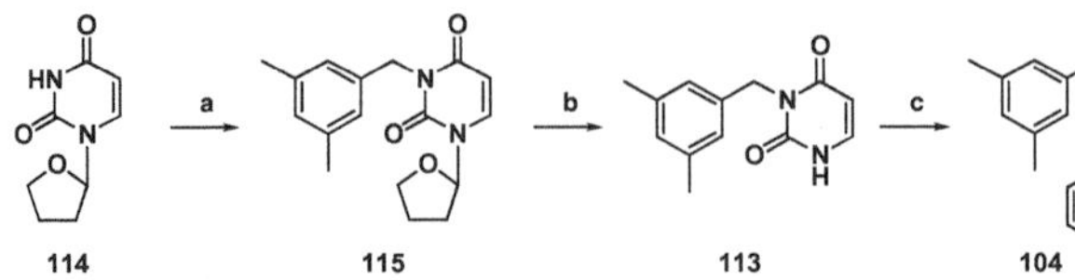

Scheme 20 Reagents and conditions: (a) THF, 3,5-dimethylbenzyl alcohol, PPh_3, DIAD, 50 °C, overnight; (b) 1 M HCl, 1,4-dioxane, reflux, 15 h; (c) 2-picolyl chloride hydrochloride, DMF, K_2CO_3; 86%.

The synthesis of 6-amino-3-(3,5-dimethylbenzyl)-1-(4-aminobenzyl) uracil **108** is shown in Scheme 21. An excess of 6-chlorouracil **116** was treated with 4-nitrobenzyl bromide in DMF in the presence of potassium carbonate, which led to 6-chloro-1-(4-nitrobenzyl)uracil **117** with 75% yield. Using the Mitsunobu reaction, compound **117** was condensed with 3,5-dimethylbenzyl alcohol in the presence of an excess of PPh_3 and *N,N,N′,N′*-tetramethylazodicarboxamide (TMAD) in THF, which gave 1-benzyl-6-chloro-3-(3,5-dimethylbenzyl)uracil **118** in 69% yield. It was treated with sodium azide in DMF at room temperature to form 6-azido derivative **119** in 90% yield. Reduction of the azido group with lithium aluminum hydride led to 6-amino derivative **120** in 71% yield, which was converted into the target 6-amino-1-(4-aminobenzyl)-3-(3,5-dimethylbenzyl)uracil **108** by White's procedure[114] with 80% yield.

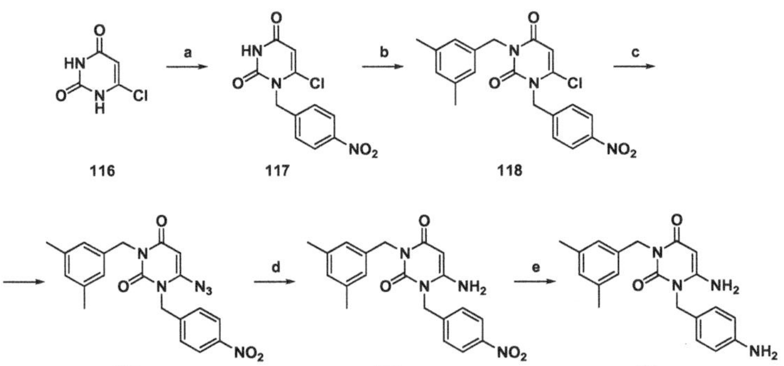

Scheme 21 Reagents and conditions: (a) 4-nitrobenzyl bromide, DMF, K_2CO_3, 70 °C, 1 h; (b) THF, 3,5-dimethylbenzyl alcohol, PPh_3, TMAD, 50 °C, 12 h; (c) NaN_3, DMF, rt., 0.5 h; (d) $LiAlH_4$, THF, 0 °C, 0.5 h; (e) $NaBH_4$, $NiCl_2$-$6H_2O$, MeOH, THF, 0 °C, 30 min.

The synthesis of 6-amino-3-(3,5-dimethylbenzyl)-1-(4-picolyl)uracil **107** described by Sakakibara et al. (Scheme 22).[108] Commercially available 6-chlorouracil **116** was treated with methoxymethyl chloride in methylene chloride in the presence of DBU to provide 6-chloro-1-methoxymethyluracil **121** in 90% yield. Condensation of uracil derivative **121** with 3,5-dimethylbenzyl alcohol under the Mitsunobu reaction conditions gave 3-(3,5-dimethylbenzyl)-derivative **122** with 93% yield. Its hydrolysis with B-bromocatecholborane in methylene chloride led to a mixture of 6-chloro- **123** and 6-bromo **124** derivatives in a 4:1 ratio and 72% yield. The mixture of compounds **123** and **124** was treated with an excess of 4-(chloromethyl)pyridine hydrochloride in DMF in the presence of potassium carbonate and then sodium azide. 6-Azido-3-(3,5-dimethylbenzyl)-1-(4-picolyl)uracil **125** is formed in 59% yield (for two stages). Its reduction with lithium aluminum hydride provided the target 6-amino derivative **107** in 74% yield.

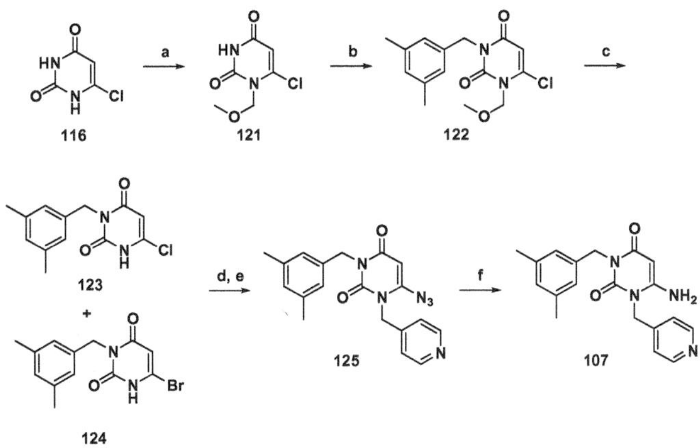

Scheme 22 Reagents and conditions: (a) CH_3OCH_2Cl, DBU, CH_2Cl_2, 0 °C, 20 min; (b) 2,6-dimethylbenzyl alcohol, TMAD, PPh_3, 50 °C, 2 h; (c) B-bromocatecholborane, CH_2Cl_2, rt., 1 h; (d) 4-(chloromethyl)pyridine hydrochloride, DMF, K_2CO_3; (e) NaN_3, DMF, rt., 30 min; (f) $LiAlH_4$, THF, 0 °C, 5 min.

The synthesis of 3-(3,5-dimethylbenzyl)-1-[2-(4-methylphenoxy)ethyl] uracil **109** included two steps (Scheme 23).[96] Condensation of equimolar amounts of bis(trimethylsilyl)uracil **126** and 1-bromo-2-(4-methylphenoxy)ethane on heating 170–180 °C without solvent for 2 h and treatment of the formed 1-[2-(4-methylphenoxy)ethyl]uracil **127** with 3,5-dibromobenzyl bromide in DMF in the presence of potassium carbonate. As a result, the target compound **109** was formed in 80% yield.

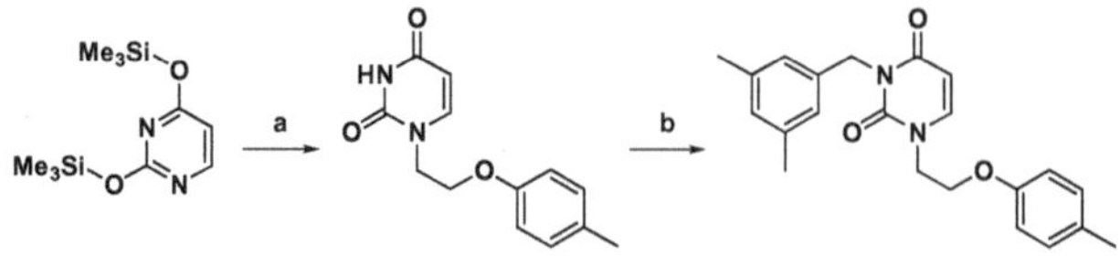

Scheme 23 Reagents and conditions: (a) 1-bromo-2-(4-methylphenoxy)ethane, 170–180 °C, 2 h; (b) 3,5-dimethylbenzyl bromide, K_2CO_3, DMF, 80 °C, 24 h.

Synthesis of 1-cinnamyl-3-(3,5-dibromobenzyl)uracil **110** also included two stages (Scheme 24). The first stage was condensation of 2,4-bis(trimethylsilyloxy)pyrimidine **126** with cinnamyl bromide. The reaction mixture was refluxed in 1,2-dichloroethane for 20 h to give **128** in a 90% yield. Next, 1-cinnamyluracil **128** was treated with the 3,5-dibromobenzyl bromide in DMF in the presence of K_2CO_3 (1.5 M excess) at room temperature. The target compound **110** was isolated in 81% yield.[111]

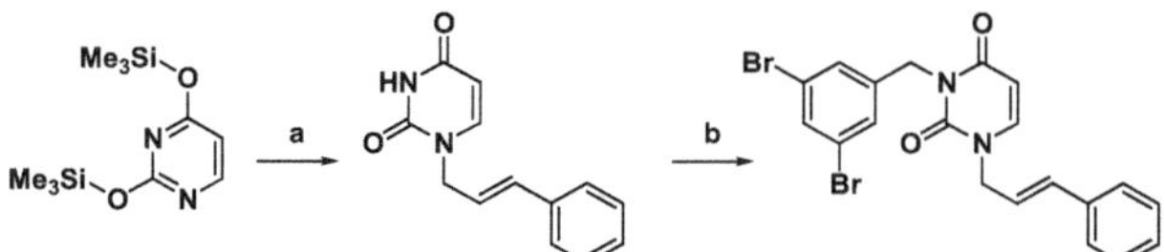

Scheme 24 Reagents and conditions: (a) cinnamylbromide, 1,2-dichloroethane, Δ, 20 h; (b) 3,5-dibromobenzyl bromide, K_2CO_3, DMF, 80 °C.

5. Nucleoside analogs as NNRTIs

Two groups of HIV NNRTIs are analogs of pyrimidine nucleosides with certain modifications of the carbohydrate fragment.

5.1 Nucleoside analogs with 3′-spiro-5″-(4″-amino-1″,2″-oxathiol-2″,2″-dioxide)-modification of the carbohydrate fragment

The first representatives of this class of HIV-1 NNRTIs were described by Camaraza et al. in 1992.[115] Structure-activity studies of a large number of derivatives showed that the presence of 3′-spiro-5″-(4″-amino-1″,2″-oxathiol-2″,2″-dioxide) group in 2′,5′-di(*tert*-butyldimethylsilyl)ribonucleosides is important for activity against HIV-1. Heterocyclic base modifications had less effect on activity than the nature of the sugar moiety.

The most active compound was [2′,5′-bis-*O*-(*tert*-butyldimethylsilyl)-*β*-D-ribofuranosyl]-3′-spiro-5″-(4″-amino-1″,2″-oxathiol-2″,2″-dioxide)thymine (TSAO-T, **129**, Fig. 16).[116] Compound **129** showed EC_{50} 0.060 μM in MT-4 cell culture infected with HIV-1, however it turned out to be quite toxic (SI 227). The mechanism of action study showed that in MT-4 and CEM cell cultures, TSAO-T can suppress the replication of not only the wild strain of HIV-1, but also virus mutants resistant to AZT. In addition, it was found that TSAO-T and its analogs inhibit the activity of HIV-1 RT in concentration 6.3–7.1 μM.[117]

The next step was the study of the antiviral properties of N3-substituted TSAO-T derivatives. It was found that the introduction of a lipophilic (alkyl or allyl) moiety in the N3 position of pyrimidine ring resulted in a significant decrease in cytotoxicity without a drop in anti-HIV-1 activity. Thus, new derivatives were reported with a selectivity index higher than TSAO-T (**129**). The most active was [1-[2′,5′-bis-*O*-(*tert*-butyldimethylsilyl)-*β*-D-ribofuranosyl]-3-*N*-methylthymine]-3′-spiro-5″-(4″-amino-1′,2′-oxathiol-2′,2′-dioxide) **130** (Fig. 16). Compound **130** inhibited HIV-1 replication in MT-4 cells at EC_{50} 0.059 μM. The CC_{50} of this compound was 240 μM and its SI exceeded 4000.[118] In addition, it was shown that **130** is able to block the dimerization of the p66 and p51 subunits of the heterodimer RT HIV-1.[119] The same anti-HIV-1 activity showed a mixture of 3-[((*E*)- and (*Z*)-4-oxopent-2-en-2-yl)oxypropyl]-TSAO-T **131** (Fig. 16) in a 2:1 ratio, EC_{50} 0.06 μM and $CC_{50} > 130$ μM.[120]

Compounds with improved antiviral properties have been obtained using computer modeling. Thus, 3-(3-hydroxypropyl)-TSAO-T **132** (Fig. 16) in the MT-4 cell culture inhibited HIV-1 replication at EC_{50} 0.03 μM and in the CEM cell culture, the EC_{50} was 0.01 μM.[121]

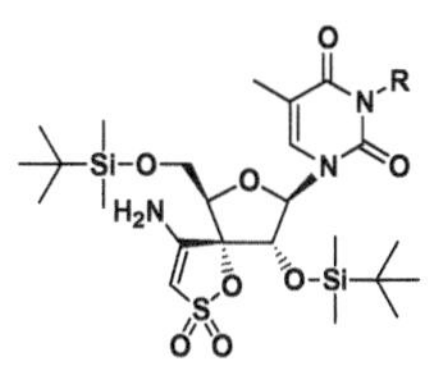

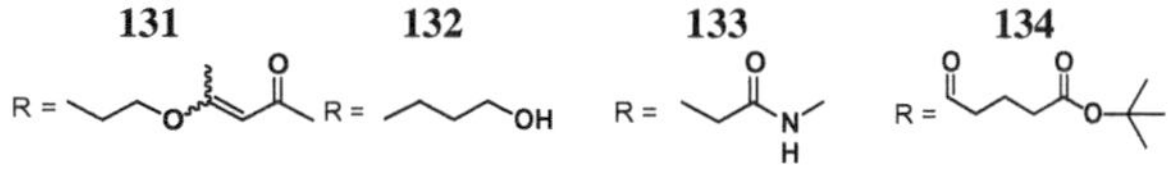

129	130	131	132	133	134
R = H TSAO-T	R = Me	R =	R =	R =	R =
EC_{50} 0.06 μM	EC_{50} 0.059 μM	EC_{50} 0.06 μM	EC_{50} 0.03 μM	EC_{50} 0.02 μM	EC_{50} 0.033 μM
CC_{50} 14 μM SI 227	CC_{50} 240 μM SI 4088	CC_{50} >130 μM SI >2160		CC_{50} ≥250 μM SI ≥12500	CC_{50} 11 μM SI 333

Fig. 16 Structures and properties of compounds **129–134**.

The next step was the introduction of a polar amide fragment, *N*-methylacetamide, into N3 position of TSAO-T. The compound **133** (Fig. 16) inhibited viral replication in MT-4 cell culture at EC_{50} 0.03 μM and in a CEM cells at EC_{50} 0.02 μM. The SI of compound **133** was 12,500.[122]

The introduction of substituents containing dicarbonyl fragments into N3 position of TSAO-T resulted to compounds with inhibitory activity against both HIV-1 and hepatitis C virus (HCV). However, if TSAO-T derivatives are active against HIV-1 in the nanomolar range, then HCV activity was observed in the micromolar range, so, at concentrations that were cytostatic for human T lymphocytes. For example, 3-(tert-butoxyglutaryl) derivative **134** (Fig. 16) inhibited HIV-1 replication in CEM cell culture at EC_{50} 0.033 μM. The cytotoxicity of the compound **134** was quite high (CC_{50} 11 μM) and the level of activity against HCV was comparable to the cytotoxicity (EC_{50} 11 μM).[123]

5.2 Synthesis of analogs with 3′-spiro-5″-(4″-amino-1″,2″-oxathiol-2″,2″-dioxide)-modification of the carbohydrate fragment

The synthesis of **129** (TSAO-T), presented in Scheme 25, included several steps. It was based on the work of Calvo-Mateo et al.,[124] which described the preparation of 3′-keto-nucleoside **135** starting from 1-(*β*-D-ribofuranozyl)thymine. The interaction of **135** with sodium cyanide in a two-phase ethyl ether/water system in the presence of sodium bicarbonate led to the formation of a mixture of two epimeric nucleoside 3′-cyanohydrins **136** and **137**. These cyanohydrins were unstable and gave the initial 3′-keto-nucleoside **135**, therefore their mixture was treated without isolation with mesyl chloride in pyridine resulting in the formation of the respective 3′-C-cyano-3′-O-mesyl-xylo- and ribo-derivatives of thymine **138** (53%) and **139** (10%), which were separated by chromatography. Treatment of 3′-O-mesylate **139** with Cs_2CO_3 afforded the spiro derivative **129** with 87% yield.

An alternative, more convenient route for obtaining **129** was proposed by Pérez-Pérez et al. (Scheme 26).[118] Starting 5-*O*-benzoyl-1,2-*O*-isopropylidene-α-D-erythro-pentofuranos-3-ulose **140**[125] was treated with sodium cyanide in a two-phase ethyl ether/water system in the presence of sodium bicarbonate resulting in the kinetically controlled ribo-cyanohydrin **141**. The cyanide ion came from the sterically less hindered β-face of the

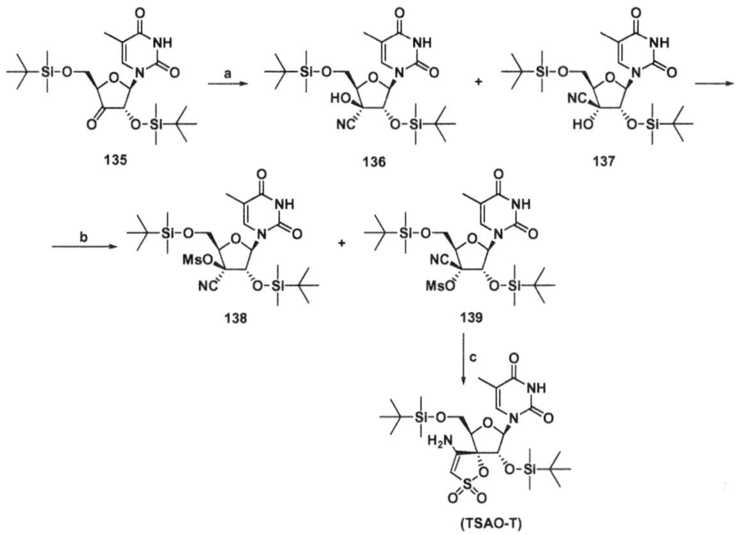

Scheme 25 Reagents and conditions: (a) NaCN, $NaHCO_3$, Et_2O-H_2O (2:1), rt., 16 h; (b) MsCl, Py, 8–10 °C, 48 h; (c) acetonitrile, Cs_2CO_3, rt., 6 h.

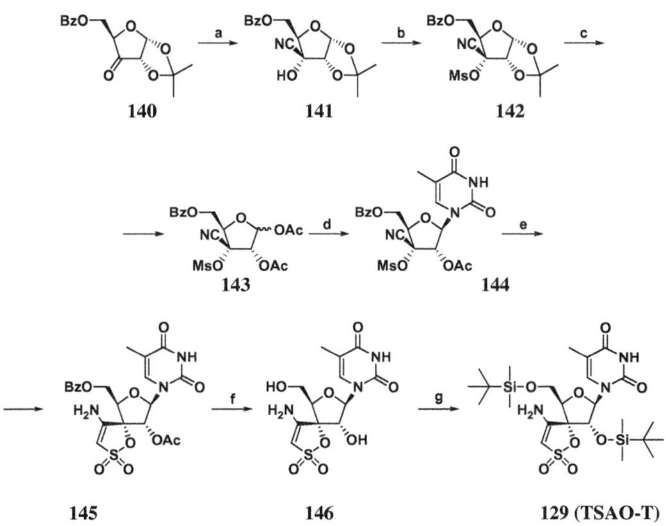

Scheme 26 Reagents and conditions: (a) NaCN, $NaHCO_3$, Et_2O-H_2O (2:1), rt., 4 h; (b) mesyl chloride, Py, 8–10 °C, 16 h; (c) (i) trifluoroacetic acid/water (9:1), rt., 4 h; (ii) acetic anhydride, Py, rt., overnight; (d) 2,4-bis(trimethylsilyloxy)-5-methylpyrimidine, acetonitrile, trimethylsilyl triflate, 2 h; (e) acetonitrile, Cs_2CO_3, rt., 3 h; (f) NH_3/MeOH, rt., overnight; (g) *tert*-butyldimethylsilyl chloride, acetonitrile, 4-(dimethylamino)pyridine, rt., 24 h.

ulose **140**, opposite to the 1,2-*O*-isopropylidene group. Without isolation and further purification **141** was transformed to the cyano mesylate **142** (78% yield after two stages). Hydrolysis of the 1,2-*O*-isopropylidene group of **142** was performed with aqueous trifluoroacetic acid, followed by reaction with acetic anhydride/pyridine. As a result, a mixture of α and β anomers (1.5:1) of the diacetate derivative **143** was obtained in 95% yield. Condensation of 1,2-bis-*O*-acetyl cyano mesylate **143** with silylated thymine using trimethylsilyl triflate as condensing reagent, as described by Vorbruggen et al.,[126] resulted in 3′-cyano mesylate nucleoside **144** in 77% yields. Due to the presence of a 2′-*O*-acyl group, only β-anomer product was obtained. Treatment of cyano mesylate **144** with Cs_2CO_3 gave the spiro derivative **145** in 65% yield. Reaction of **145** with saturated ammonia in methanol gave the fully deprotected nucleoside **146** in 66% yield. Reaction of **146** with *tert*-butyldimethylsilyl chloride yielded the 2′,5′-bis-*O*-silylated nucleoside **129**. The last was treated with methyl iodide in the presence of potassium carbonate to give 3-methyl derivative **130** in 55% yield.

The synthesis of N3-substituted TSAO-T derivatives is shown in Scheme 27.[118,120–123] Compounds **130–133** were obtained in 55%–78% yields by alkylation of **129** in the presence of K_2CO_3 in acetone. The final

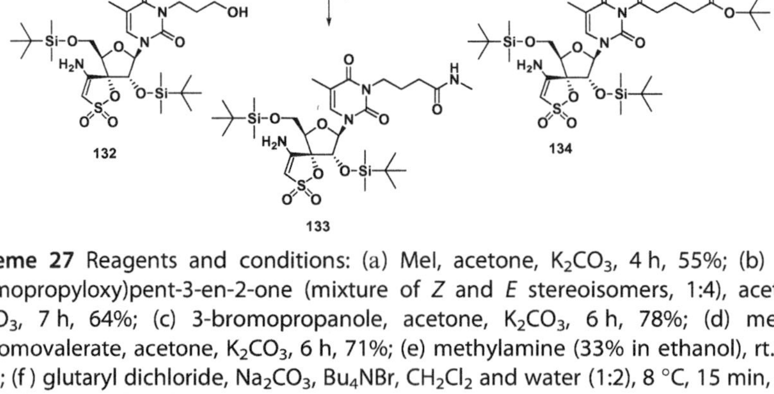

Scheme 27 Reagents and conditions: (a) MeI, acetone, K_2CO_3, 4 h, 55%; (b) 4-(3-bromopropyloxy)pent-3-en-2-one (mixture of *Z* and *E* stereoisomers, 1:4), acetone, K_2CO_3, 7 h, 64%; (c) 3-bromopropanole, acetone, K_2CO_3, 6 h, 78%; (d) methyl-5-bromovalerate, acetone, K_2CO_3, 6 h, 71%; (e) methylamine (33% in ethanol), rt., 2 h, 50%; (f) glutaryl dichloride, Na_2CO_3, Bu_4NBr, CH_2Cl_2 and water (1:2), 8 °C, 15 min, then t-BuOH, rt., 15 min, 62%.

stage in the case of derivative **133** was treatment with methylamide solution in ethanol.[122] The 3-N-(tert-butoxyglutaryl)-substituted analog **134** was obtained by the reaction of TSAO-T **129** with glutaryl dichloride and the subsequent formation of tert-butyl ether in 62% overall yield.[123]

5.3 5′-Norcarbocyclic uracil derivatives

1-(4′-Hydroxy-2′-cyclopenten-1′-yl)uracil derivatives substituted at the 4′-hydroxyl group and 3 position of the heterocyclic base represent another group of HIV-1 NNRTI. Compounds **135–138** (Fig. 17) showed activity on the isolated HIV-1 RT enzyme in 5–19 μM concentrations. These compounds were also studied on mutant forms of HIV-1 RT, carrying L100I and K103N amino acids substitutions. The activity of carbocyclic analogs on L100I mutant form of enzyme was about an order of magnitude higher than on the wild type RT HIV-1 virus with Ki values 1.2–2.4 μM, and on the enzyme with K103N mutation, the activity decreased by only 2–3 times (Ki 8.8–55.4 μM).[127,128]

The synthesis was carried out starting from 1-(4′-hydroxy-2′-cyclopentenyl)uracil **139** (Scheme 28), which was obtained earlier.[129] Compound **139** was acylated with benzoyl-, 3,5-dimethylbenzoyl- or 3,5-dichlorobenzoyl chloride in pyridine at 36°C for 1 h to form 1–4′-(benzoyloxy)-, 4′-(3, 5-dimethylbenzoyloxy)- and 4′-(3,5-dichloro-benzoyloxy)-2′-cyclopenten-1′-yl)uracil derivatives **140–142**. Their treatment with the appropriate benzyl bromide gave 3-benzyl derivatives **135–138** in 65%–85% yield.[127,128]

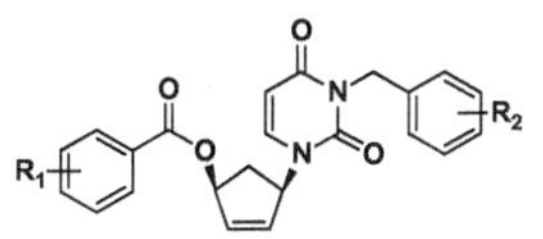

135: $R_1 = R_2 = H$ **136:** $R_1 = H$, $R_2 = 4\text{-Me}$ **137:** $R_1 = 3,5\text{-Me}_2$, $R_2 = H$ **138:** $R_1 = 3,5\text{-Cl}_2$, $R_2 = H$

Fig. 17 Structures of compounds **135–138.**

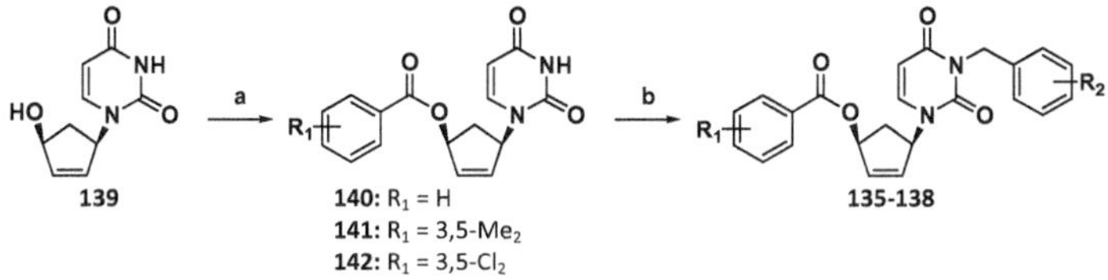

Scheme 28 Reagents and conditions: (a) benzoyl-, 3,5-dimethylbenzoyl- or 3,5-dichlorobenzoyl chloride, Py, rt.; (b) benzyl bromide or 4-methylbenzyl bromide, K_2CO_3, DMF.

6. Uracil derivatives with dual activity against HIV

HIV-1 reverse transcriptase has several catalytic functions: (1) synthesis of proviral DNA on a genomic RNA template with the formation of an RNA-DNA complex (RNA-dependent DNA polymerization), (2) hydrolysis of the RNA-DNA complex and release of the DNA strand (ribonuclease H activity), (3) synthesis of the second strand of DNA and the formation of double-stranded proviral DNA (DNA-dependent DNA polymerization). The active site of RT polymerase activity is localized at the foot of the "thumb" domain, the active center responsible for ribonuclease H activity is located in the "connection" domain.[130] The distance between these two centers is estimated at approximately 18 base pairs. It was found that for the implementation of ribonuclease H activity, the active center must include two metal cations Mg^{2+} or Mn^{2+}.[131] It is these cations that can serve as a good target for RT inhibitors, which can form various chelates with them.

IN, which is responsible for the incorporation of proviral DNA into the DNA of the host cell, is another excellent target for chemotherapeutic effects on HIV-1. Integration is one of the key stages in the life cycle of a virus. The active site of integrase contains a metal cation Mg^{2+}, which plays an important role in the conformation of the enzyme and its interaction with cellular DNA, forming a stable IN-metal-DNA complex.[132] The integrase active site contains a catalytic triad of three highly conserved residues of one glutamic and two aspartic acids. It was found that Mg^{2+} cations are coordinated by two conserved aspartic acid residues of the catalytic triad, mediating their catalytic activity.[133] Mg^{2+} cations can serve as a good target for IN inhibitors, which can form various chelates with them. For this reason, chelating compounds can have both the ability to inhibit ribonuclease H activity and IN.

6.1 RT and IN inhibitors

There is a group of bifunctional chimeric inhibitors, which are based on the structures of HIV-1 NNRTI (HEPT derivatives) and integrase (IN) inhibitors (diketo acids and quinolones).[134,135] Biochemical studies have revealed activity against RT and IN in nanomolar and micromolar concentrations, respectively. Compound **143** (Fig. 18) showed the best inhibitory properties (IC_{50} 24 nM against RT and 4.4 μM against IN, EC_{50} 10 nM against HIV-1) from this group of chimeric compounds.[134]

The introduction of N-hydroxyl group into the structure of HEPT analogs led to the new type of HIV-1 replication inhibitors with activity against both RT and IN. N3-Hydroxylation of the pyrimidine ring in HEPT derivatives afforded the appearance of a triad chelating Mg^{2+} cations in the molecule, which, along with the benzyl group, apparently satisfied the main structural requirements for IN inhibition. This modification did not greatly affect the ability of compounds to inhibit RT. A structure–activity relationship study shows that the presence of the N3-OH group is sufficient for inhibition of IN, and benzyl group in the side chain at N1 is more important for suppressing IN than the benzyl group at C6 of the uracil residue. The best properties were shown by compound **144** (Fig. 18), which suppressed HIV-1 replication in the CEM-SS cells at EC_{50} 0.024 μM and inhibited viral enzymes at IC_{50} 0.17 μM (RT) and 3.5 μM (IN).[136,137]

N3-Hydroxylated derivatives of 6-benzoyl-uracil analogues, which were also originally designed as NNRTIs, have demonstrated dual inhibitory activity for both RT polymerase and IN in the micromolar range. The most active was compound **145** (Fig. 18), which inhibited HIV-1 replication in the CEM-SS cell culture at EC_{50} 0.061 μM and inhibited IN at IC_{50} 0.41 μM.[138]

6.2 RNase H and IN inhibitors

Substitution of a benzyl group for a cyclohexylmethyl moiety led to a new class of HIV-1 RT inhibitors. Such compounds did not influence IN activity, however, inhibited RNase H activity. Compound **146** (Fig. 18) was found to be the most active among these compounds. It suppressed HIV-1 replication in CEM-SS cell culture at EC_{50} 0.026 μM, inhibited RNase H activity at IC_{50} of 5.2 μM and RT polymerase activity at IC_{50} 0.45 μM.[139]

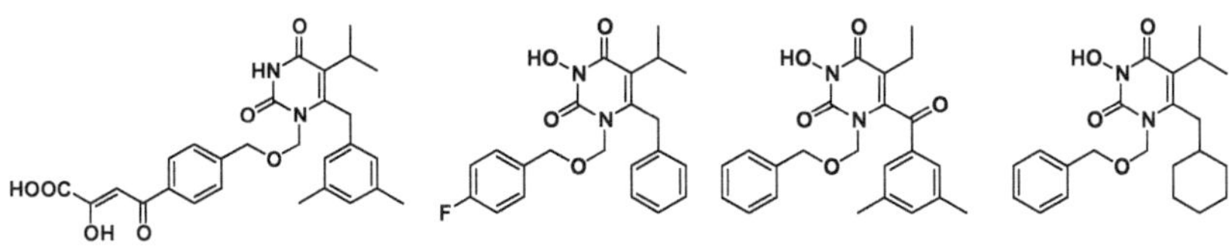

143
EC_{50} 0.0097 μM
IC_{50} 0.024 μM (RT)
IC_{50} 4.4 μM (IN)
CC_{50} >10 μM

144
EC_{50} 0.024 μM
IC_{50} 0.17 μM (RT)
IC_{50} 3.5 μM (IN)
CC_{50} >20 μM

145
EC_{50} 0.061 μM
IC_{50} 41 μM (IN)
CC_{50} 33 μM

146
EC_{50} 0.026 μM
IC_{50} 5.2 μM (RNase H)
IC_{50} 0.45 μM (Pol)
CC_{50} >100 μM

Fig. 18 Structures and properties of compounds **143–146**.

Another class of RNase H activity inhibitors contained N3-OH group, which responsible for chelating Mg^{2+} cations, and diaryl fragment that is linked to the uracil residue through a short bridge (NH, O, S, CH_2, or CO). The most interesting compound among the series of 6-(4-phenylphenylamino)-derivatives was **147** (Fig. 19), which inhibited the RNase H activity of RT at IC_{50} 2.0 μM and blocked the polymerase activity of RT at IC_{50} 0.45 μM. At the same time, the compound was inactive against integrase.[140]

The next modification was associated with the replacement of the NH-bridge, linking the diaryl fragment with the uracil residue, by a sulfur atom. This sufficiently enhanced the inhibitory properties of the compounds. The most promising from this group was compound **148** (Fig. 19), which inhibited RT RNase H activity at IC_{50} 0.005 μM and RT polymerase activity at IC_{50} 10 μM. In addition, the compound inhibited IN activity at IC_{50} 4.0 μM and suppressed HIV-1 replication in a CEM-SS cell culture at EC_{50} 0.061 μM.[141] Replacing a sulfur atom with a methylene or carbonyl group markedly decreased the antiviral effect.[142,143]

One more interesting derivative was 6-amino-3-hydroxyisoorotic acid **149** (Fig. 19). It inhibited HIV-1 replication in CEM-SS cell culture at EC_{50} 0.015 μM, RT RNase activity at IC_{50} 0.005 μM and the IN activity at IC_{50} 0.021 μM. The compound also suppressed the replication of HIV-1 strains, resistant to the action of reltegravir.[144]

6.3 Synthesis of uracil derivatives with dual activity against HIV

The synthesis of compound **143** included three stages (Scheme 29).[134] Trimethylsilyl derivative of 5-isopropyl-6-(3,5-dimethylbenzyl)uracil **150** was condensed with compound **151**, which led to the formation of ketone **152** with 76% yield. Reaction of ketone **152** with diethyl oxolate gave the ketoester, which was hydrolyzed to give the target **143** with 79% yield.[134]

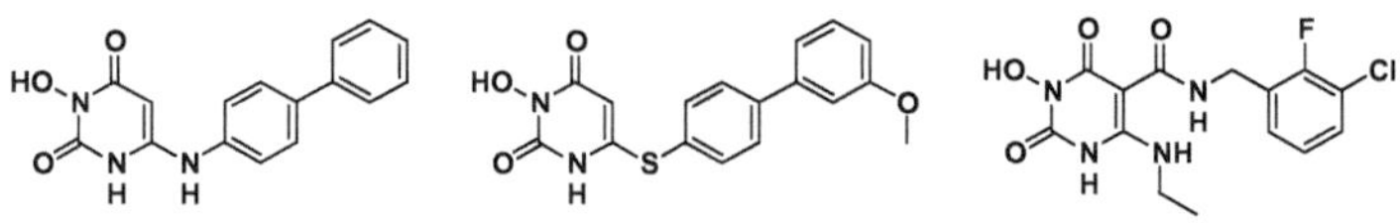

147	**148**	**149**
IC_{50} 2.0 μM (RNase H)	EC_{50} 7.7 μM	EC_{50} 0.015 μM
IC_{50} 0.45 μM (Pol)	IC_{50} 0.005 μM (RNase H)	IC_{50} 0.005 μM (RNase H)
CC_{50} >25 μM	IC_{50} 10 μM (Pol)	IC_{50} 0.021 μM (IN)
	IC_{50} 4.0 μM (IN)	CC_{50} 18 μM
	CC_{50} > 100 μM	

Fig. 19 Structures and properties of compounds **147–149**.

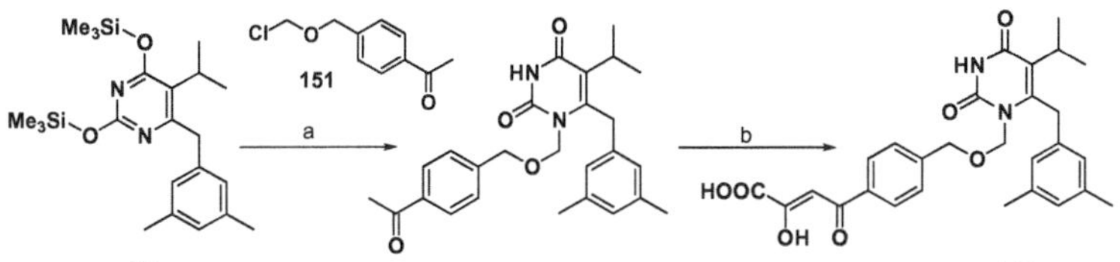

Scheme 29 Reagents and conditions: (a) **151**, tetrabutylammonium iodide, CH_2Cl_2, rt.; (b) (i) diethyl oxalate, (ii) NaOH (1 N), EtOH/CH_2Cl_2, 79%.

The synthesis of compounds **144–146** (Scheme 30) started in a similar manner. The condensation of trimethylsilyl derivatives of uracil **153** with benzyloxymethyl chloride **154** gave 1-benzyloxymethyl derivative **155** with 83%–87% yield. Subsequent N3-hydroxylation was achieved by base-mediated NaH oxidation of m-CPBA to give final compounds **144–146** in the range of 53%–70%.[135–139]

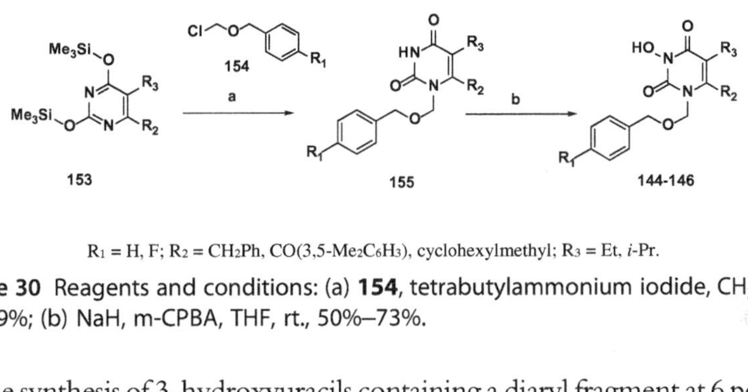

R_1 = H, F; R_2 = CH_2Ph, CO(3,5-$Me_2C_6H_3$), cyclohexylmethyl; R_3 = Et, *i*-Pr.

Scheme 30 Reagents and conditions: (a) **154**, tetrabutylammonium iodide, CH_2Cl_2, rt., 51%–89%; (b) NaH, m-CPBA, THF, rt., 50%–73%.

The synthesis of 3-hydroxyuracils containing a diaryl fragment at 6 position is shown in Scheme 31. Thus, 3-benzyloxy-6-chlorouracil was condensed with 4-phenylaniline in DMF in the presence of *N,N*-dimethylaniline at 170 °C, which led to the formation of 4-phenylphenylamino derivative **156** with the yield in the range of 54%–85%. Debenzylation was carried out in methanol in the presence of Pd/C in H_2 atmosphere. Target compound **147** was obtained in 82%–90% yield.[140]

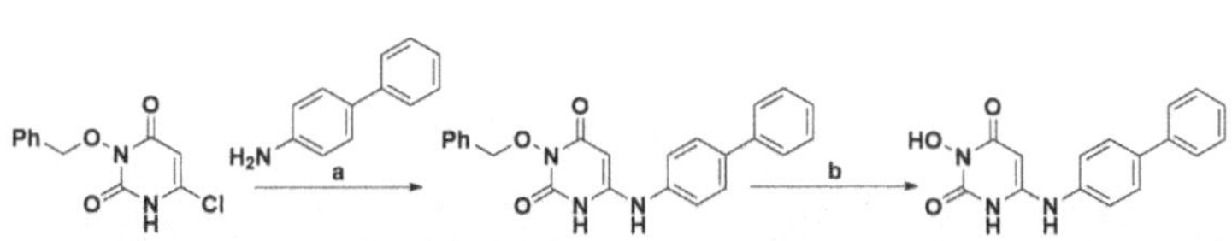

Scheme 31 Reagents and conditions: (a) DMF, *N,N*-dimethylaniline, MW, 170 °C, 30–40 min; (b) Pd/C, H_2, MeOH, 3–4 h.

The synthesis of 3-hydroxy-6-[3′-methoxy-(1,10-biphenyl)-4-yl)sulfanyl] uracil **148** is presented in Scheme 32. 3-Benzyloxy-6-(4-bromophenylsulfanyl)uracil **157**, obtained in accordance with a published method,[145] was condensed under the Suzuki reaction conditions with 3-methoxyphenylboronic acid in the presence of a $Pd(PPh_3)_4$ to give 3-methoxyphenyl derivative **158**, which was hydrolyzed with TFA. The target compound **148** was formed in 57% yield.[141]

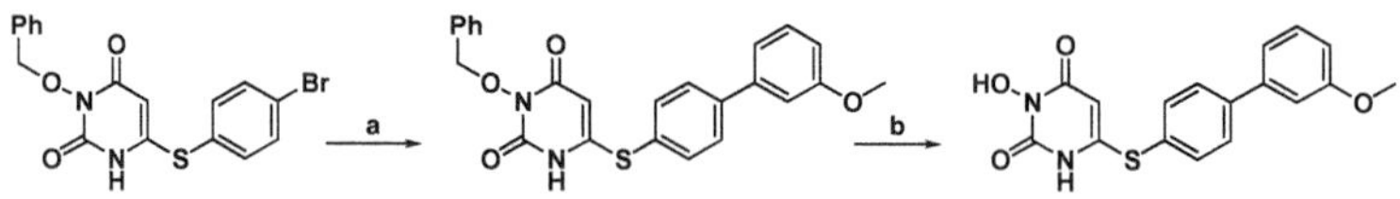

Scheme 32 Reagents and conditions: (a) arylboronic acid, $Pd(PPh_3)_4$, K_2CO_3, MeOH/H_2O, microwave, 120 °C, 30–50 min, 40%–79%; (b) TFA, microwave, 120 °C, 40–50 min, 39%–90%.

7. Uracil derivatives as non-nucleoside inhibitors of various viruses

7.1 Human cytomegalovirus (HCMV)

Human cytomegalovirus (HCMV) is a member of the subfamily Betaherpesvirinae; it is prevalent in nearly 90% of the adult population.[146] The virus does not manifest itself in any way in a healthy body, but it can be deadly for people with an impaired immune system: for HIV-infected people, transplant recipients, newborns.[146] HCMV is the most common cause of congenital viral infections in newborns,[147] it is able to penetrate the placenta and infect the fetus, leading to stillbirths and innate deformities.[148]

Maruyama et al. found that some 1,3-disubstituted uracil derivatives, in addition to anti-HIV activity, are active against HCMV. The highest anti-HCMV activity among HIV-1 NNRTIs of this type was demonstrated by derivative **102** (Fig. 20), which inhibited viral replication (strain AD-169) at EC_{50} 1.0 μM. Other 1,3-disubstituted uracil derivatives were less active or completely inactive against HCMV.[104] Another example of 1,3-disubstituted uracil with significant activity against HCMV is 1-cinnamyl-3-(3,5-dichloro benzyl)uracil **159** (Fig. 20).[111] The synthesis of compounds relative to **102** and **159** is described above (see Section 4.2).

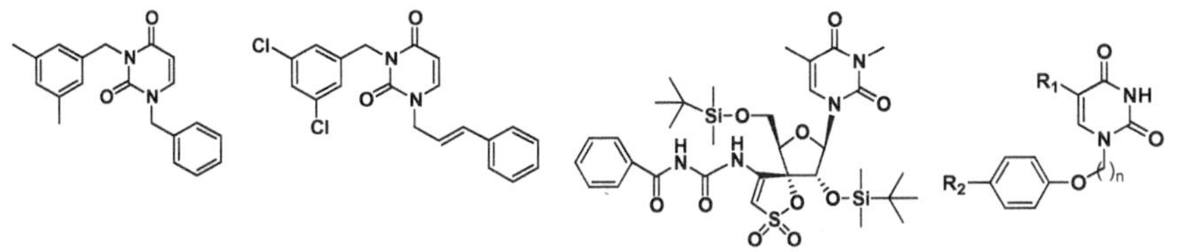

102

EC_{50} 1μM

CC_{50} > 200μM

159

EC_{50} 9.4 μM (AD-169)
EC_{50} 12 μM (Davis)

160: $R_1 = R_2$ = TBDMS;
161: R_1 =TDS, R_2 = TBDMS;
162: $R_1 = R_2$ = TDS

(**160**): EC_{50} 1.1 μM (AD-169)
(**160**): EC_{50} 1.4 μM (Davis)
(**160**): CC_{50} > 200μM

163: n = 5, R_1= H, R_2= CN
164: n = 5, R_1= H, R_2= Br
165: n = 5, R_1= Me, R_2= Br
166: n = 8, R_1= H, R_2= Br
EC_{50} 5.5–12 μM (AD-169)
EC_{50} 5.5–12 μM (Davis)
CC_{50} ≥ 100μM

Fig. 20 Structures and properties of compounds **102, 159–166.**

3-Benzoylureido-substituted analogue **160** (Fig. 20), is a derivative of 1-[2′,5′-bis-*O*-(*tert*-butyldimethylsilyl)-*β*-D-ribofuranosyl]-3-N-(methyl) thymine]-3′-spiro-5″-[4″-(3-benzoylureido)-1″,2″-oxathiole-2″,2″-dioxide **130**, which proved to be an effective NNRTI of HIV.[149] In contrast to TSAO-T derivatives **129–134** described above, compound **160** did not have significant anti-HIV activity, but was able to inhibit HCMV at concentrations in the lower μM range (EC_{50} 1.4 μM (Davis), 1.1 μM (AD-169)) and without cytotoxicity in human embryonic lung (HEL) cells (CC_{50} > 200 μM). Related compounds where one of the *tert*-butyldimethylsilyl groups **161** or both **162** (Fig. 20) are replaced by TDS (thexyldimethylsilyl) exhibited similar anti-HCMV activity: in HEL cell cultures (EC_{50} 1.4–1.5 μM (Davis), 1.1–1.2 μM (AD-169), but appeared to be more toxic (CC_{50} > 50). Replacement of the *tert*-butyldimethylsilyl group in the 5′-position of the molecule to benzoyl group also led to a higher cytotoxicity while maintaining antiviral activity. Removal of 2′-TBDMS, as well as replacement with an acetyl or benzyl residue, lead to complete loss of activity.[150] HCMV strains with mutations in DNA polymerase or viral protein kinase responsible for phosphorylation of nucleoside drugs remained completely sensitive to the inhibitory activity of compounds **161** and **162**. This indicates that mechanism of action is different from that of clinically used nucleoside analogs. Derivatives **161** and **162** also showed high activity against mouse cytomegalovirus (MCMV) with EC_{50} values 0.66 and 0.54 μM, respectively. A lower activity of compounds **161** and **162** was recorded against varicella-zoster virus (VZV) (EC_{50} 11–16 μM, on wild-type and thymidine kinase-deficient strains). There was no activity against herpes simplex virus types 1 and 2, vaccinia virus and other RNA viruses.

Another group of compounds with pronounced activity against HCMV (EC_{50} 5.5–12 μM, Davis and AD-169 strains; $CC_{50} \geq 100$ μM, HEL cell culture) are 1-[ω-(phenoxy)alkyl]uracil derivatives **163–166**.[151] Compounds were obtained in 73%–88% yield by condensation of equimolar amounts of the corresponding 2,4-bis(trimethylsilyl)pyrimidine and ω-(phenoxy)alkyl bromide at 160–170 °C without solvent (Scheme 33).[151]

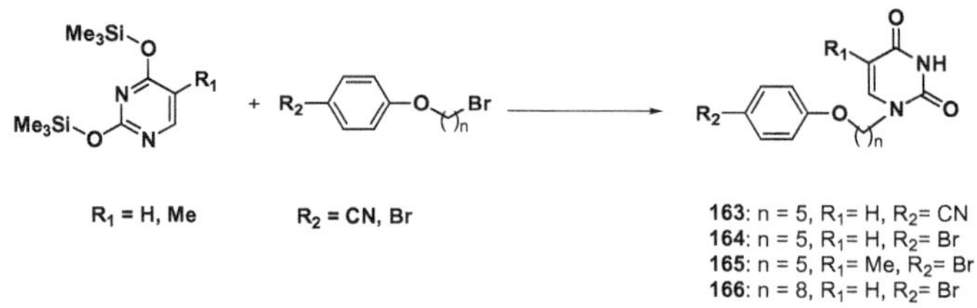

Scheme 33

7.2 Hepatitis C virus (HCV)

Hepatitis C virus (HCV) is a (+)-stranded RNA virus of the Flaviviridae family that was first identified in 1989.[152] The long-term natural course of HCV infection is diverse. Liver involvement can range from minimal histological changes to extensive fibrosis and cirrhosis with hepatocellular carcinoma. Currently, there are several highly effective treatment regimens for HCV infection, which include various combinations of inhibitors of NS3 protease and interferon-resistant NS5A protein, nucleoside and non-nucleoside inhibitors of RNA-dependent RNA polymerase (NS5B polymerase). HCV NS5B polymerase inhibitors are among the most clinically useful agents for HCV infection treatment. Nucleoside inhibitors bind to the catalytic site of NS5B as their triphosphate metabolites and function by interrupting the genetic message as RNA chain terminators. Non-nucleoside inhibitors (NNIs) bind to one of four allosteric sites in NS5B and interfere with enzyme catalytic function by affecting its conformation.[153]

A series of N1-aryl uracil analogs **167–170** was described[154,155] as a novel structural class of non-nucleoside inhibitors of HCV NS5B polymerase with selective activity (EC_{50} 1–27 nM, HCV replicons) limited to genotypes 1a and 1b (Fig. 21.).

171 (ABT-075)[153] and **172** (Dasabuvir, ABT-333)[156] (Fig. 22) are relative compounds, their preclinical studies also have shown high antiviral activity in infected cell cultures and tissues.[153,156–158] The results of the study of pharmacokinetic parameters in rats, dogs and rabbits, as well as the absence of acute and chronic toxicity for laboratory animals, made it possible to recommend both compounds **171** and **172** for clinical trials.[153,159]

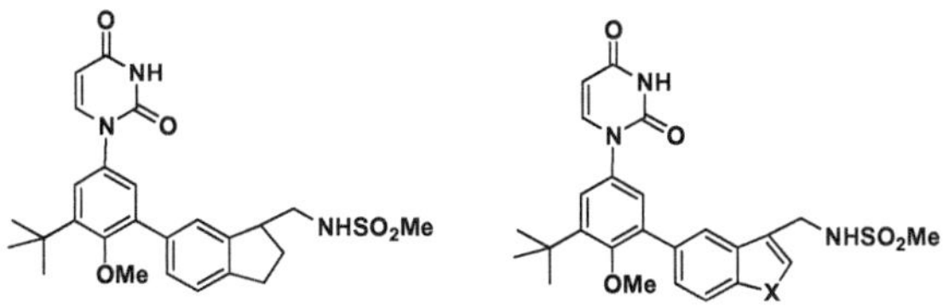

167
EC_{50} 20 nM (GT 1a)
EC_{50} 3 nM (GT 1b)

168a: X = CH_2,
EC_{50} 14 nM (GT 1a)
EC_{50} 3 nM (GT 1b)

168b: X = S
EC_{50} 27 nM (GT 1a)
EC_{50} 3 nM (GT 1b)

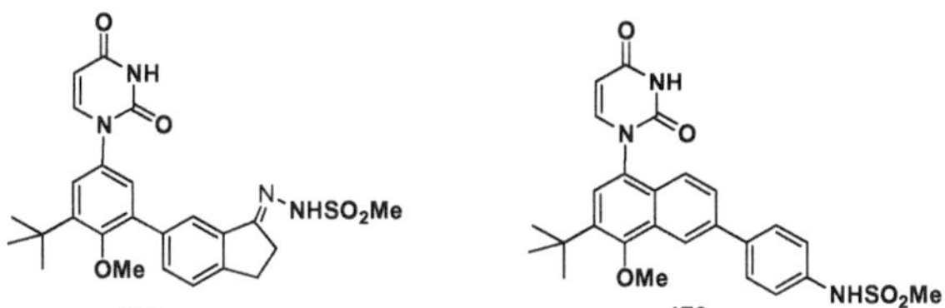

169
EC_{50} 2 nM (GT 1a)
EC_{50} 1 nM (GT 1b)

170
EC_{50} 4 nM (GT 1a)
EC_{50} 1 nM (GT 1b)

Fig. 21 Structures and properties of compounds **167–170**.

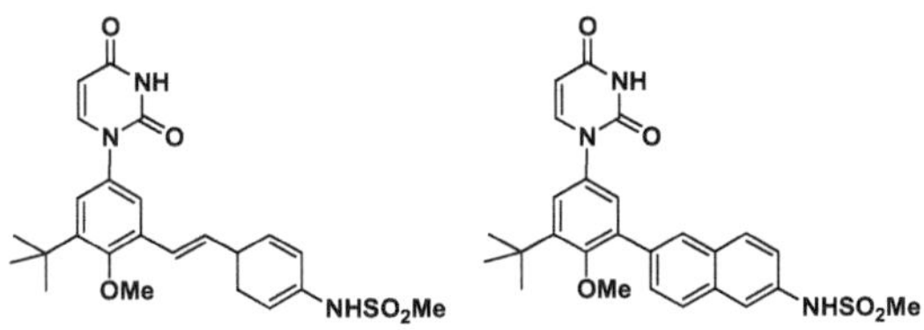

171 (ABT-072)

IC_{50} 1.1 nM (HCV pol GT 1a)
IC_{50} 0.3 nM (HCV pol GT 1b)

EC_{50} 19 nM (HCV replicons, GT 1a)
EC_{50} 2.4 nM (HCV replicons, GT 1b)

172 (ABT-333, Dasabuvir)

IC_{50} 2.4–4.2 nM (HCV pol GT 1a)
IC_{50} 2.2–10.7 nM (HCV pol GT 1b)
Human polymerases >100nM
EC_{50} 99 nM (HCV replicons, GT 1a)
EC_{50} 21 nM (HCV replicons, GT 1b)
CC_{50} > 10360 nM

Fig. 22 Structures and properties of compounds **171** and **172**.

Clinical trials have confirmed the efficacy of dasabuvir (**172**) for the treatment of chronic HCV genotypes 1a and 1b. This drug is considered safe and does not cause serious side effects. Dasabuvir has been shown to be synergistic in combination therapy with other drugs, although it exhibits moderate antiviral efficacy when used alone. Dasabuvir is currently licensed and is prescribed at a dose of 250 mg twice daily in combination with paritaprevir/ritonavir, ombitasvir and ribavirin.[160] This drug combination, along with high activity, demonstrates an acceptable safety profile and is well tolerated by patients, which maintains adherence to treatment and slows down emergence of resistance.[160]

The current method for the synthesis of dasabuvir[161] is shown in Schemes 34 and 35. Reaction of picolinic acid with toluenesulfonyl chloride (Scheme 34) led to mixed anhydride **173**, which reacted in situ with anthranilonitrile **174** to give ligand **175** in 85% yield. Treatment of 2-tert-butylphenol with sodium iodide and sodium hypochlorite (bleach) and subsequent methylation (NaOH, MeI, acetone) led to the formation of diiodanisole **176** in high yield. The reaction of uracil with diiodanisole **176** in the presence of copper iodide and ligand **175** in DMSO gave N1 substituted uracil **177**. The presence of 70 mol% water relative to **176** did not significantly affect the yield of product **177**. Optimal reaction conditions included use of powdered potassium phosphate as a base, 10 mol% of copper (I) iodide, and 12 mol% of ligand **175** (Scheme 34). The reaction mixture was washed with water to remove $Cu(OAc)_2$. Product **177** was isolated by crystallization from i-PrOAc/heptanes. Recrystallization from 2-propanol afforded the product with a purity of more than 97.8%. The yield of intermediate **177** was 63%–64% in the synthesis scaled up to hundreds of kilograms based on the starting 2-tert-butylphenol (Scheme 34).

The Suzuki cross-coupling reaction of N1-substituted uracil **177** with hydroxynaphthylboronic acid in the presence of Pd_2dba_3 and ligand **178** led to naphthol **179** (Scheme 35). The complete conversion was reached in 1.5 h, providing a high yield of naphthol **179** with more than 90% of purity. The synthesis of nonaflate **180** from naphthol **179** was carried out under the action of perfluorobutanesulfonyl fluoride in the presence of K_2CO_3 in a mixture of DMF/MeCN solvents. The final sulfonamidation reaction catalyzed by Pd using t-Bu-BrettPhos **181** or VincePhos **182** gave dasabuvir **172** in 90% yield with a high purity (>99.5%) and comparable efficiency (Scheme 35).

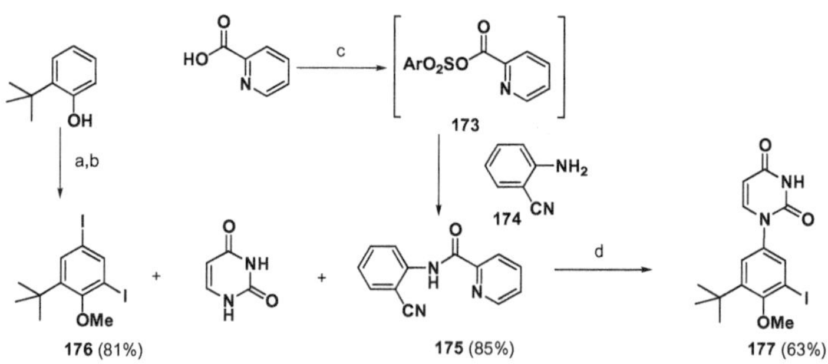

Scheme 34 Reagents and conditions: (a) NaI, NaOCl, MeOH, H_2O; (b) NaOH, MeI, acetone; (c) TsCl, Et_3N, CH_2Cl_2; (d) CuI, K_3PO_4, DMSO, 60 °C

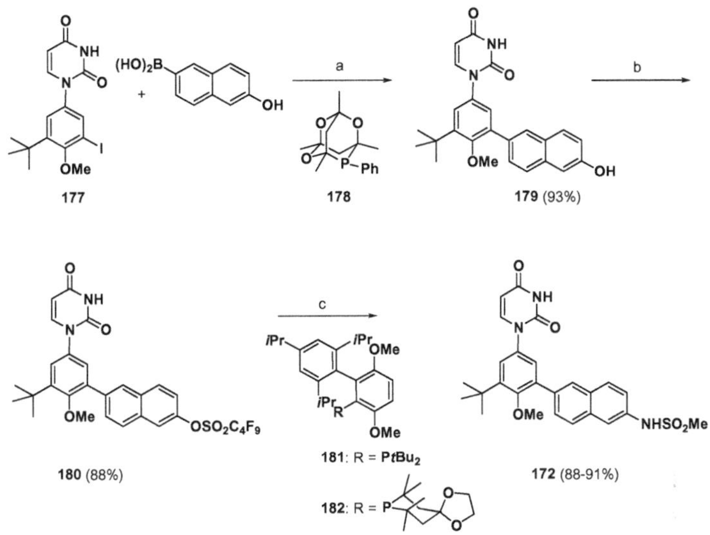

Scheme 35 Reagents and conditions: (a) Pd_2dba_3, **178**, K_3PO_4, THF/H_2O, 60–65 °C; (b) $FSO_2C_4F_9$, K_2CO_3, MeCN/DMF, 30 °C; (c) Pd_2dba_3, K_3PO_4, ethyl acetate, 90 °C, **181** or **182**

Another uracil-containing HCV inhibitor **183** (Z401, Fig. 23) was obtained[162] among other N1,N3-substituted uracils. Compound **183** showed the best in the series the ratio of antiviral activity (IC_{50} 0.092 μM) and cytotoxicity (CC_{50} 22.6 μM) with SI exceeding 244 on Huh7-J20 cells infected with HCV.[162] It was shown that compound **183** suppresses the replication of various HCV genotypes, 1b, 2a, and 3a at early stages, with IC_{50} values of 2.2, 0.092, and 0.69 μM, respectively.

7.3 Hepatitis B virus (HBV)

Today, around 350–400 million people worldwide are chronically infected with the hepatitis B virus (HBV).[163–165] This disease can lead to cirrhosis,

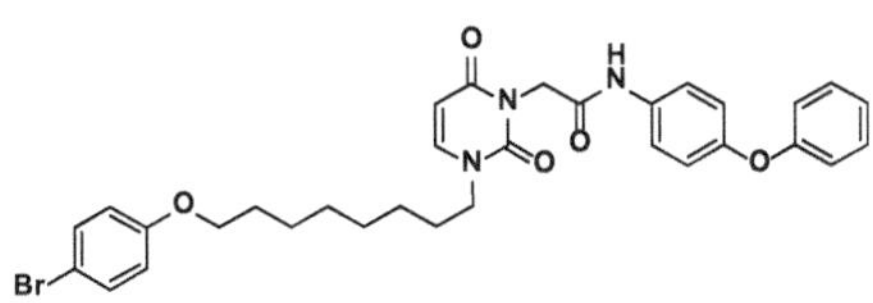

183 (Z401)

IC_{50} 0.092 μM
CC_{50} 22.6 μM

Fig. 23 Structure and properties of compound **183**.

hepatocellular carcinoma and death. Current HBV treatments include certain nucleoside reverse transcriptase inhibitors (NRTIs) and an innate immune modulator, interferon alpha. Interferon has limited efficacy in patients and has many side effects. NRTIs can effectively control the production, replication of the virus and the development of the disease, but they do not cure the infection. Accordingly, people with hepatitis B infection now require lifelong treatment, which is often costly and can cause side effects due to prolonged drug use.[165,166]

RNase H of HBV is an attractive therapeutic target because of its important role in viral replication. RNase H inhibitors can help control HBV infections more effectively. A group of 52 compounds was tested in HepAD38 cell culture and some of them showed ability to inhibit the activity of RNase H. The most active among them, compounds **184** and **185**, inhibited HBV replication at low micromolar concentrations (Fig. 24).[167]

7.4 Respiratory syncytial virus (RSV)

Respiratory syncytial virus (RSV) is a major cause of lower respiratory tract infections without available effective treatment currently. The search for new inhibitors of RSV is an important first step towards the development of an effective therapy for RSV. Compound **186** (Fig. 25) is a non-nucleoside inhibitor of RSV polymerase.[168] A structure of **186** is different from the previously described series YM-53403/AZ-27 and BI,[169–171] however, it is likely to have a similar mechanism of action.[168] Inhibitor **186** was identified during the screening of 1 million compounds from the AstraZeneca collection on the RSV replicon, it acts selectively on RSV and is not active against other viruses. The selection of RSV strains resistant to **186** made it possible to identify mutations mapped to the RSV L protein, RNA-dependent RNA polymerase (Y1631C and I1413T). Recombinant RSV containing the L substitution Y1631C, resistant to the action of compound **186**, suggests that RSV polymerase is the target of this inhibitor.[168]

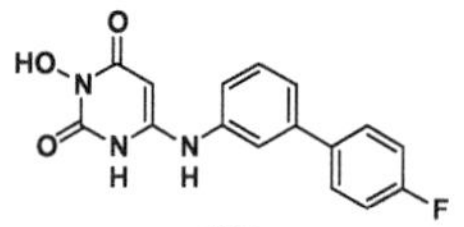

184
EC_{50} 5.5 μM
CC_{50} >100 μM

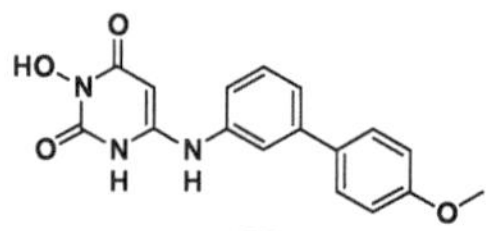

185
EC_{50} 8.0 μM
CC_{50} >100 μM

Fig. 24 Structures and properties of compounds **184** and **185**.

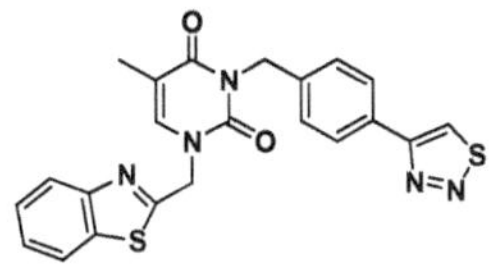

186

EC_{50} 1.6 μM (RVS (A2) in HEp2 cells)
CC_{50} >100 μM (HEp2 cells)

Fig. 25 Structure and properties of compound **186**.

7.5 Human adenovirus (HAdV)

Human adenoviruses (HAdVs) are non-enveloped viruses and have a linear, unsegmented double-stranded DNA as a genome.[172] Adenovirus infections are widespread and highly infectious, most commonly affecting the mucous membranes of the respiratory tract, eyes, gastrointestinal tract, and genito-urinary system.[173–178] In humans with weak immune systems, adenoviral infections can lead to the development of acute diseases leading to death.[179] Currently, there are no specific effective drugs against adenovirus infections.

As a result of screening a number of 5-amino-derivatives of 1-substituted uracils in a HEK293 cell culture, it was found that 1-[4-(phenoxy)benzyl]-5-(morpholino)uracil **187** (Fig. 26) inhibits the replication of a recombinant adenovirus (strain HAdV5-eGFP) at EC_{50} 0.5 μM and exhibits cytotoxicity at 47.6 μM. The effect of compound **187** on the infectivity of the adenoviral progeny of HAdV5-eGFP was also evaluated. A decrease in progeny virus titer was observed under the influence of **187**. Thus, it can be assumed that the mechanism of action of **187** is associated with inhibition of key factors of replication of human adenoviruses, such as viral DNA polymerase and E1A gene products.[178]

7.6 SARS-CoV-2

The pandemic of COVID-19 caused by the β-coronavirus SARS-CoV-2 has claimed about 4 million lives during first 1.5 years.[180] Despite the

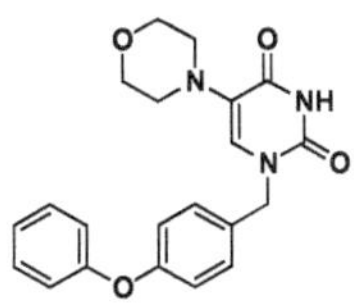

187

EC_{50} 0.5 μM
CC_{50} 47.6 μM

Fig. 26 Structure and properties of compound **187**.

development and use of vaccines against COVID-19, the disease remains the main global public health problem in 2021. As part of the repurposing of existing drugs, intensive research is underway on drugs approved for the treatment of other viral, bacterial and parasitic infections, autoimmune, oncological and other diseases in order to create an effective therapeutic agent.

Tipiracil (**188**) in combination with trifluridine approved by the FDA for the treatment of colorectal cancer.[181] It inhibits the enzyme thymidine phosphorylase, which metabolizes trifluridine. It was shown that, **188** (Fig. 27) also inhibits the activity of Nsp15 endoribonuclease NendoU from SARS-CoV-2.[182] This enzyme is highly conserved in coronaviruses and is a uridine-specific endoribonuclease with a C-terminal catalytic domain belonging to the EndoU family. Tipiracil (**188**) inhibits Nsp15 by interacting with the uridine binding pocket in the active site of the enzyme, as confirmed by crystallography. Reaction catalyzed by SARS-CoV-2 Nsp15 endoribonuclease is decreased by 50% in the presence of 7.5 μM **188**. S antigen ELISA assay in A549 cells and CoV-2 virus replication assay using qRT-PCR were performed. Tipiracil did not affect viability of cells but the inhibition of virus was modest in the concentration range 1–50 μM. The activity of Tipiracil is not high enough for an effective drug, but it is a promising lead compound for the design of new highly active inhibitors of viral proliferation.[182]

In addition, virtual screening of USA FDA approved drugs by computer-aided tools has led to the identification of Tipiracil (**188**) among other potential inhibitors of SARS-CoV-2 main protease (M^{pro}).[183]

7.7 Rigid amphipathic fusion inhibitors: HCV, HSV-1, HSV-2, TBEV (tick-borne encephalitis virus)

Rigid amphipathic fusion inhibitors (RAFIs) **189–192** are potent antiviral drugs.[184–187] They are synthetic amphiphilic molecules based on a nucleoside-linked perylene core and their shape similar to that of phospholipids. RAFIs **189** and **190** (Fig. 28) inhibit the infectivity of several unrelated enveloped viruses[185,188] with no cytotoxic or cytostatic effects (SI > 3000).[186] RAFIs' mechanism of action is biophysical, not biochemical.

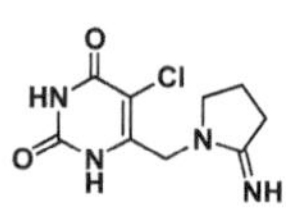

188 (Tipiracil)

Fig. 27 Structure of compound **188**.

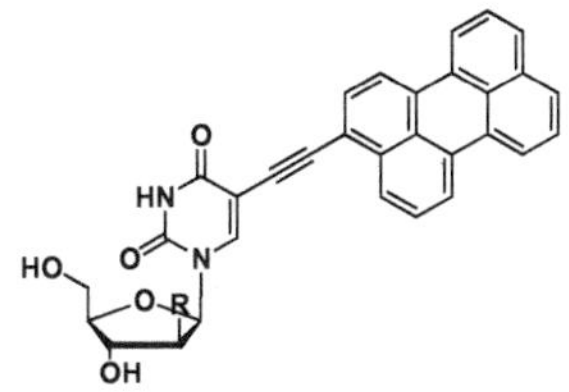

189: R = H
190: R = OH

Virus	189	190
	IC_{50}, μM	IC_{50}, μM
Influenza (A/PR/8/34; A/USSR/90/77, A/Aichi/2/68, A/PC/1/73)	0.097-0.187	0.078-0.283
HCV (JFH-1)	0.107	0.187
VSV	0.002	0.005
HSV-1 (KOS)	0.048	0.131
HSV-2 (186)	0.052	0.031
mCMV	0.013	0.037
TBEV [188]	0.024	0.018

Fig. 28 Structures and properties of compounds **189** and **190**.

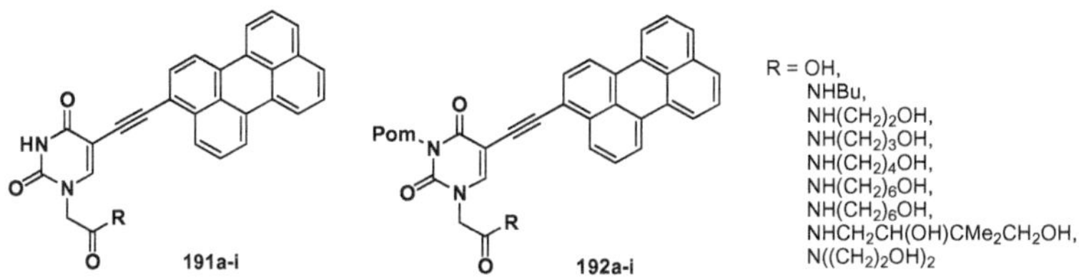

EC_{50} 0.003-0.05μM (TBEV strain Absettarov) CC_{50} >50 μM (PEK cells)

EC_{50} 0.0004-0.05μM (TBEV strain Absettarov) CC_{50} >50 μM (PEK cells)

Fig. 29 Structures and properties of compounds **191** and **192**.

They interact with virion envelope lipids and prevent fusion between viral and cellular membranes by inhibiting of the increased negative curvature required for the initial stages of fusion.

5-(Perylen-3-ylethynyl)uracil-1-acetic acid **191a** and its amides **191b–i** (Fig. 29) have low toxicity to porcine embryonic kidney cells (PEK) and high activity against tick-borne encephalitis virus (TBEV). It turned out that synthetic precursors **191a–i** protected by 3-pivaloyloxymethyl (Pom) **192a–i** exhibited even higher activity (Fig. 29).[187]

8. Dimers of NRTI and NNRTI

In an attempt to increase antiviral efficacy by combining nucleoside and non-nucleoside inhibitors of HIV RT in one molecule, different types of heterodimers have been obtained. There are several arguments for creating NRTI-NNRTI heterodimers. The use of NRTIs and NNRTIs in combination therapy supports the hypothesis that the synergistic activity of drugs leads to greater efficacy and slows down the emergence of resistance to individual agents.[189] Cooperative interaction between the NNRTI binding pocket and the substrate binding site suggests that a heterodimer combining the functions of NNRTI and NRTI, can very strongly and specifically bind to HIV RT.[190] Calculations of the additivity of binding energies for multicomponent systems show that the affinity of RT for the heterodimer molecule may be greater than for individual components.[191] These arguments led to the emergence of an alternative approach to combination therapy using "mixed site inhibitors," namely heterodimeric molecules that bind simultaneously in both sites of the RT.

The first type of dimers contained TSAO-T and AZT **193** or d4T **194** (Fig. 30) linked at the N3-position of the heterocyclic bases by an alkyl linker.[192,193] Dimers of this type have activity against HIV-1, but lower than original inhibitors. In addition, these dimeric derivatives are ineffective against HIV-2, indicating their excessive stability and inability to release HIV NRTIs outside and inside cells. The antiviral activity of the dimers can be attributed to a portion of the TSAO-T molecule without any significant contribution of the nucleoside moiety (AZT or d4T).

Another type of dimer consists of a HEPT derivative as an NNRTI fragment and a nucleoside analog: AZT, ddC, or d4T.[194] The authors studied the dependence of the activity from the nature of the linker and the position of its attachment to the pyrimidine base of the nucleoside analog. The AZT-HEPT dimers **195** (attachment of the linker to N3-position) and **196** (attachment of the linker to C5-position) (Fig. 31) exhibited moderate anti-HIV-1 activity in different cell cultures, but do not affect the replication

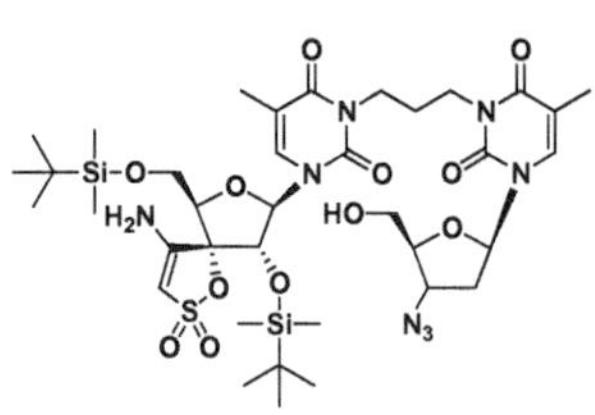

193
EC_{50} 0.38 μM (MT-4, HIV-1)
CC_{50} >100 μM (MT-4)
SI >263

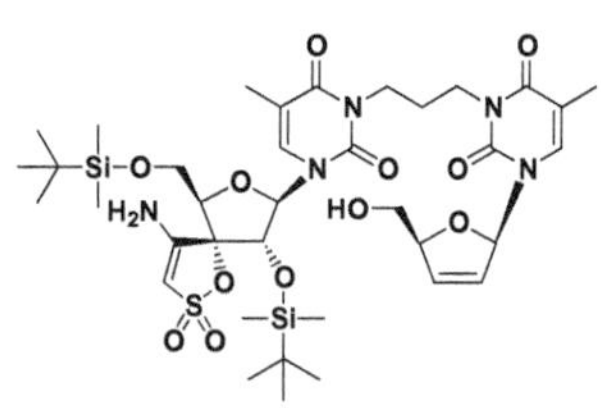

194
EC_{50} 0.04 μM (MT-4, HIV-1)
CC_{50} 16 μM (MT-4)
SI 400

Fig. 30 Structures and properties of compounds **193** and **194**.

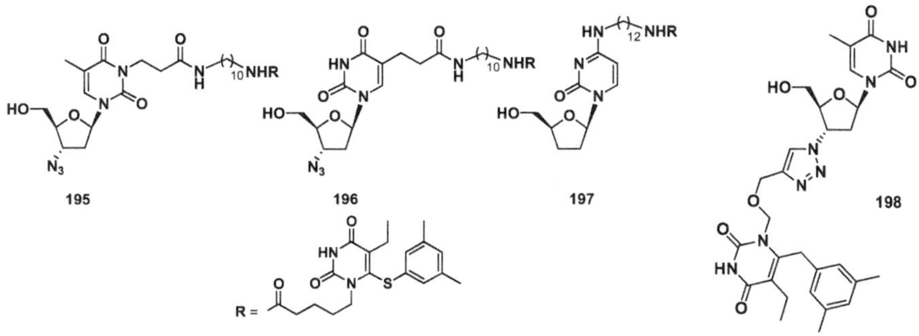

195	196	197	198
EC_{50} 2.8 μM (CEM-SS, HIV-1 LAI) CC_{50} >25 μM (CEM-SS)	EC_{50} 1.7 μM (CEM-SS, HIV-1 LAI) CC_{50} >25 μM (CEM-SS)	EC_{50} 0.43 μM (CEM-SS, HIV-1 LAI) CC_{50} >25 μM (CEM-SS)	EC_{50} 0.08 μM (MT-4, HIV-1 wt) CC_{50} >100 μM (MT-4)

Fig. 31 Structures and properties of compounds **195–198.**

of HIV-2 or the HIV-1 strain with the Y181C mutation. The ddC-HEPT **197** (Fig. 31) is equally effective against HIV-1 (wild-type strain and resistant to nevirapine Y181C) and HIV-2 in cell culture. The lack of a synergistic effect suggests that the two separate components of the dimer do not bind simultaneously at their respective sites.

Interestingly, the dimers AZT-HEPT (**195** and **196**) and ddC-HEPT **197** inhibited reverse transcriptase (RT) in the opposite way. The authors suggested that compounds **195** and **196** preferentially interacted with the hydrophobic pocket of RT, while **197** (after possible triphosphorylation) inhibited RT through binding at the catalytic site.[194] Obvious advantages of dimer **198**, in which the NNRTI fragment (analog of MKC-442) was attached to the 3′-azido group of AZT by the "click" chemistry method,[195] compared to the individual inhibitors, was not revealed. Detailed studies of the HIV RT structure showed that, although the distance between the catalytic site and the hydrophobic binding pocket is relatively small (10–15 Å),[196] these two sites are different and the probability that the heterodimer molecule can bind to both sites is extremely small. This led to a decline in interest in non-hydrolysable heterodimers.

With the development of the depot forms and prodrugs strategy, the idea of co-drugs, dimers capable of releasing two active components that inhibit HIV through different mechanisms, appeared. Such heterodimers are prodrugs that provide certain advantages, for example, improved solubility, gradual release of active components leading to reduced toxicity, improved pharmacokinetic parameters, and delayed onset of resistance. Examples of successful implementation of this approach are compounds **199–201** (Fig. 32).[197]

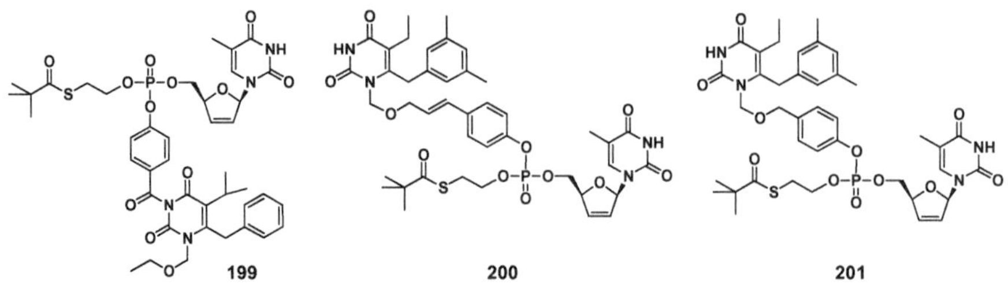

199
EC$_{50}$ 0.03 μM (HIV-1, IIIB)
EC$_{50}$ 0.002 μM (HIV-2)
CC$_{50}$ 32 μM (CEM)

200
EC$_{50}$ 0.01 μM (HIV-1, IIIB)
EC$_{50}$ 0.1 μM (HIV-2)
CC$_{50}$ 32 μM (CEM)

201
EC$_{50}$ 0.003 μM (HIV-1)
EC$_{50}$ 4.0 μM (HIV-2)
CC$_{50}$ 35 μM (MT-4)

Fig. 32 Structures and properties of compounds **199–201**.

In compounds **199–201**, the monophosphate of the nucleoside reverse transcriptase inhibitor d4T contained S-acyl-2-thioethyl (SATE) as phosphoester groups and an aromatic linker binding the NNRTI HIV. Dimer **199** was a hybrid of d4T monophosphate and the well-known NNRTI MKC-442 (**26**, see Fig. 4, Section 2.2.1), attached at the N3-position by a cleavable p-hydroxybenzoyl protecting group. In dimers **200** and **201** the nucleoside and non-nucleoside fragments were linked through phenol, which is part of these NNRTIs as a substituent at the N1-position. Dimers **199–201** had good activity against wild-type HIV-1, Y181C mutant, as well as against HIV-2 strain in MT-4, CEM and CEM TK-cell cultures, which confirms their ability to release d4T monophosphate inside cells.[197]

Heterodimers **202a-f** and **203**[198] (Fig. 33) included AZT or 3TC, as typical HIV-1 NRTIs, and 1-[ω-(phenoxy)alkyl]uracil derivatives[151] (**164**, **166** or related, see Fig. 20, Section 7.1), which are non-nucleoside inhibitors of HCMV. Such heterodimers are able to release active antiviral components under the action of hydrolyzing enzymes (esterases). The obtained compounds demonstrated antiviral activity in vitro against HCMV (ID$_{50}$ 3–12 μM) and HIV-1 (ID$_{50}$ 0.19–0.83 μM) without cytotoxicity (CD$_{50}$ 170–600 μM in a culture of human lung embryo diploid fibroblasts, CD$_{50}$ > 100 μM on MT-4 cells). Experiments on the human tissue system *ex vivo* simultaneously infected with HIV-1 (LAI.04) and HCMV confirmed the ability of the compounds to completely suppress the replication of both viruses in concentration of 10 μM, with an absence of cytotoxicity at this concentration.[198]

Another heterodimers containing an acetic acid moiety as a binding linker were compounds **204** and **205**.[199] They consisted of a non-nucleoside inhibitor of HIV (**66** or related compounds) and a nucleoside inhibitor of herpes viruses (acyclovir or 5′-noraristeromycin). Unfortunately, compounds **204**

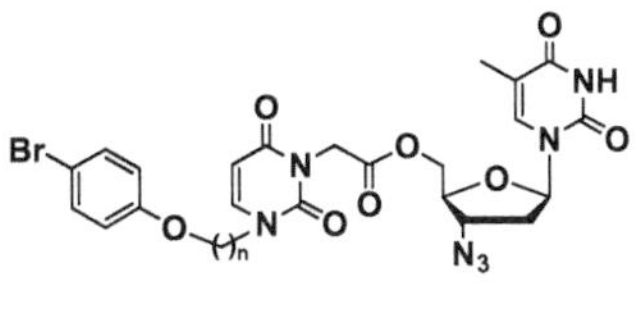

202a-f: n=3, 5, 6, 8, 10,12

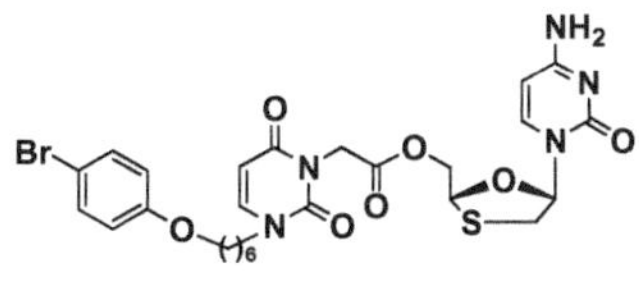

203

Fig. 33 Structures of compounds **202** and **203**.

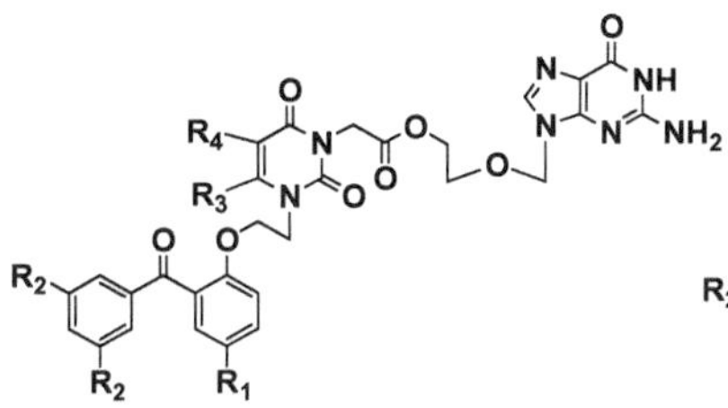

204

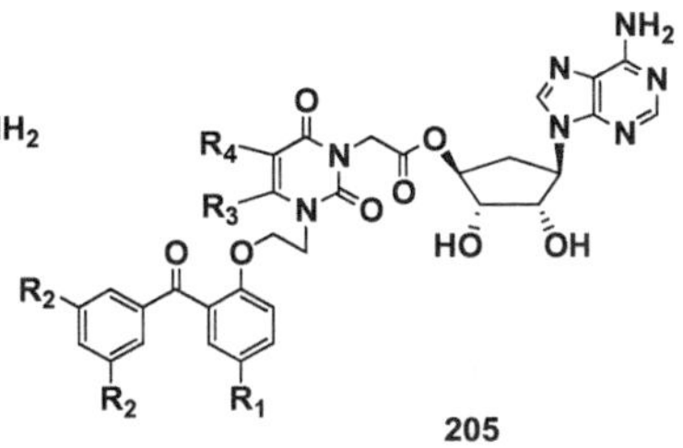

205

Fig. 34 Structures of compounds **204** and **205**.

and **205** did not show any advantages over the individual active agents they include. The lack of significant antiviral activity for the heterodimers **204** and **205**[199] (Fig. 34) suggests that the hydrolysis of the linker and the release of nucleoside analogs under cellular experiment conditions goes too slowly to effectively deliver the active components. Thus, the use of acetic acid as the linker, which proved successful in previously studied and structurally similar heterodimers,[198] did not give the desired results for this type of compounds.[199]

9. Conclusion

The widespread use of antiviral agents in the 20th century has made it possible to significantly alleviate the course of infectious diseases and reduce mortality. However, at present, most of the pathogenic viruses have developed resistance to the drugs used in the clinic. In addition, in recent years, new viruses such as influenza H5N1 and H1N1, SARS, MERS, Ebola and Zika have appeared. Currently, there is a Covid-19 pandemic caused by the SARS-CoV-2 coronavirus. In this regard, the development of new antiviral agents remains an extremely urgent task for medicinal chemists. It is known that considerable attention of researchers is paid to the chemistry of heterocyclic compounds containing nitrogen atoms and uracil derivatives - this is one of the most studied and productive areas of medicinal chemistry.

Uracil derivatives exhibit a wide range of biological activities, including antitumor, antibacterial, antiparasitic, antiviral, *etc.* In this review we introduced for readers a wide range of uracil derivatives, which are non-nucleoside inhibitors of various DNA and RNA viruses. We also described conditions for the synthesis of antiviral agents, based on uracil scaffold, and issues related to the structure-antiviral activity relationship of the presented uracil derivatives. We sincerely hope that this review will stimulate further research into the synthesis of uracil derivatives and encourage researchers to develop new approaches to the creation of new highly effective antiviral agents.

Acknowledgment

The work was supported by the Russian Foundation for Basic Research Project № 18-29-08010.

References

1. Jordheim, L. P.; Durantel, D.; Zoulim, F.; Dumontet, C. *Nat. Rev. Drug Discov.* **2013**, *12*, 447–464.
2. Seley-Radtke, K. L.; Yates, M. K. *Antiviral Res.* **2018**, *154*, 66–86.
3. Yates, M. K.; Seley-Radtke, K. L. *Antiviral Res.* **2019**, *162*, 5–21.
4. Longley, D. B.; Harkin, D. P.; Johnston, P. G. *Nat. Rev. Cancer* **2003**, *3*, 330–338.
5. Palasz, A.; Ciez, D. *Eur. J. Med. Chem.* **2015**, *97*, 582–611.
6. Thomson, J. M.; Lamont, I. L. *Front. Microbiol.* **2019**, *10*, 952.
7. Ali, J. A.; Tagoe, D. N.; Munday, J. C.; Donachie, A.; Morrison, L. J.; de Koning, H. P. *PLoS One* **2013**, *8*, e58034.
8. Campagnaro, G. D.; de Koning, H. P. *Med. Res. Rev.* **2020**, *40*, 1679–1714.
9. De Clercq, E. *Chem. Asian J.* **2019**, *14*, 3962–3968.
10. De Clercq, E.; Li, G. *Clin. Microbiol. Rev.* **2016**, *29*, 695–747.
11. Tyrrell, D. A. *Trans. Ophthalmol. Soc. U. K.* **1963**, *83*, 573–581.
12. Pomerantz, R. J.; Schooley, R. T. *Clin. Lab. Med.* **1987**, 7, 793–813.
13. Balzarini, J. *Pharm. World Sci.* **1994**, *16*, 113–126.
14. Schalm, S. W.; de Man, R. A.; Heijtink, R. A.; Niesters, H. G. *J. Hepatol.* **1995**, *22*, 52–56.
15. Sun, S.; Yang, Q.; Sheng, Y.; Fu, Y.; Sun, C.; Deng, C. *Exp. Ther. Med.* **2021**, *21*, 35.
16. Richman, D. D. *Antivir. Ther.* **2001**, *6*, 83–88.
17. Hirsch, M. S.; Schooley, R. T. *N. Engl. J. Med.* **1983**, *309*, 963–970.
18. Lea, A. P.; Bryson, H. M. *Drugs* **1996**, *52*, 225–230 (discussion 231).
19. D'Cruz, O. J.; Uckun, F. M. *J. Enzyme Inhib. Med. Chem.* **2006**, *21*, 329–350.
20. Chen, W.; Zhan, P.; Wu, J.; Li, Z.; Liu, X. *Curr. Pharm. Des.* **2012**, *18*, 4165–4186.
21. Zhuang, C.; Pannecouque, C.; De Clercq, E.; Chen, F. *Acta Pharm. Sin. B* **2020**, *10*, 961–978.
22. Song, Y.; Fang, Z.; Zhan, P.; Liu, X. *Curr. Med. Chem.* **2014**, *21*, 329–355.
23. Zhan, P.; Liu, X.; Li, Z. *Curr. Med. Chem.* **2009**, *16*, 2876–2889.
24. Famiglini, V.; Silvestri, R. *Molecules* **2016**, *21*.
25. Miyasaka, T.; Tanaka, H.; Baba, M.; Hayakawa, H.; Walker, R. T.; Balzarini, J.; Declercq, E. *J. Med. Chem.* **1989**, *32*, 2507–2509.

26. Baba, M.; Tanaka, H.; Declercq, E.; Pauwels, R.; Balzarini, J.; Schols, D.; Nakashima, H.; Perno, C. F.; Walker, R. T.; Miyasaka, T. *Biochem. Biophys. Res. Commun.* **1989**, *165*, 1375–1381.
27. Baba, M.; Declercq, E.; Tanaka, H.; Ubasawa, M.; Takashima, H.; Sekiya, K.; Nitta, I.; Umezu, K.; Nakashima, H.; Mori, S.; Shigeta, S.; Walker, R. T.; Miyasaka, T. *Proc. Natl. Acad. Sci. U. S. A.* **1991**, *88*, 2356–2360.
28. Tanaka, H.; Baba, M.; Saito, S.; Miyasaka, T.; Takashima, H.; Sekiya, K.; Ubasawa, M.; Nitta, I.; Walker, R. T.; Nakashima, H.; Declercq, E. *J. Med. Chem.* **1991**, *34*, 1508–1511.
29. Tanaka, H.; Takashima, H.; Ubasawa, M.; Sekiya, K.; Nitta, I.; Baba, M.; Shigeta, S.; Walker, R. T.; Declercq, E.; Miyasaka, T. *J. Med. Chem.* **1992**, *35*, 4713–4719.
30. Baba, M.; Declercq, E.; Tanaka, H.; Ubasawa, M.; Takashima, H.; Sekiya, K.; Nitta, I.; Umezu, K.; Walker, R. T.; Mori, S.; Ito, M.; Shigeta, S.; Miyasaka, T. *Mol. Pharmacol.* **1991**, *39*, 805–810.
31. Balzarini, J.; Velazquez, S.; Sanfelix, A.; Karlsson, A.; Perezperez, M. J.; Camarasa, M. J.; Declercq, E. *Mol. Pharmacol.* **1993**, *43*, 109–114.
32. Pan, B. C.; Chen, H. C.; Piras, G.; Dutschman, G. E.; Rowe, E. C.; Cheng, Y. C.; Chu, S. H. *J. Heterocyclic Chem.* **1994**, *31*, 177–185.
33. Hopkins, A. L.; Ren, J. S.; Tanaka, H.; Baba, M.; Okamato, M.; Stuart, D. I.; Stammers, D. K. *J. Med. Chem.* **1999**, *42*, 4500–4505.
34. Sun, G. F.; Chen, X. X.; Chen, F. E.; Wang, Y. P.; De Clercq, E.; Balzarini, J.; Pannecouque, C. *Chem. Pharm. Bull.(Tokyo)* **2005**, *53*, 886–892.
35. Goudgaon, N. M.; Mcmillan, A.; Schinazi, R. F. *Antivir. Chem. Chemother.* **1992**, *3*, 263–266.
36. Goudgaon, N. M.; Schinazi, R. F. *J. Med. Chem.* **1991**, *34*, 3305–3309.
37. Kim, D. K.; Kim, Y. W.; Gam, J. S.; Kim, G. Y.; Lim, J. S.; Lee, N. Y.; Kim, H. T.; Kim, K. H. *J. Heterocyclic Chem.* **1996**, *33*, 1275–1283.
38. Nguyen, M. H.; Schinazi, R. F.; Shi, C. F.; Goudgaon, N. M.; Mckenna, P. M.; Mellors, J. W. *Antimicrob. Agents Chemother.* **1994**, *38*, 2409–2414.
39. Kim, D. K.; Kim, H. T.; Lim, J. S.; Gam, J.; Kim, Y. W.; Kim, K. H.; Shin, Y. O. *J. Heterocyclic Chem.* **1996**, *33*, 885–894.
40. Tanaka, H.; Baba, M.; Hayakawa, H.; Sakamaki, T.; Miyasaka, T.; Ubasawa, M.; Takashima, H.; Sekiya, K.; Nitta, I.; Shigeta, S.; Walker, R. T.; Balzarini, J.; Declercq, E. *J. Med. Chem.* **1991**, *34*, 349–357.
41. Tanaka, H.; Takashima, H.; Ubasawa, M.; Sekiya, K.; Nitta, I.; Baba, M.; Shigeta, S.; Walker, R. T.; Declercq, E.; Miyasaka, T. *J. Med. Chem.* **1992**, *35*, 337–345.
42. Novikov, M. S.; Geisman, A. N. *Chem. Heterocycl. Compd.* **2014**, *49*, 1426–1450.
43. Baba, M.; Shigeta, S.; Yuasa, S.; Takashima, H.; Sekiya, K.; Ubasawa, M.; Tanaka, H.; Miyasaka, T.; Walker, R. T.; Declercq, E. *Antimicrob. Agents Chemother.* **1994**, *38*, 688–692.
44. Meng, G.; Chen, F. E.; De Clercq, E.; Balzarini, J.; Pannecouque, C. *Chem. Pharm. Bull.* **2003**, *51*, 779–789.
45. Tanaka, H.; Takashima, H.; Ubasawa, M.; Sekiya, K.; Inouye, N.; Baba, M.; Shigeta, S.; Walker, R. T.; Declercq, E.; Miyasaka, T. *J. Med. Chem.* **1995**, *38*, 2860–2865.
46. El-Brollosy, N. R.; Jorgensen, P. T.; Dahan, B.; Boel, A. M.; Pedersen, E. B.; Nielsen, C. *J. Med. Chem.* **2002**, *45*, 5721–5726.
47. El-Brollosy, N. R.; Pedersen, E. B.; Nielsen, C. *Arch. Pharm.* **2003**, *336*, 236–241.
48. El-Brollosy, N. R.; Loddo, R. *Drug Res. (Stuttg)* **2016**, *66*, 181–188.
49. Wamberg, M.; Pedersen, E. B.; El-Brollosy, N. R.; Nielsen, C. *Bioorg. Med. Chem.* **2004**, *12*, 1141–1149.
50. Loksha, Y. M.; Pedersen, E. B.; Loddo, R.; La Colla, P. *Arch. Pharm.* **2009**, *342*, 501–506.

51. Mitchell, M. L.; Son, J. C.; Guo, H. Y.; Im, Y. A.; Cho, E. J.; Wang, J. H.; Hayes, J.; Wang, M.; Paul, A.; Lansdon, E. B.; Chen, J. M.; Graupe, D.; Rhodes, G.; He, G. X.; Geleziunas, R.; Xu, L. H.; Kim, C. U. *Bioorg. Med. Chem. Lett.* **2010**, *20*, 1589–1592.
52. Aly, Y. L.; Pedersen, E. B.; La Colla, P.; Loddo, R. *J. Heterocyclic Chem.* **2007**, *44*, 381–387.
53. Ji, L.; Chen, F. E.; Feng, X. Q.; De Clercq, E.; Balzarini, J.; Pannecouque, C. *Chem. Pharm. Bull.* **2006**, *54*, 1248–1253.
54. Ji, L.; Chen, F. E.; Xie, B.; De Clercq, E.; Balzarini, J.; Pannecouque, C. *Eur. J. Med. Chem.* **2007**, *42*, 198–204.
55. Petersen, L.; Jessen, C. H.; Pedersen, E. B.; Nielsen, C. *Org. Biomol. Chem.* **2003**, *1*, 3541–3545.
56. Son, J.-C.; Lee, I.-Y.; Bae, B.-I.; Han, J.-S.; Choi, J.-K.; Chae, Y. B. *Antiviral 2,4-pyrimidinedione derivatives*; 1998.
57. Son, J.-C.; Shin, S.-S.; Kim, S.-K.; Lee, C.-K.; Kim, H.-S. *Antiviral 2,4-pyrimidinedione derivatives and process for the preparation thereof*; 2004.
58. Elshehry, M. F.; Balzarini, J.; Meier, C. *Synthesis-Stuttgart* **2009**, 841–847.
59. Geisman, A. N.; Valuev-Elliston, V. T.; Ozerov, A. A.; Khandazhinskaya, A. L.; Chizhov, A. O.; Kochetkov, S. N.; Pannecouque, C.; Naesens, L.; Seley-Radtke, K. L.; Novikov, M. S. *Bioorg. Med. Chem.* **2016**, *24*, 2476–2485.
60. Nazaretyan, A. K.; Torosyan, G. O.; Babayan, A. T. *J. Appl. Chem. USSR* **1985**, *58*, 2396–2400.
61. Tanaka, H.; Baba, M.; Takahashi, E.; Matsumoto, K.; Kittaka, A.; Walker, R. T.; De Clercq, E.; Miyasaka, T. *Nucleosides Nucleotides* **1994**, *13*, 155–162.
62. Boudet, N.; Knochel, P. *Org. Lett.* **2006**, *8*, 3737–3740.
63. Danel, K.; Larsen, E.; Pedersen, E. B. *Synthesis-Stuttgart* **1995**, *8*, 934–936.
64. El-Brollosy, N. R.; Al-Deeb, O. A.; El-Emam, A. A.; Pedersen, E. B.; La Colla, P.; Collu, G.; Sanna, G.; Loddo, R. *Arch. Pharm. (Weinheim, Ger.)* **2009**, *342*, 663–670.
65. Mitchell, M. L.; Son, J. C.; Lee, I. Y.; Lee, C. K.; Kim, H. S.; Guo, H. Y.; Wang, J. H.; Hayes, J.; Wang, M.; Paul, A.; Lansdon, E. B.; Chen, J. M.; Eisenberg, G.; Geleziunas, R.; Xu, L. H.; Kim, C. U. *Bioorg. Med. Chem. Lett.* **2010**, *20*, 1585–1588.
66. Buckheit, R. W.; Hartman, T. L.; Watson, K. M.; Chung, S. G.; Cho, E. H. *Antimicrob. Agents Chemother.* **2008**, *52*, 225–236.
67. Buckheit, R. W., Jr.; Hartman, T. L.; Watson, K. M.; Kwon, H. S.; Lee, S. H.; Lee, J. W.; Kang, D. W.; Chung, S. G.; Cho, E. H. *Antivir. Chem. Chemother.* **2007**, *18*, 259–275.
68. Watson Buckheit, K.; Yang, L.; Buckheit, R. W., Jr. *Antimicrob. Agents Chemother.* **2011**, *55*, 5243–5254.
69. Buckheit, R. W., Jr.; Watson Buckheit, K.; Sturdevant, C. B.; Buckheit, R. W., 3rd. *Antiviral Res.* **2013**, *100*, 382–391.
70. Buckheit, R. W.; Watson, K.; Fliakas-Boltz, V.; Russell, J.; Loftus, T. L.; Osterling, M. C.; Turpin, J. A.; Pallansch, L. A.; White, E. L.; Lee, J. W.; Lee, S. H.; Oh, J. W.; Kwon, H. S.; Chung, S. G.; Cho, E. H. *Antimicrob. Agents Chemother.* **2001**, *45*, 393–400.
71. Mahalingam, A.; Simmons, A. P.; Ugaonkar, S. R.; Watson, K. M.; Dezzutti, C. S.; Rohan, L. C.; Buckheit, R. W., Jr.; Kiser, P. F. *Antimicrob. Agents Chemother.* **2011**, *55*, 1650–1660.
72. Pereira, L. E.; Mesquita, P. M. M.; Ham, A.; Singletary, T.; McNicholl, J.; Buckheit, K. W.; Buckheit, R. W.; Smith, J. *J. Med. Primatol.* **2015**, *44*, 320.
73. Pereira, L. E.; Singletary, T.; Martin, A.; Dinh, C. T.; Deyounks, F.; Holder, A.; McNicholl, J.; Buckheit, K. W.; Buckheit, R. W.; Ham, A.; Katz, D. F.; Smith, J. M. *Drug Deliv. Transl. Res.* **2018**, *8*, 1180–1190.
74. Ham, A. S.; Rohan, L. C.; Boczar, A.; Yang, L.; Buckheit, K. W.; Buckheit, R. W. *Pharm. Res-Dordr.* **2012**, *29*, 1897–1907.

75. Srinivasan, P.; Zhang, J. N.; Martin, A.; Kelley, K.; McNicholl, J. M.; Buckheit, R. W.; Smith, J. M.; Ham, A. S. *Antimicrob. Agents Chemother.* **2016**, *60*, 4140–4150.
76. Al-Khouja, A.; Shieh, E.; Fuchs, E. J.; Marzinke, M. A.; Bakshi, R. P.; Hummert, P.; Ham, A. S.; Buckheit, K. W.; Breakey, J.; Weld, E. D.; Chen, H.; Caffo, B. S.; Buckheit, R. W.; Hendrix, C. W. *AIDS Res. Hum. Retroviruses* **2021**.
77. Danel, K. *Synthesis-Stuttgart* **1997**, *9*, 1021–1026.
78. Loksha, Y. M.; Pedersen, E. B.; Loddo, R.; Sanna, G.; Collu, G.; Giliberti, G.; La Colla, P. *J. Med. Chem.* **2014**, *57*, 5169–5178.
79. Bardagi, J. I.; Rossi, R. A. *J. Org. Chem.* **2008**, *73*, 4491–4495.
80. Nichols, S. E.; Domaoal, R. A.; Thakur, V. V.; Tirado-Rives, J.; Anderson, K. S.; Jorgensen, W. L. *J. Chem. Inf. Model.* **2009**, *49*, 1272–1279.
81. Bollini, M.; Domaoal, R. A.; Thakur, V. V.; Gallardo-Macias, R.; Spasov, K. A.; Anderson, K. S.; Jorgensen, W. L. *J. Med. Chem.* **2011**, *54*, 8582–8591.
82. Chan, J. H.; Freeman, G. A.; Tidwell, J. H.; Romines, K. R.; Schaller, L. T.; Cowan, J. R.; Gonzales, S. S.; Lowell, G. S.; Andrews, C. W.; Reynolds, D. J.; St Clair, M.; Hazen, R. J.; Ferris, R. G.; Creech, K. L.; Roberts, G. B.; Short, S. A.; Weaver, K.; Koszalka, G. W.; Boone, L. R. *J. Med. Chem.* **2004**, *47*, 1175–1182.
83. Novikov, M. S.; Ivanova, O. N.; Ivanov, A. V.; Ozerov, A. A.; Valuev-Elliston, V. T.; Temburnikar, K.; Gurskaya, G. V.; Kochetkov, S. N.; Pannecouque, C.; Balzarini, J.; Seley-Radtke, K. L. *Bioorg. Med. Chem.* **2011**, *19*, 5794–5802.
84. Leporati, A.; Novikov, M. S.; Valuev-Elliston, V. T.; Korolev, S. P.; Khandazhinskaya, A. L.; Kochetkov, S. N.; Gupta, S.; Goding, J.; Bolotin, E.; Gottikh, M. B.; Bogdanov, A. A. *Nanosci. Nanotechnol.* **2016**, *12*, 2405–2413.
85. Janssen, P. A. J.; Lewi, P. J.; Arnold, E.; Daeyaert, F.; de Jonge, M.; Heeres, J.; Koymans, L.; Vinkers, M.; Guillemont, J.; Pasquier, E.; Kukla, M.; Ludovici, D.; Andries, K.; de Bethune, M. P.; Pauwels, R.; Das, K.; Clark, A. D.; Frenkel, Y. V.; Hughes, S. H.; Medaer, B.; De Knaep, F.; Bohets, H.; De Clerck, F.; Lampo, A.; Williams, P.; Stoffels, P. *J. Med. Chem.* **2005**, *48*, 1901–1909.
86. Lee, W. G.; Gallardo-Macias, R.; Frey, K. M.; Spasov, K. A.; Bollini, M.; Anderson, K. S.; Jorgensen, W. L. *J. Am. Chem. Soc.* **2013**, *135*, 16705–16713.
87. Lee, W. G.; Frey, K. M.; Gallardo-Macias, R.; Spasov, K. A.; Bollini, M.; Anderson, K. S.; Jorgensen, W. L. *ACS Med. Chem. Lett.* **2014**, *5*, 1259–1262.
88. Lee, W. G.; Chan, A. H.; Spasov, K. A.; Anderson, K. S.; Jorgensen, W. L. *ACS Med. Chem. Lett.* **2016**, 7, 1156–1160.
89. Kudalkar, S. N.; Beloor, J.; Chan, A. H.; Lee, W. G.; Jorgensen, W. L.; Kumar, P.; Anderson, K. S. *Mol. Pharmacol.* **2017**, *91*, 383–U168.
90. Kudalkar, S. N.; Beloor, J.; Quijano, E.; Spasov, K. A.; Lee, W. G.; Cisneros, J. A.; Saltzman, W. M.; Kumar, P.; Jorgensen, W. L.; Anderson, K. S. *Proc. Natl. Acad. Sci. U. S. A.* **2018**, *115*, E802–E811.
91. Sweeney, Z. K.; Kennedy-Smith, J. J.; Wu, J.; Arora, N.; Billedeau, J. R.; Davidson, J. R.; Fretland, J.; Hang, J. Q.; Heilek, G. M.; Harris, S. F.; Hirschfeld, D.; Inbar, P.; Javanbakht, H.; Jernelius, J. A.; Jin, Q. W.; Li, Y.; Liang, W. L.; Roetz, R.; Sarma, K.; Smith, M.; Stefanidis, D.; Su, G. P.; Suh, J. M.; Villasenor, A. G.; Welch, M.; Zhang, F. J.; Klumpp, K. *ChemMedChem* **2009**, *4*, 88–99.
92. Zhang, H.; Tian, Y.; Kang, D. W.; Huo, Z. P.; Zhou, Z. X.; Liu, H. Q.; De Clercq, E.; Pannecouque, C.; Zhan, P.; Liu, X. Y. *Eur. J. Med. Chem.* **2017**, *130*, 209–222.
93. Claisen, L.; Kremers, F.; Roth, F.; Tietze, E. *Justus Liebigs Ann. Chem.* **1925**, *442*, 210–245.
94. McKay, A. F.; Baker, H. A.; Gaudry, R.; Garmaise, D. L.; Ranz, R. J. *J. Med. Chem.* **1963**, *6*, 816–817.
95. Kikumoto, R.; Tobe, A.; Fukami, H.; Egawa, M. *J. Med. Chem.* **1983**, *26*, 246–250.

96. Novikov, M. S.; Ozerov, A. A. *Chem. Heterocycl. Compd.* **2005**, *41*, 905–908.
97. Frieden, M.; Giraud, M.; Reese, C. B.; Song, Q. L. *J. Chem. Soc. Perkin Trans.* **1998**, *1*, 2827–2832.
98. Bode, M. L.; Kaye, P. T. *J. Chem. Soc. Perkin Trans.* **1993**, *1*, 1809–1813.
99. Gourves, J. P.; Ruzziconi, R.; Vilarroig, L. *J. Org. Chem.* **2001**, *66*, 617–619.
100. Hartmann, R. W.; Palusczak, A.; Lacan, F.; Ricci, G.; Ruzziconi, R. *J. Enzyme Inhib. Med. Chem.* **2004**, *19*, 145–155.
101. Liu, Z. Q.; Chen, W. M.; Zhan, P.; De Clercq, E.; Pannecouque, C.; Liu, X. Y. *Eur. J. Med. Chem.* **2014**, *87*, 52–62.
102. Mejias, X.; Feliu, L.; Planas, M.; Bardaji, E. *Tetrahedron Lett.* **2006**, *47*, 8069–8071.
103. Li, X.; Chen, W. M.; Tian, Y.; Liu, H. Q.; Zhan, P.; De Clercq, E.; Pannecouque, C.; Balzarini, J.; Liu, X. Y. *Eur. J. Med. Chem.* **2014**, *80*, 112–121.
104. Maruyama, T.; Kozai, S.; Yamasaki, T.; Witvrouw, M.; Pannecouque, C.; Balzarini, J.; Snoeck, R.; Andrei, G.; De Clercq, E. *Antivir. Chem. Chemother.* **2003**, *14*, 271–279.
105. Maruyama, T.; Kozai, S.; Demizu, Y.; Witvrouw, M.; Pannecouque, C.; Balzarini, J.; Snoecks, R.; Andrei, G.; De Clercq, E. *Chem. Pharm. Bull.(Tokyo)* **2006**, *54*, 325–333.
106. Isono, Y.; Sakakibara, N.; Ordonez, P.; Hamasaki, T.; Baba, M.; Ikejiri, M.; Maruyama, T. *Antivir. Chem. Chemother.* **2011**, *22*, 57–65.
107. Ordonez, P.; Hamasaki, T.; Isono, Y.; Sakakibara, N.; Ikejiri, M.; Maruyama, T.; Baba, M. *Antimicrob. Agents Chemother.* **2012**, *56*, 2581–2589.
108. Sakakibara, N.; Hamasaki, T.; Baba, M.; Demizu, Y.; Kurihara, M.; Irie, K.; Iwai, M.; Asada, E.; Kato, Y.; Maruyama, T. *Bioorg. Med. Chem.* **2013**, *21*, 5900–5906.
109. Sakakibara, N.; Baba, M.; Okamoto, M.; Toyama, M.; Demizu, Y.; Misawa, T.; Kurihara, M.; Irie, K.; Kato, Y.; Maruyama, T. *Antivir. Chem. Chemother.* **2015**, *24*, 3–18.
110. Babkov, D. A.; Valuev-Elliston, V. T.; Paramonova, M. P.; Ozerov, A. A.; Ivanov, A. V.; Chizhov, A. O.; Khandazhinskaya, A. L.; Kochetkov, S. N.; Balzarini, J.; Daelemans, D.; Pannecouque, C.; Seley-Radtke, K. L.; Novikov, M. S. *Bioorg. Med. Chem.* **2015**, *23*, 1069–1081.
111. Novikov, M. S.; Valuev-Elliston, V. T.; Babkov, D. A.; Paramonova, M. P.; Ivanov, A. V.; Gavryushov, S. A.; Khandazhinskaya, A. L.; Kochetkov, S. N.; Pannecouque, C.; Andrei, G.; Snoeck, R.; Balzarini, J.; Seley-Radtke, K. L. *Bioorg. Med. Chem.* **2013**, *21*, 1150–1158.
112. Singh, H.; Aggarwal, P.; Kumar, S. *Synthesis-Stuttgart* **1980**, *6*, 520–522.
113. Noell, C. W.; Cheng, C. *J. Heterocyclic Chem.* **1968**, *5*, 25–28.
114. White, R. D.; Wood, J. L. *Org. Lett.* **2001**, *3*, 1825–1827.
115. Camarasa, M. J.; Perezperez, M. J.; Sanfelix, A.; Balzarini, J.; Declercq, E. *J. Med. Chem.* **1992**, *35*, 2721–2727.
116. Das, K.; Bauman, J. D.; Rim, A. S.; Dharia, C.; Clark, A. D.; Camarasa, M. J.; Balzarini, J.; Arnold, E. *J. Med. Chem.* **2011**, *54*, 2727–2737.
117. Balzarini, J.; Perezperez, M. J.; Sanfelix, A.; Schols, D.; Perno, C. F.; Vandamme, A. M.; Camarasa, M. J.; Declercq, E. *Proc. Natl. Acad. Sci. U. S. A.* **1992**, *89*, 4392–4396.
118. Perez-Perez, M. J.; San-Felix, A.; Balzarini, J.; De Clercq, E.; Camarasa, M. J. *J. Med. Chem.* **1992**, *35*, 2988–2995.
119. Camarasa, M. J.; Velazquez, S.; San-Felix, A.; Perez-Perez, M. J. *Antivir. Chem. Chemother.* **2005**, *16*, 147–153.
120. Chamorro, C.; Camarasa, M. J.; Perez-Perez, M. J.; de Clercq, E.; Balzarini, J.; San Felix, A. *Antivir. Chem. Chemother.* **1998**, *9*, 413–422.
121. Rodriguez-Barrios, F.; Perez, C.; Lobaton, E.; Velazquez, S.; Chamorro, C.; San-Felix, A.; Perez-Perez, M. J.; Camarasa, M. J.; Pelemans, H.; Balzarini, J.; Gago, F. *J. Med. Chem.* **2001**, *44*, 1853–1865.

122. Bonache, M. C.; Chamorro, C.; Velazquez, S.; De Clercq, E.; Balzarini, J.; Barrios, F. R.; Gago, F.; Camarasa, M. J.; San-Felix, A. *J. Med. Chem.* **2005**, *48*, 6653–6660.
123. Moura, M.; Josse, S.; Albert, N. V. N.; Fournier, C.; Duverlie, G.; Castelain, S.; Soriano, E.; Marco-Contelles, J.; Balzarini, J.; Postel, D. *Eur. J. Med. Chem.* **2011**, *46*, 5046–5056.
124. Calvo-Mateo, A.; Camarasa, M. J.; Diazortiz, A.; Heras, F. G. D.; Alemany, A. *Tetrahedron* **1988**, *44*, 4895–4903.
125. Hollemberg, D.H.; Klein, R. S.; Fox, J. J. *J. Carbohydr. Rev.* **1978**, *67*, 491–494.
126. Vorbruggen, H.; Krolikiewicz, K.; Bennua, B. *Chem. Ber./Recl.* **1981**, *114*, 1234–1255.
127. Matyugina, E. S.; Valuev-Elliston, V. T.; Babkov, D. A.; Novikov, M. S.; Ivanov, A. V.; Kochetkov, S. N.; Balzarini, J.; Seley-Radtke, K. L.; Khandazhinskaya, A. L. *Medchemcomm* **2013**, *4*, 741–748.
128. Matyugina, E. S.; Valuev-Elliston, V. T.; Geisman, A. N.; Novikov, M. S.; Chizhov, A. O.; Kochetkov, S. N.; Seley-Radtke, K. L.; Khandazhinskaya, A. L. *Medchemcomm* **2013**, *4*, 1443–1451.
129. Matyugina, E.; Khandazhinskaya, A.; Chernousova, L.; Andreevskaya, S.; Smirnova, T.; Chizhov, A.; Karpenko, I.; Kochetkov, S.; Alexandrova, L. *Bioorg. Med. Chem.* **2012**, *20*, 6680–6686.
130. Tantillo, C.; Ding, J.; Jacobo-Molina, A.; Nanni, R. G.; Boyer, P. L.; Hughes, S. H.; Pauwels, R.; Andries, K.; Janssen, P. A.; Arnold, E. *J. Mol. Biol.* **1994**, *243*, 369–387.
131. Tramontano, E.; Di Santo, R. *Curr. Med. Chem.* **2010**, *17*, 2837–2853.
132. Asante-Appiah, E.; Skalka, A. M. *Adv. Virus Res.* **1999**, *52*, 351–369.
133. Hindmarsh, P.; Leis, J. *Microbiol. Mol. Biol. Rev.* **1999**, *63*, 836–843. Table of Contents.
134. Wang, Z.; Bennett, E. M.; Wilson, D. J.; Salomon, C.; Vince, R. *J. Med. Chem.* **2007**, *50*, 3416–3419.
135. Wang, Z.; Vince, R. *Bioorg. Med. Chem. Lett.* **2008**, *18*, 1293–1296.
136. Tang, J.; Maddali, K.; Dreis, C. D.; Sham, Y. Y.; Vince, R.; Pommier, Y.; Wang, Z. *ACS Med. Chem. Lett.* **2011**, *2*, 63–67.
137. Tang, J.; Maddali, K.; Metifiot, M.; Sham, Y. Y.; Vince, R.; Pommier, Y.; Wang, Z. *J. Med. Chem.* **2011**, *54*, 2282–2292.
138. Tang, J.; Maddali, K.; Dreis, C. D.; Sham, Y. Y.; Vince, R.; Pommier, Y.; Wang, Z. *Bioorg. Med. Chem. Lett.* **2011**, *21*, 2400–2402.
139. Tang, J.; Kirby, K. A.; Huber, A. D.; Casey, M. C.; Ji, J.; Wilson, D. J.; Sarafianos, S. G.; Wang, Z. *Eur. J. Med. Chem.* **2017**, *128*, 168–179.
140. Tang, J.; Liu, F.; Nagy, E.; Miller, L.; Kirby, K. A.; Wilson, D. J.; Wu, B.; Sarafianos, S. G.; Parniak, M. A.; Wang, Z. *J. Med. Chem.* **2016**, *59*, 2648–2659.
141. Wang, L.; Tang, J.; Huber, A. D.; Casey, M. C.; Kirby, K. A.; Wilson, D. J.; Kankanala, J.; Xie, J.; Parniak, M. A.; Sarafianos, S. G.; Wang, Z. *Eur. J. Med. Chem.* **2018**, *156*, 652–665.
142. Tang, J.; Do, H. T.; Huber, A. D.; Casey, M. C.; Kirby, K. A.; Wilson, D. J.; Kankanala, J.; Parniak, M. A.; Sarafianos, S. G.; Wang, Z. *Eur. J. Med. Chem.* **2019**, *166*, 390–399.
143. Wang, L.; Tang, J.; Huber, A. D.; Casey, M. C.; Kirby, K. A.; Wilson, D. J.; Kankanala, J.; Parniak, M. A.; Sarafianos, S. G.; Wang, Z. *Eur. J. Med. Chem.* **2018**, *156*, 680–691.
144. Wu, B.; Tang, J.; Wilson, D. J.; Huber, A. D.; Casey, M. C.; Ji, J.; Kankanala, J.; Xie, J.; Sarafianos, S. G.; Wang, Z. *J. Med. Chem.* **2016**, *59*, 6136–6148.
145. Yoneda, F.; Tsukuda, K.; Kawazoe, M.; Sone, A. *J. Heterocyclic Chem.* **1981**, *18*, 1329–1334.
146. Griffiths, P. D.; Walter, S. *Curr. Opin. Infect. Dis.* **2005**, *18*, 241–245.

147. Bialas, K. M.; Swamy, G. K.; Permar, S. R. *Clin. Perinatol.* **2015**, *42*, 61–75 (viii).
148. Swanson, E. C.; Schleiss, M. R. *Pediatr. Clin. North Am.* **2013**, *60*, 335–349.
149. de Castro, S.; Lobaton, E.; Perez-Perez, M. J.; San-Felix, A.; Cordeiro, A.; Andrei, G.; Snoeck, R.; De Clercq, E.; Balzarini, J.; Camarasa, M. J.; Velazquez, S. *J. Med. Chem.* **2005**, *48*, 1158–1168.
150. de Castro, S.; Peromingo, M. T.; Naesens, L.; Andrei, G.; Snoeck, R.; Balzarini, J.; Velazquez, S.; Camarasa, M. J. *J. Med. Chem.* **2008**, *51*, 5823–5832.
151. Novikov, M. S.; Babkov, D. A.; Paramonova, M. P.; Khandazhinskaya, A. L.; Ozerov, A. A.; Chizhov, A. O.; Andrei, G.; Snoeck, R.; Balzarini, J.; Seley-Radtke, K. L. *Bioorg. Med. Chem.* **2013**, *21*, 4151–4157.
152. Choo, Q. L.; Kuo, G.; Weiner, A. J.; Overby, L. R.; Bradley, D. W.; Houghton, M. *Science* **1989**, *244*, 359–362.
153. Randolph, J. T.; Krueger, A. C.; Donner, P. L.; Pratt, J. K.; Liu, D.; Motter, C. E.; Rockway, T. W.; Tufano, M. D.; Wagner, R.; Lim, H. B.; Beyer, J. M.; Mondal, R.; Panchal, N. S.; Colletti, L.; Liu, Y.; Koev, G.; Kati, W. M.; Hernandez, L. E.; Beno, D. W. A.; Longenecker, K. L.; Stewart, K. D.; Dumas, E. O.; Molla, A.; Maring, C. J. *J. Med. Chem.* **2018**, *61*, 1153–1163.
154. Krueger, A. C.; Randolph, J. T.; DeGoey, D. A.; Donner, P. L.; Flentge, C. A.; Hutchinson, D. K.; Liu, D.; Motter, C. E.; Rockway, T. W.; Wagner, R.; Beno, D. W.; Koev, G.; Lim, H. B.; Beyer, J. M.; Mondal, R.; Liu, Y.; Kati, W. M.; Longenecker, K. L.; Molla, A.; Stewart, K. D.; Maring, C. J. *Bioorg. Med. Chem. Lett.* **2013**, *23*, 3487–3490.
155. Talamas, F. X.; Abbot, S. C.; Anand, S.; Brameld, K. A.; Carter, D. S.; Chen, J.; Davis, D.; de Vicente, J.; Fung, A. D.; Gong, L.; Harris, S. F.; Inbar, P.; Labadie, S. S.; Lee, E. K.; Lemoine, R.; Le Pogam, S.; Leveque, V.; Li, J.; McIntosh, J.; Najera, I.; Park, J.; Railkar, A.; Rajyaguru, S.; Sangi, M.; Schoenfeld, R. C.; Staben, L. R.; Tan, Y.; Taygerly, J. P.; Villasenor, A. G.; Weller, P. E. *J. Med. Chem.* **2014**, *57*, 1914–1931.
156. Kati, W.; Koev, G.; Irvin, M.; Beyer, J.; Liu, Y.; Krishnan, P.; Reisch, T.; Mondal, R.; Wagner, R.; Molla, A.; Maring, C.; Collins, C. *Antimicrob. Agents Chemother.* **2015**, *59*, 1505–1511.
157. Kwo, P. Y.; Badshah, M. B. *Curr. Opin. Organ Transplant.* **2015**, *20*, 235–241.
158. Trivella, J. P.; Gutierrez, J.; Martin, P. *Expert Opin. Pharmacother.* **2015**, *16*, 617–624.
159. El Kassas, M.; Elbaz, T.; Hafez, E.; Wifi, M. N.; Esmat, G. *Expert Opin. Drug Discovery* **2017**, *12*, 635–642.
160. Martinello, M.; Bajis, S.; Dore, G. J. *Gastroenterol. Clin. North Am.* **2020**, *49*, 253–277.
161. Barnes, D. M.; Shekhar, S.; Dunn, T. B.; Barkalow, J. H.; Chan, V. S.; Franczyk, T. S.; Haight, A. R.; Hengeveld, J. E.; Kolaczkowski, L.; Kotecki, B. J.; Liang, G.; Marek, J. C.; McLaughlin, M. A.; Montavon, D. K.; Napier, J. J. *J. Org. Chem.* **2019**, *84*, 4873–4892.
162. Magri, A.; Ozerov, A. A.; Tunitskaya, V. L.; Valuev-Elliston, V. T.; Wahid, A.; Pirisi, M.; Simmonds, P.; Ivanov, A. V.; Novikov, M. S.; Patel, A. H. *Sci. Rep.* **2016**, *6*, 29487.
163. Cox, N.; Tillmann, H. *Expert Opin. Emerg. Drugs* **2011**, *16*, 713–729.
164. Lai, C. L.; Yuen, M. F. *N. Engl. J. Med.* **2008**, *359*, 2488–2491.
165. Tang, C. M.; Yau, T. O.; Yu, J. *World J. Gastroenterol.* **2014**, *20*, 6262–6278.
166. Xie, Y. H.; Hong, R.; Liu, W.; Liu, J.; Zhai, J. W. *Virol. Sin.* **2010**, *25*, 294–300.
167. Huber, A. D.; Michailidis, E.; Tang, J.; Puray-Chavez, M. N.; Boftsi, M.; Wolf, J. J.; Boschert, K. N.; Sheridan, M. A.; Leslie, M. D.; Kirby, K. A.; Singh, K.; Mitsuya, H.; Parniak, M. A.; Wang, Z.; Sarafianos, S. G. *Antimicrob. Agents Chemother.* **2017**, *61*.

168. Laganas, V. A.; Dunn, E. F.; McLaughlin, R. E.; Tiong-Yip, C. L.; Yuzhakov, O.; Isabella, V. M.; Hill, P.; Yu, Q. *Antiviral Res.* **2015**, *115*, 71–74.
169. Liuzzi, M.; Mason, S. W.; Cartier, M.; Lawetz, C.; McCollum, R. S.; Dansereau, N.; Bolger, G.; Lapeyre, N.; Gaudette, Y.; Lagace, L.; Massariol, M. J.; Do, F.; Whitehead, P.; Lamarre, L.; Scouten, E.; Bordeleau, J.; Landry, S.; Rancourt, J.; Fazal, G.; Simoneau, B. *J. Virol.* **2005**, *79*, 13105–13115.
170. Mason, S. W.; Lawetz, C.; Gaudette, Y.; Do, F.; Scouten, E.; Lagace, L.; Simoneau, B.; Liuzzi, M. *Nucleic Acids Res.* **2004**, *32*, 4758–4767.
171. Sudo, K.; Miyazaki, Y.; Kojima, N.; Kobayashi, M.; Suzuki, H.; Shintani, M.; Shimizu, Y. *Antiviral Res.* **2005**, *65*, 125–131.
172. Ghebremedhin, B. *Eur. J. Microbiol. Immunol. (Bp)* **2014**, *4*, 26–33.
173. Eckardt, A. J.; Baumgart, D. C. *Recent Pat. Antiinfect. Drug Discov.* **2011**, *6*, 54–63.
174. Gonzalez-Lopez, J. J.; Morcillo-Laiz, R.; Munoz-Negrete, F. J. *Arch. Soc. Esp. Oftalmol.* **2013**, *88*, 108–115.
175. Gupta, P.; Tobias, J. D.; Goyal, S.; Hervie, P.; Harris, J. B.; Sadot, E.; Noviski, N. *J. Intensive Care Med.* **2011**, *26*, 267–272.
176. Lu, X.; Trujillo-Lopez, E.; Lott, L.; Erdman, D. D. *J. Clin. Microbiol.* **2013**, *51*, 1089–1093.
177. Lynch, J. P., 3rd; Fishbein, M.; Echavarria, M. *Semin. Respir. Crit. Care Med.* **2011**, *32*, 494–511.
178. Nikitenko, N. A.; Gureeva, E. S.; Ozerov, A. A.; Tukhvatulin, A. I.; Izhaeva, F. M.; Prassolov, V. S.; Deryabin, P. G.; Novikov, M. S.; Logunov, D. Y. *Acta Nat.* **2018**, *10*, 58–64.
179. Robinson, C. M.; Seto, D.; Jones, M. S.; Dyer, D. W.; Chodosh, J. *Infect. Genet. Evol.* **2011**, *11*, 1208–1217.
180. https://covid19.who.int/.
181. Kish, T.; Uppal, P. *P T* **2016**, *41*, 314–325.
182. Kim, Y.; Wower, J.; Maltseva, N.; Chang, C.; Jedrzejczak, R.; Wilamowski, M.; Kang, S.; Nicolaescu, V.; Randall, G.; Michalska, K.; Joachimiak, A. *Commun. Biol.* **2021**, *4*, 193.
183. Baby, K.; Maity, S.; Mehta, C. H.; Suresh, A.; Nayak, U. Y.; Nayak, Y. *Arch. Med. Res.* **2021**, *52*, 38–47.
184. Hakobyan, A.; Galindo, I.; Nanez, A.; Arabyan, E.; Karalyan, Z.; Chistov, A. A.; Streshnev, P. P.; Korshun, V. A.; Alonso, C.; Zakaryan, H. *J. Gen. Virol.* **2018**, *99*, 148–156.
185. Colpitts, C. C.; Ustinov, A. V.; Epand, R. F.; Epand, R. M.; Korshun, V. A.; Schang, L. M. *J. Virol.* **2013**, *87*, 3640–3654.
186. St Vincent, M. R.; Colpitts, C. C.; Ustinov, A. V.; Muqadas, M.; Joyce, M. A.; Barsby, N. L.; Epand, R. F.; Epand, R. M.; Khramyshev, S. A.; Valueva, O. A.; Korshun, V. A.; Tyrrell, D. L.; Schang, L. M. *Proc. Natl. Acad. Sci. U. S. A.* **2010**, *107*, 17339–17344.
187. Chistov, A. A.; Orlov, A. A.; Streshnev, P. P.; Slesarchuk, N. A.; Aparin, I. O.; Rathi, B.; Brylev, V. A.; Kutyakov, S. V.; Mikhura, I. V.; Ustinov, A. V.; Westman, G.; Palyulin, V. A.; Jain, N.; Osolodkin, D. I.; Kozlovskaya, L. I.; Korshun, V. A. *Eur. J. Med. Chem.* **2019**, *171*, 93–103.
188. Orlov, A. A.; Chistov, A. A.; Kozlovskaya, L. I.; Ustinov, A. V.; Korshun, V. A.; Karganova, G. G.; Osolodkin, D. I. *Medchemcomm* **2016**, 7, 563.
189. Vandamme, A. M.; Van Vaerenbergh, K.; De Clercq, E. *Antivir. Chem. Chemother.* **1998**, *9*, 187–203.
190. Spence, R. A.; Kati, W. M.; Anderson, K. S.; Johnson, K. A. *Science* **1995**, *267*, 988–993.

191. Rittinger, K.; Divita, G.; Goody, R. S. *Proc. Natl. Acad. Sci. U. S. A.* **1995**, *92*, 8046–8049.
192. Velazquez, S.; Alvarez, R.; San-Felix, A.; Jimeno, M. L.; De Clercq, E.; Balzarini, J.; Camarasa, M. J. *J. Med. Chem.* **1995**, *38*, 1641–1649.
193. Velazquez, S.; Tunon, V.; Jimeno, M. L.; Chamorro, C.; De Clercq, E.; Balzarini, J.; Camarasa, M. J. *J. Med. Chem.* **1999**, *42*, 5188–5196.
194. Pontikis, R.; Dolle, V.; Guillaumel, J.; Dechaux, E.; Note, R.; Nguyen, C. H.; Legraverend, M.; Bisagni, E.; Aubertin, A. M.; Grierson, D. S.; Monneret, C. *J. Med. Chem.* **2000**, *43*, 1927–1939.
195. Danel, K.; Larsen, L. M.; Pedersen, E. B.; Sanna, G.; La Colla, P.; Loddo, R. *Bioorg. Med. Chem.* **2008**, *16*, 511–517.
196. Shuker, S. B.; Hajduk, P. J.; Meadows, R. P.; Fesik, S. W. *Science* **1996**, *274*, 1531–1534.
197. Petersen, L.; Jorgensen, P. T.; Nielsen, C.; Hansen, T. H.; Nielsen, J.; Pedersen, E. B. *J. Med. Chem.* **2005**, *48*, 1211–1220.
198. Khandazhinskaya, A. L.; Mercurio, V.; Maslova, A. A.; Palomino, R. A.Ñ.; Novikov, M. S.; Matyugina, E. S.; Paramonova, M. P.; Kukhanova, M. K.; Fedorova, N. E.; Yurlov, K. I.; Kushch, A. A.; Tarasova, O.; Margolis, L.; Kochetkov, S. N.; Vanpouille, C. *Biochimie* **2021**.
199. Maslova, A. A.; Matyugina, E. S.; Snoeck, R.; Andrei, G.; Kochetkov, S. N.; Khandazhinskaya, A. L.; Novikov, M. S. *Molecules* **2020**, 25.